I dedicate this work to two extraordinary women: Jody Swann, Data Facility Manager at the United States Geological Survey (USGS), Branch of Astrogeology, for her invaluable and irreplaceable collaboration, and Patricia Conklin, talented planetary cartographer at JPL-Nasa, author of the wonderful 'Equatorial Photomosaic' of Mars.

Viking Orbiter 1 Mars Mosaic

It has been more than 15 years since the Viking mission spacecraft first approached Mars. There were four spacecraft that made the journey to the red planet. The Viking 1 and 2 landers entered Mars' thin atmosphere and, by using parachutes and then breaking rockets, came to rest on the surface. Viking Orbiters 1 and 2 remained in orbit. Although all Viking spacecraft are now silent, the data collected by them is still providing scientists with new insights about this solar system neighbor.

Using the Planetary Image Cartography System (PICS) developed at the U.S. Geological Survey, Flagstaff, Arizona, Tammy Becker, Alfred S. McEwen, and Larry Soderblom recently processed 102 Viking Orbiter 1 images taken of Mars in 1980 to form this dramatic mosaic of nearly a full hemisphere of the planet. PICS, a computer-based system, permitted each image to be aligned with the others in a manner like fitting together the pieces of a jigsaw puzzle. Image distortions were adjusted by PICS to provide perfect border matches. To bring out detail, color variations of the dark materials were enhanced by a factor of about two and brightness variations of the images adjusted to provide consistency across the hemisphere. The view is comparable to what would be seen from a spacecraft orbiting 2,500 kilometers above the surface of the planet.

The center of the Mars mosaic shows the entire Valles Marineris, a canyon system that stretches over 3,000 kilometers in length and is up to 8 kilometers deep. The Grand Canyon of Earth compares to just one of the tributary canyons of this giant system. Layers of bright material in the eastern and central region of the canyon may be carbonate-rich sediments deposited in an ancient lake. Huge ancient river channels, such as Kasai Valles, stretch northward from the central and western regions of the canyon.

South of Valles Marineris is very ancient terrain covered by many impact craters. North of the eastern end of the Valles Marineris is the Chryse Planitia. The Viking 1 spacecraft landed in the place indicated on the sketch map.

To the west of Valles Marineris lie three of Mars' huge volcanoes. The Tharsis volcanoes, as they are known, appear as dark reddish spots. Each volcano is about 25 kilometers high, over 350 kilometers in diameter, and has a central crater at its summit. The most famous member of the Tharsis volcanoes, Olympus Mons, is not visible in the mosaic. It lies approximately 1,100 kilometers west of Ascraeus Mons.

The next important step for Mars exploration is the Mars Observer (MO) mission scheduled for launch in September 1992. MO will arrive at Mars in September 1993 and, after a few months of orbiting maneuvers, will begin its 2-year mapping mission. MO will make a global scientific map showing surface chemistry, mineral composition, landforms, atmospheric characteristics, and magnetic and gravitational fields. The map produced by MO will offer significant improvements over Viking in terms of image resolution and detail.

For the Classroom

1. Research the Latin roots of surface feature names on Mars such as abyrinthus, mons, planitia, and valles.
2. Learn about the life and accomplishments of Percival Lowell.
3. To visualize the size of the Martian features found on this mosaic, draw an outline of the United States on a separate piece of paper to the same scale as the mosaic. Cut out the outline and place it on the mosaic for comparison. (The distance between the central craters of Ascraeus Mons and Pavonis Mons is approximately 700 km.)

References

Carr, M. H., ed., (1984), *Geology of the Terrestrial Planets*, NASA SP-469, NASA Scientific and Technical Information Branch, Washington, DC.

Greeley, R. & Batson, R. M., ed., (1991), *Planetary Mapping*, Cambridge University Press, New York.

NASA, (1978), *The Martian Landscape*, NASA SP-425, NASA Scientific and Technical Information Branch, Washington, DC.

Spitzer, G. R., ed., (1980), *Viking Orbiter Views of Mars*, NASA SP-441, NASA Scientific and Technical Information Branch, Washington, DC.

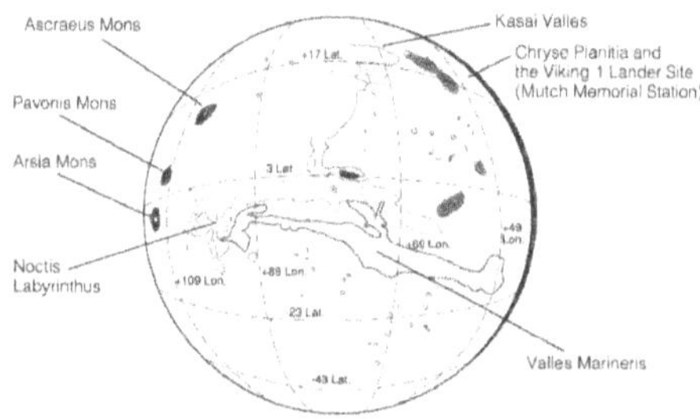

Cover Photo Caption: 1) Above, the caption of the Nasa image (Viking Orbiter 1 Mission); 2) Statue of the Goddess "Flora of Spring" – Acireale (Sicily) – Piazza Lionardo Vigo; sculptor Francesco Contarino (1925-2019); photo by the artist photographer Mr Mario Bella of Acireale.

CONTACT INFORMATION

Gianni Viola – Germinal280@libero.it – 3425597743
Paolo Fienga – paolocf1963@gmail.com - 3383634362

Gianni Viola

THE SCIENTIFIC DISCOVERY OF LIFE ON MARS

Handbook of Planethography for the study of planets and the search for life in the Solar System

Translation of Maria Grazia Lucrezia Leotta

Paolo C. Fienga
Scientific Advisor

IF PRESS

Contributors: Paolo C. Fienga (Italy), Maria Grazia Lucrezia Leotta (United Kindom), Yulia Bondarenko (Russia), Melania Cassisi (Italy); Elena Salemi (Italy); *NASA* (USA - headquarters): Joseph L. Headle, David J. Bohlin, Cynthia M. O'Carrol, Bert Ulrich; *Goddard Space Flight Center* (NASA-USA): Joseph King, David A. Williams, Barbara Pope; *Jet Propulsion Laboratory* (NASA-USA): John Bluth, Elizabeth Moorthy, Amy Culver; *United States Geological Survey* (USA): Jody Swann, Adrienne Wassermann, Kathy Hoyt, Karen Zanter; *Arizona State University* (USA): Charles Hewett, Daniel Ball, Sara Santiago, Veronica Ann Zabala-Aliberto; *Regional Planetary Image Facility* (USA): Susanne Pieth (Germany), Daniel Gaspard (France); Livia Giacomini (Italy); *Lunar and Planetary Institute* (NASA): Mary Noel-Black; *National Air and Space Museum* (USA): Diana Barbe; *Armagh Planetarium* (Ireland): Ian Griffin; *Geoscience Node* (USA): Susan Slavney, Margo Muller; *Agenzia Spaziale Europea – Istituto Europeo di Ricerche Spaziali (ESA-ESRIN-Italy)*: Angelo Bodini; *Consiglio Nazionale delle Ricerche* (Italy): Ruggero Casacchia; *Istituto di Astrofisica e Planetologia Spaziali* (IAPS – INAF Italy): Marco Feroci.

IF Press srl - Roma, Italia
info@if-press.com - www.if-press.com

ISBN 978-88-6788-426-1

ABOUTH THE AUTHOR

Gianni Viola studied sociology at the Carlo Bo University of Urbino (Italy). Photo reporter for *Stampa Alternativa* in Rome; he published *Police* (Lewis&Cann, London, 1976) and *Police Chronicles and Documents of the Repression in Italy 1860–1977* (Bertani, Verona, 1978). He contributed to the *New Police and Reformation of the State*, Rome (Union for the Police Workers).

He worked on international politics and served as a political-military advisor in military embassy offices.

He published *The Civilisation on Mars*, Mediterranean Rome, 2002. This is a scientific work based on precise references, experimental data, proven facts, and rereuevable documents.

He co-operates with *L'Astronomia*, an important astronomy magazine in Italy. In this periodical, between 2005 and 2006, several articles by the author appeared concerning the neo-discipline called "Planetography", of which he is the founder and promoter. The founding elements of "Planetography", as a scientific discipline were officially registered in 1994 at the Società Autori ed Editori in Rome.

For many years, the author collaborated with NASA's photographic library and the Regional Planetary Image Facility (Research Area Roma 2 in Tor Vergata, Institute of Astrophysics and Space Planetology).

He is a member of "The Society for Planetary SETI Research" (USA).

THE NASA INTERNATIONAL PHOTO LIBRARY NETWORK

"REGIONAL PLANETARY IMAGE FACILITY"

The need to promote the quantitative and qualitative analysis of the photographic data obtained during the many NASA's space missions led to the creation, in USA, year 1977, as well as in Europe, during the eighties, of the Regional Planetary Image Facilities (RPIF). The seeds of the RPIF network were first planted in the 1971 with the first facility opening at the United States geological Survey in Flagstaff, Arizona.

These organisations, which operated on the basis of a 'Memorandum of Agreement' signed between the NASA Planetary Geology Program and the host institutions, had as one of their founding purposes the essential one of enabling a wider consultation of archived photographic data.

The various RPIF centres were libraries that preserved and made available to the public satellite images taken by spacecraft exploring the Solar System, as well as planetary data and maps of the planets, their satellites, and collections of books and other documents from NASA's planetary missions. Many facilities possessed unique collections, data and documents saved from local research programs and also had an area of specialisation.

The RPIF network charter provided access to visiting scientists and the public, assistance to users in understanding the heritage of the facilities, assistance in acquiring data from the National Space Science Data Centre (NSSDC) (NASA's Central Archive) and NASA's Planetary Data System (PDS) archives, and communication and training of the community on relevant topics (see Acknowledgement).

Over the decades, the facilities evolved in terms of their capabilities with regard to digital imagery, while continuing to provide access to NASA's initial planetary data for helping and training users.

The Regional Planetary Image Facility (RPIF) network was operated in part by NASA until 2020.

To this day, several US centres of the former RPIF Network continue to function as NASA planetary data centres at their home institutions and continue NASA's original mandate to facilitate access to planetary mission data while expanding their planetary data and document collections.

Over time, many former photo libraries have informally become an important part of NASA's planetary data ecosystem, with facilities that meet the need for physical and online access to planetary data and supporting documentation, and also provide expert assistance and training on planetary data and its use.

Mention should be made here of the 'data centres' at Arizona State University, Cornell University, and the United States Geological Survey (USGS) Astrogeology Science Center in Flagstaff, Arizona, which began conducting a public planetary data laboratory in 2023.

The RPIF facilities were geographically distributed to increase accessibility and were located at institutions with long-standing groups of planetary scientists using the collections.

There were seventeen active photo libraries worldwide, nine of which were in the USA and the rest in Asia and Europe (including one based in Rome).

The first personal contact with the RPIFs, was with the Lunar and Planetary Institute (Centre for Information and Research Services – Houston, TX), through Mary Noel-Black (manager), who provided the initial information regarding the entire network of the various RPIF locations.

Among the photo libraries located within the United States, mention must be made of the United States Geological Survey – USGS (Branch of Astrogeology – Department of the Interior – Flagstaff, Arizona), and, in particular, with the notable contribution of Miss Jody Swann (in charge) for her precious, irreplaceable and fundamental collaboration; Adrienne Wassermann, Kathy Hoyt and Karen M. Zanter (the latter, also in charge of the Earth Resources Observation and Science EROS Center). With regard to the photo library housed at the Jet Propulsion Laboratory in Pasadena (California), it should be mentioned that it provided a series of maps of Mars and Venus, as well as fundamental information on some satellite images of Mars, and a valuable volume of data reconversions of the Mariner 9 mission.

In the photo library housed at the Space Photography Laboratory (ASU-EDU, NASA Space Grant Intern, JPL Solar System Ambassador, Arizona State University, Department of Geological Sciences - Planetary Geology Group - Tempe AZ), Charles Hewett (director), Daniel Ball (manager), Sara Santiago and Veronica Ann Zabala-Aliberto played an important role, as they sent in photographic material, in particular that referring to the Mariner 9 mission.

The photo library housed at the National Air and Space Museum (Center for Earth and Planetary Studies – Washington DC), in the person of Diane Barbe (volunteer – press officer) provided important planetographic material.

Of the international photo libraries (outside the USA) we mention: the Déutsches Zentrum für Luft und Raumfahrt and V. (German Aerospace Center, Institute of Planetary Research, RPIF – Berlin, Germany), with Miss Susanne Pieth (in charge), who provided some useful data and information for further contacts; the Photothéque Planétaire d'Orsay (Université Paris-Sud – Department des Sciences de la Terre – Orsay – France), with Miss Daniéle Gaspard (in charge) who provided fundamental information regarding some photomosaics of the Mariner 9 and Viking 1 missions.

A special mention must go to the Italian section of RPIF, Nasa's Southern Europe Regional Planetary Imaging Facility in Rome, with which a collaboration was established from 2006 to 2019 (Tabs.**I,II**).

This facility began its activities in 1981 and was inaugurated by NASA in May 1983, at the Institute of Space Astrophysics (IAS) of the National Research Council, then located at the Research Area of Frascati (Rome); in the early 1990s the archive was transferred to Rome (Viale dell'Università, 11) and in 1996 the facility was relocated to the new Research Area Roma 2 of Tor Vergata-ARTOV (Via del Fosso del Cavaliere, 100). In the meantime, the host facility, the Institute of Space Astrophysics, changed its name to the Institute of Space Astrophysics and Cosmic Physics - IASF, becoming part of the INAF (National Institute of Astrophysics).

Administratively, until 2011 the photo library was managed jointly by the Istituto di Astrofisica Spaziale e Fisica Cosmica di Roma (IASF) and the Istituto di Fisica dello Spazio Interplanetario (IFSI), then, in December 2011, the two institutes merged into the IAPS (Istituto di Astrofisica e Planetologia Spaziali).

In 2014, the photo library was dedicated to the planetologist Angioletta Coradini, who died prematurely in 2011, an inspiration for many space instruments and Director of the INAF-IFSI.

In terms of investments, the Photo Library devoted a large part of the funding provided by the CNR, both through the Frascati Research Area and through the Institute's Directorate (IASF), to the acquisition of materials for the archiving of photographic data, sent on several occasions by NASA.

For the processing of planetary photographic data and, in particular, for their transformation into cartographic projections and photomosaics (as well as the subsequent analysis of the products obtained), the RPIF-IRPS prototype system was used, developed with the contribution of all the RPIFs (thus including the Photo Library of Rome) and jointly developed by the McDonnell Center for the Space Science, Washington University of St. Louis in Missouri (USA).

This system allowed researchers to access information, both on planetary data to be processed and on data that had already been processed, to conduct detailed research on mission-derived images, photomosaics, maps and finally to conduct local data processing.

In view of the considerable amount of data stored in the photo library, and in order to facilitate the research of visitors and researchers, in 1989 the people in charge of the structure elaborated and published an extremely useful general consultation guide, with the title "Information and consultation guide" **(Tab.III)** and in 1992 "Planetary cartography", a descriptive guide to the available planetary maps **(Tab.IV)**.

In 1995, the writer published a CD - distributed free of charge - in which the various programmes for decoding satellite images of Mars and Venus from the Mariner (Mariner 6,7,9,10) and Viking (Viking 1&2) missions were collected. The CD also contained the complete list of the images from the Mariner 9 and Viking Orbiter 1&2 missions, including the readings from the descent modules, Viking Lander 1&2 **(Tab.V)**.

Over the years, the Archives of Rome assumed a truly remarkable importance, offering one of the richest collections of paper and digital images and data concerning all the planets of the Solar System, most of their satellites and some minor bodies (comets and asteroids).

The material included: photographs, photomosaics, microfiches and films obtained from the most important space missions carried out by NASA; popular slides on astronomical, terrestrial, astrophysical and planetary subjects; videotapes, CDs, atlases, publications, both popular and historical, some of them very rare; geological and topographical maps of the surfaces of the planets and satellites of the Solar System.

All these items were available to the European scientific community interested in planetary research, through which they represented a point of reference for the documentation, updating and dissemination of planetary science itself.

In the facility, there was a wealth of equipment necessary for data consultation, research, and image display (electronic databases, CDs, analogue optical laser discs, etc.).

These instruments allowed the projection of audiovisual material, through which introductory seminars were held on the Solar System, as well as on data acquisition techniques during space missions. The facility was also characterised as a reference centre for supporting material for popular lectures and educational activities.

In this area, the photo library was used by scholars working in the field of space research. Alongside this branch of institutional activity, a popularisation and information service activity developed over time, which touched the most diverse subjects: from science teachers to students interested in preparing research papers and baccalaureate dissertations; from astrophiles who collected material for exhibitions and conferences on popularising science, to producers of science documentaries for theatre shows and television programmes. The populariser Piero Angela collected a lot of material for the programme 'Journey to the Cosmos'.

Among the staff of this institute, I think it is only right to mention, in particular, my friend and researcher Livia Giacomini, who facilitated the meticulous planetographic research that, over time, has made the development and evolution of this work possible, and with whom I shared projects in 2007 and 2012 regarding the management of the Photo Library, of which she is, in effect, the data manager.

As Ms. Tricia Talbert (Senior Multimedia Designer - NASA Headquarters, at www.nasa.gov/solar-system/nasa-commemorates-more-than-40-years-of-regional-planetary-image-facility-excellence/) aptly wrote on 27 August 2020, commemorating the 40th anniversary of the founding of RPIFs, "The images that RPIFs hold helped form the foundations of modern planetary science".

Tab. I	The Italian branch of the "NASA Photo Library" of the Regional Planetary Image Facility, at the Institute of Space Astrophysics and Planetology in Rome (entrance).

CONSIGLIO NAZIONALE DELLE RICERCHE

ISTITUTO DI ASTROFISICA SPAZIALE - AREA DELLA RICERCA DI FRASCATI

Southern-European Regional Planetary Image Facility

INFORMATION AND CONSULTATION GUIDE

R. Bianchi (*), R. Casacchia (*), A. Picchiotti (*)
M. Poscolieri (*), R. Salvatori (+)

(*)C.N.R. - Istituto di Astrofisica Spaziale
Frascati (Roma), Italy

(+) C.N.R. Area della ricerca di Frascati

October 1989

Tab. III	The guide-consultation of the Photo Library of Rome (1989).

LA CARTOGRAFIA PLANETARIA

A. Picchiotti, R. Salvatori, R. Casacchia

ARF-92-12 Settembre 1992

AREA RICERCA FRASCATI

CONSIGLIO NAZIONALE DELLE RICERCHE

VIA G. GALILEI - 00044 FRASCATI (ITALIA)

Tab. IV	A. Picchiotti, R. Salvatori, R. Casacchia - La Cartografia Planetaria (The Planetary Cartography), Area Ricerca Frascati, Roma 1992.

Tab. V	"NASA Image Programs" (edited by Gianni Viola) - CD containing decoding programs for Mariner 6,7,10 (m67v1m01), Mariner 9 (MarinerView203), Viking Orbiter 1&2 (Viking NASAView;Viking Orbiter Image Viewer; Viking Orbiter View), Viking Lander 1&2 (Viking Lander 1-2 EDR Image Browser), as well as the complete lists of Mariner 9 images (Mariner9-complete lists; Complete Index of Mariner 9 images) and Viking Orbiter 1-2 (Image Index), Viking Lander 1&2 (Viking Lander Image Data Dictionary). Distribution out of print, Rome 1995.

CHRONOLOGY OF THE PLANET MARS (1609-2025)

1609 – First telescopic observations of Mars by Galileo Galilei (Italy,1564-1642). Giovanni Kepler (Germany, 1571-1630) published his great 'Martian work', as he called the work *Astronomia Nova.*

1630 – Observations by Giovan Battista Zupi (Italy, 1590-1650).

1636 – First drawing of Mars by Francesco Fontana (Italy, 1585-1656), a spot discovered on the Red Planet.

1640 – Observations by Niccolò Zucchi (Italy, 1586-1670).

1644-1645 – Observations by Father Daniello Bartoli (Italy, 1608-1685). Discovery of two distinct spots.

1659-1672 – Christian Huygens (Netherlands,1629-1695) made the earliest drawing representing a defined aspect of the surface of Mars. It is the present Syrtis Major Planum (a plain).

1666 – Cassini Giandomenico (Italy,1625-1712) calculated the rotation of Mars. The results were published in the *Jourmal de Savants* of 31 May 1666. The calculation of the Martian day gave 24 hours and 40 minutes (less than 3 minutes longer than today's measurements). Cassini observes one of the channels detected two centuries later by Schiaparelli. It is the *Gehon,* a natural formation near the Mare Acidalium (Acidalia Planitia). in 1672 he made important observations of the planet.

1698 – Huygens is convinced that Mars hosts an intelligent civilisation. He reaches this conclusion by applying the analogical method, comparing other celestial bodies with the Earth.

1704 – Giacomo Filippo Maraldi (Italy, 1709-1788), grandson of Cassini, redetermines Mars' rotation period and observes other spots.

1719 – Observations by Francesco Bianchini (Italy, 1662-1729). in some specific cases, Bianchini's topographical precision precedes the formidable descriptions made by Schiaparelli.

1777-1779-1881-1883 – Observations by William Herschel (Germany,1738-1822). He predicted a true *geography of Mars.*

1785-1802 – Observations by Wolfang Hieronimus Schroter (Germany,1745-1816). Studies of Martian geography.

1834 – Johann Heinrich von Madler (Germany,1794-1874) and Wilhem Beer (Germany, 1797-1850). Drafting of a map of Mars with fixation of the *zero meridian*, later called *Meridiani Sinus*.

1859 – Father Angelo Secchi (Italy,1818-1878), astronomer of the Specola Vaticana, drew one of the first maps of Mars, very detailed, including the polar caps.

1864 – Observations by Frederik Kaiser (Netherlands, 1808-1872). Compilation of an important cartographic document of Mars; Observations by William Rutter Dawes (United Kingdom,1799-1868).

1867 – William Huggins (UK, 1824-1910) used spectrographic studies and discovered the presence of water in the atmosphere of Mars.

1867-1869 -Map of Mars by Richard Anthony Proctor (UK,1837-1888). He was the first to match the names of famous scholars of the planet to features on the surface of Mars, attempting to create an areographic language.

1877 – First contemporary observations of Mars by Giovanni Schiaparelli (Italy, 1835-1910). First toponymy of Mars (**Table 18**).

1892 – Observations by William Henry Pickering (USA, 1858-1938). According to Camille Flammarion (France – 1846-1936), of all the potentially habitable planets, Mars was the one of greatest interest.

1895 – Stimulated by Schiaparelli's studies, American Percival Lowell (USA – 1855-1916) began personal studies of the surface of Mars. He was in constant contact with Schiaparelli (**Tables 8,9**).

1931 – Desiderius Papp (Germany – 1895-1993). in his work 'Was lebt auf den Stermen?' he stated that the presence of intelligent, higher life on Mars was certain.

1939 – Observations by Mentore Maggini (Italy – 1890-1941). Publication of the text 'The Planet Mars' (1941); Observations by Georges Fournier (France – 1881-1954).

1954 – Earl C. Slipher (USA – 1883-1964) produced the first colour photograph of Mars.

1956 – Observations by Gerard Peter Kuiper (Netherlands – 1905-1973).

1957 – Glauco De Mottoni (Italy – 1901-1988). From his photographic maps of Mars he drew a global map that was adopted by the *International Astronomical Union.*

1965 – First satellite surveys of Mars by the Mariner 4 probe (21 ½ images, taken at a geometric resolution of 1,250 metres). The surface retracted was approximately 1%. (**Tables 4A,59**).

1966 – The text 'Intelligent Life on Universe' by J.S. Slovsky (USSR, 1916-1985), and Carl Sagan (USA. 1934-1996) was published. It is an extensive work on the planet Mars from a perspective free of the typical prejudices of astronomers in the years following Schiaparelli's death, when it was considered useless to study Mars directly (observational studies).

1969 – More in-depth satellite surveys of Mars by the Mariner 6 and 7 probes. Mariner 6 took 428 images and Mariner 7 took 749 photos, with a geometric resolution of 1,000 metres (**Table 4B**).

1971 – Guido Ruggeri (Italy – 1913-1976). Publication of the text 'The Discovery of the Planet Mars' where the author stigmatises the anti-scientific behaviour of astronomers against Schiaparelli.

1971-1972 – First almost complete survey of the planet by the Mariner 9 probe (Table 61). A total of 7,329 satellite surveys were carried out, of which 3,820 images were taken with the wide-angle camera (Camera A) (Tables 5A,19) and a maximum geometric resolution of 430 metres; the other 3,509 images were taken with the telephoto camera (Camera B) and a maximum resolution of 37 metres (**Tables 59,50**). Construction of the first photomosaic of Mars (**Tables 66,67,68,70A, 70B**).

1971 – First detection of artificial structures on Mars by the Mariner 9 probe (23 December 1971). It is the photo Das 4402165 with a geometric resolution of 43.90 metres. (**Table 39**).

1976-1980 – Automatic space exploration by the Viking 1&2 missions. 50,028 satellite surveys were carried out with a maximum resolution of 8 metres (**Tables 5BCD;23,24,25,27,28,29,30,44,45,48,59,60**). Reconfirmation and redefinition of the structures detected by Mariner 9 (1972), by the orbital missions Viking 1 & 2 (1976) (**Tables 42,44,48, 133,134,135,144,145**).

1976 – First detection of artificial structures on Mars by the Viking Lander 1 probe (landing module): image 12A235 (Tav.47) of 19 August 1976, with a geometric resolution of 13.3 metres. All the measurements made by the descent modules were 3,983 (**Tables 16,22,26,47,59,60**).

1976 – In order to ascertain the presence of elementary life on Mars, the first biological experiments were conducted in 1976 by the descent modules of the Viking probes, launched in August-September 1975, which arrived on Mars the following summer. The two landers landed on the planet's surface and performed a series of experiments to analyse soil samples **(Table 6).** The conclusions of Gilbert Levin – the technician proposed to direct the experiments – were that: "*They all were in favour of the presence of life*".

1976-2004 – Following satellite surveys of the Viking missions, in frame 35A72 (**Table 45** – with 47.13 metres of geometric resolution), an anthropomorphic-looking structure was noticed, later named the 'Face on Mars'. It was immediately established that the image was an optical illusion (Table 49A). This statement would be correct, since what is seen is not what exists in reality, however, it would be a *relative* optical illusion, where the latter represents a *deformed image of an existing reality*. The image in question is deformed due to the Sun being too low on the horizon (80°), but in two subsequent images, the structure will be detected in a redefined manner.

1) in 1976 (code 70A13, from the Viking 1 mission), with a resolution increase of 7.87% and a decrease in the solar incidence angle of 21.78%. 2) in 2004 (code V12445004, from the Mars Odyssey mission), with a resolution increase of 19.73% and a decrease in the solar incidence angle of 15.56% (**Tables 45,46,58**). With this, the question is technically resolved in a positive sense and it can be stated beyond any reasonable doubt that the structure actually corresponds to a real construction.

The discoverers of the peculiar structure were two Italian-American researchers, Vincent Di Pietro and Gregor Molenaar, who presented their findings on 1st May 1980 at a press conference in Lanham, Maryland. On the following 16th June they planned another meeting at the Annual Convention of the American Astronomical Society. Between May and December 1980, the news was revealed to the world and was published in many newspapers (see 1981).

The question of Mars in general and the 'Face on Mars' in particular – in the following years – were covered by journalists, researchers and popularisers. Among the others there are Jeff Greenwald (1985), Mark Carlotto (1988,1997), Erol Torun (1988-89), Vladimir Avinsky (1984), Brian Hurtak and James Crowley (1989), Allen Harold (1997), René Anaya (2005), Frissel Bob (1994), Richard Grossinger (1993), Hatcher David Childress (1994), Richard Hoagland (1987,1992), C. N. Mahr (2007), McDaniel Stanley (1994,1998), Moore Patrick (2000), Pozos Randolfo Rafael (1986), Rux Bruce (1996), Brandenburg J.E. (1991), Lambert Dolphon (1983), Tom Ruthemberg (1983), Merton Davis (1983), Gene Cordell (1983), C. West Churchmann (1984), Horace W. Crater (1993).

1977 - The need to promote the quantitative and qualitative analysis of photographic data obtained during NASA's numerous space missions led to the creation of the Regional Planetary Image Facilities (RPIF) in the United States in 1977 and in Europe in the 1980s (see infra The Nasa International Photo Library Network "Regional planetary Image Facility") (**Tables I,II,III,IV,V,187**).

1981 - The Roman branch of the RPIF began operations in 1981 and was inaugurated by NASA in May 1983, at the Institute of Space Astrophysics (IAS) of the National Research Council, then located at the Research Area of Frascati (Rome); in the early 1990s the archive was transferred to Rome (Viale dell'Università, 11) and in 1996 the facility was moved to the new Research Area Roma 2 of Tor Vergata-ARTOV (Via del Fosso del Cavaliere, 100,**Table 187-IV of cover**).

1981 – Vincent Di Pietro and Gregor Molenaar published the results of their work up to that time, in "Unusual Mars Surface Features".

1993 - Beginning of personal research conducted by Gianni Viola at Nasa's Southern Europe Regional Planetary Imaging Facility in Rome, with which a collaboration was also established from 2006 to 2019 (**Tables I,II,III,IV,V,59,60,64,65,67, 81,82,83,84,85A,186,187**).

1994 - First registration with the SIAE of the work 'Mars' by Gianni Viola (containing the first discoveries of artificial structures on Mars, made by automatic space exploration) (**Table 185**).

1997 - The booklet 'The Discovery of Mars Civilisation' by Gianni Viola is published (**Table 183**).

1998 - Interview with the author by journalist Giovanna Crisà: 'Viola, a scholar with his eye on Mars' - La Sicilia, 20 February 1998.

1999 - First satellite surveys of current vegetation and lake structures on Mars by the Mars Global Surveyor mission (**Table 6**).

2002 - Unpublished work by the author: 'Comparative testing - in the aerial photogrammetric analysis of Mars'; Printing of the work "La Civiltà di Marte" (by Gianni Viola) (**Table 184**).

2004 - Article by the author: 'The search for life in the Solar System' in the Rome daily newspaper 'Rinascita' (7 December 2004). Start of the Author's collaboration with the 'NASA Photo Library' in Rome. The collaboration ended in 2019, following the closure of the photo library; Reconfirmation and redefinition - by the Mars Odyssey probe - of some artificial structures detected in 1976 by the Viking 1 probe (**Table 46**).

2005 – Reconfirmation and redefinition – by the Mars Global Surveyor probe – of some artificial structures detected in the 1972 Mariner 9 probe (**Table 114**). First report on Planetography by Gianni Viola published in the magazine *L'Astronomia* di Milano. (**Table 10**).

2006 – Second report on Planetography by Gianni Viola published in the magazine *L 'Astronomia* di Milano. (**Table 13**).

2010 – In No. 4, April-May 2010 of the periodical "The Youth Time magazine" (Czech Republic), part of the Youth Time International Movement, published an interview with the author by journalist Angela Curatolo entitled "The Other Truth About the Red Planet", which appeared in both Russian and English. The article was among the 30 winners of the Gagarin 2011 competition in Baikonur (Kazakhstan), a former Soviet missile base.

2011 - The author's article 'Il cammino della scienza nella scoperta di Marte' (The path of science in the discovery of Mars) is published in 'Punto Zero' (Nexus New Time Magazine); September 2011.

2012 - Presentation of the 'Operational Project - Planetography Laboratory' at the IAPS in Rome; 22 June 2012 (Gianni Viola).

2025 - The author is accepted as a full member of 'The Society for Planetary SETI Research' (SPSR - USA). (10 February 2025) – (**Table 185B**).

TABLE OF CONTENTS

List of the Photographic Tables

Part Three – Analysis of Areographic Maps

THE CARTOGRAPHY OF MARS

MC.10: LUNAE PALUS

Chap. 1: ECHUS FOSSAE (MC.10.1306.SW)

Chap. 2: NANEDI VALLES (MC.10.1307.SE)

MC.18: COPRATES

Chap. 3: OPHIR PLANUM (MC.18.1208.NE)

Chap. 4: AURORAE PLANUM (MC.18.1208.NE)

MC.19: MARGARITIFER SINUS

Chap. 5: CAPRI CHASMA (MC.19.1381.NW)

MC.23: AEOLIS

Chap.7: AEOLIS MENSAE Mc.23.1213.NW)

MC.26: ARGYRE

Chap. 8: NIRGAL VALLIS (MC.26.1191.NW-1192.NC)

Biographical Notes

Table 182: Author (p. 348).
Table 183: Gianni Viola, La scoperta della civiltà di Marte, Pubblicato a cura del "Centro Studi Biologia Sociale" di Acireale (Sicilia). – Acireale, 1997 **(p, 349).**
Table 184: Author's text "La Civiltà di Marte" (Mediterranean Editions, Rome 2002) **(p. 350).**
Table 184A – La Scienza di Cartapesta (The Science of Papier-Mâché) of Gianni Viola (2025); (**p. 383**).
Table 185: Certificate of deposit of the work "Marte" by Gianni Viola, dated 28 April 1994, with the "Società Italiana Autori ed Editori" (S.I.A.E.) – Sezione OLAF – Servizio deposito opere inedite **(p. 351).**
Table 185A - The boss checks over report with Girl Friday. Gene Shoemaker, researcher on the Ranger and Surveyor programs, and secretary Jody Loman Swann in new office (USGS). From: Arizona Daily Sun of 1965. **(p. 352).**
Table 185B – Author's membership card for "The Society for Planetary SETI Research – SPSR" (USA) **(p. 353).**

Bibliography

Table 186: The fundamental scientific texts, published by NASA, summarising the space exploration of Mars, in the period 1969-1983: Atlas of Mars (1979) – Mariner 9 and Viking 1&2 Missions; The Martian Landscape (1978) – Viking Lander 1&2 Missions; The New Mars (1974) – Mariner 9 Mission; Mars as viewed by Mariner 9 (1974); Viking Obiter views of Mars (1980); The Mariner 6 and 7 pictures of Mars (1971). **(p. 358).**

III of cover

Table 186A: Historical Map of Mars (Gianni Viola).

Back Cover

Table 187: Headquarters of the Institute for Space Astrophysics and Planetology (IAPS) located in the Roma 2 Tor Vergata Research Area. The same building houses the NASA Photo Library in Rome, the Italian section of the RPIF, the Southern Europe Regional Planetary Imaging Facility **(IV of cover).**

Note of the Translator

When I first met Gianni Viola, I was a student of Modern Foreign Languages and Literature at the University of Catania.

I often travelled with my father who worked at the Magistrates' Court and one morning I met Gianni on our train. He introduced himself with the kindness and impeccability that characterise him and we started a discussion. It took me less than 5 minutes to realise we had a lot in common, especially our passion for culture and the interest in extraterrestrial life. Gianni told me a lot about the Red Planet and showed me very interesting pictures of Mars.

We became friends very soon and he held me in high esteem and I reciprocated until for some years, life got in the way and we lost touch. I took my degree; I married, had children and moved to the United Kingdom. Thanks to Facebook (socials are not to demonise completely if they are used appropriately), I found Gianni and we started from the point we had left, like time had never passed. I realised that his interest in Mars had grown to the point that my friend was writing a book about the planet. Meanwhile, I had taken a MA Translation Studies at the University of Sheffield and Gianni asked me to translate his book into English.

I asked him to send me the Italian version to have a look and I was dazzled!

The book was a real masterpiece not only for the content in itself but also for the style and the gentle and impeccable way to "accompany" the reader through the Martian adventure thank to the irreplaceable work of the astronomer Giovanni Virginio Schiaparelli and the crowd of detractors who denigrated the patient and precious contributions of the scientist. What struck me was the structure of Gianni's book: well-dosed, balanced, precise and detailed but above all readable also by those who are not experts or are approaching the topic for the very first time.

In my opinion, this is Gianni's best quality; to make simple and pleasant a topic as complex and articulate as the possibility of life on Mars. I read the book greedily and I accepted the Author's invitation to translate it with great enthusiasm. It was not a job for me but a very pleasant experience that made my months in lockdown an excuse to enjoy the reading and the translation of a book that in my humble opinion is amongst the most intriguing about the topic.

Thanks to this translation, I also had the possibility to improve my knowledge about a planet that has already fascinated mankind's imagination and has inspired sci-fi films and books.

However, Gianni Viola's work is not sci-fi ma pure, unbiased science, free from the prejudices of the academic science that I like to define "convenient science" a typical behaviour that tends to reject every new contribution that can rock the system and implies a new investigation that is a synonym for hard work and humility.

I wish Gianni the success he deserves for all these years of resilience and perseverance. I appreciate the man, the writer and the friend.

Good luck Gianni and *ad majora, semper*!

Maria Grazia Lucrezia Leotta
(*York, United Kingdom*)

Author's Note

I cannot understand why so many astronomers, astrophysicists and planetologists scramble to answer the question of whether 'we are alone in the Universe'. They play it and sing it. The studies they conduct do not involve the morphological analysis of planets, this being the sole pertinence of planetography, the only discipline capable of identifying and examining the 'modified' landscape of a planet. Yes, because a modified landscape (meaning not by natural factors) is an indication of the presence of intelligent life. Good. All this stuff is far removed, indeed totally absent, from the cultural horizons of the aforementioned scholars. What follows is that these scholars continue to speak inaccuracies because they refuse to use the only material relevant to the study of the surface of planets. They do not study planets and when they pretend to study them, they would never consider satellite images. But they are experts, to such an extent that in over 50 years of space exploration of the Solar System, they have seen nothing, although since 1971 the presence of intelligent life, at least, on the planet Mars has been established beyond reasonable doubt.

When Schiaparelli watched Mars using a telescope, he did it as a geographer, not as an astronomer! This was his main strength, his method and his philosophical guide. He was an astronomer when he watched the stars, the nebulae and even when he had an intuition about other celestial bodies that could not be visible just by using a telescope.

His observations were quite different from those of his predecessors – even apart from the fact that he operated with more powerful instruments than in the past. While his colleagues emphasised colours and hues, he was attentive to lines, which he translated into a description based on geometry and calculation, analogous to the earth's surface. With this, Schiaparelli applied the first Copernican principle that every physical law experienced on Earth has equal validity at all points in the Universe. This was the first time that the methodologies already applied for the study of the Earth's surface and the compilation of its cartography were introduced for planetary observations.

After Giovanni Schiaparelli's death (1910), many astronomers, with Antoniadi in the lead, carried out the systematic *destruction of the geometry of*

Mars, on the specious and vulgar grounds that someone – in this case the populariser Lowell (but not Schiaparelli!) – 'had attributed it to the Martians'.

The reality is that academic science, in that case, began a *process of decontextualisation, fragmentation and break-up of the logical and scientific organicity* within the study of the planet Mars, effectively blocking scientific research, at the point that half a century later, the brilliant results achieved by the satellite survey carried out from 1971 onwards were totally disregarded.

The circumstance or better yet: the enterprise according to which, nowadays, the astronomy wants to deal with the geographical planetary surface features (such as the so-called *planetography*) was begun by Schiaparelli, means that a century after his death, he remains misunderstood.

In effect, nobody wanted to carry on with the geographical achievement by Schiaparelli examining the visual material as the satellite images of Mars that are qualitatively superior compared with those that Schiaparelli could observe with a telescope.

This lack of appreciation regarding Schiaparelli's effort has been coded in academic terms so that today an astrophysicist (so a planetary scientist) admitted shamelessly that he had no time to examine the images of Mars because he was waiting for better ones!

At this point, in front of a similar declaration, we should consider a thing.The images of the Viking missions (1975–1980) caught, among other things, surfacing hills (*structures*) with dimensions going from approx. 10 metres of extension (through orbital photograms) to just a few centimetres (through photograms obtained on the surface from two landers—or *descent stages*) and *this guy* (as his colleagues) would expect better images?

We could say that the planet's surface was not their *thing*, but in the attempt to hide their *lack of will and skills*, these *scientists* grasped to very childish and ridiculous excuses. It seems that every researcher follows a precise *rule*, a *guideline* that is not an order but a sort of faith to an alleged *scientific method* that is opposite to the *experimental scientific method.*

This "rule/guideline" should oblige the researcher to ignore the satellite images but in this way, a direct study on Mars becomes impossible, including the one related to the possible past and present existence of life on the "Red" planet.

At the same time, it is not possible to "deeply" study its chemistry without carrying out *biological experiments in loco.*

The above, if we want to be honest, represents some sort of a logical continuity of the studies related to the physics of Mars (i.e.: *Spectroscopic Analysis of the Red Planet*, started in the 19th century by Schiaparelli and piteously destroyed by Antoniadi & Co.) and not only about its shape—that can change depending on the technical means used—but also in its substance, so therefore, going against the principle whereby a planet, (not just

Mars!) must be observed from the morphological point of view by using the best available tools existing at the time of the research is carried out.

All the above, in my humble opinion, clearly shows the substance of many facts and problems that, unfortunately, are still existing.

The rest is just contemporary history.

For instance, we could mention some footages executed by students and researchers that, in this context, represent the prototype of the study's typology of all those who did not want to apply the scientific method to know (or, at least, to better understand) the planet (in other words and talking in a *Copernican sense*, the extension/application of the physics laws experimented on Earth must also be tested/verified in the rest of universe, if and when it is possible).

These individuals dealt—in a way—with Schiaparelli, they did mention the exploration of Mars by Mariner 4, but *skipping* the relevant detections and eventually they spoke directly and inopportunely of the biological experiments carried out in 1976. Obviously, this operation—well protected by academics—includes the omission of the detailed discussion of the Mariner 9 (1971–72) and Viking 1 and 2 (1976–1983) missions.

And this *detail* (just in a manner of speech) represents the *original sin* of today's science!

A black spot that produced endless disasters and whose ostracism Schiaparelli himself had to suffer, since, as we see written in a recent publication[1]: "To undertake topographical studies of a distant planet was (...) for Schiaparelli an act of intellectual courage. For an astronomer at the time, only the description of the movement of the planets of the Solar System, their satellites, comets (...) was considered scientifically correct"(...). "He (...) was the first astronomer to carry out observations aimed at studying the physical nature of the planets of the Solar System, following a rigorous method and relying on the Earth Sciences: Geodesy, Geophysics, Meteorology, for the interpretation of the data collected. A method that orthodox astronomers considered alien to the research tradition of their discipline". And it should be noted that, apart from a few exceptions, of which I am happy to quote the above statements (by Prof. Pasquale Tucci, of the Brera Astronomical Observatory), the situation has remained completely unchanged for over a century.

When the Author proposed (2005) an article about Martian Planetography for the magazine "L'Astronomie", (official organ of the Association Astronomique de France in Paris), his proposal was rejected because (and it was correct after all) the article was not exactly about Astronomy but about

[1] Pasquale Tucci, on: Giovanni Virginio Schiaparelli, La Vita sul Pianeta Marte – Tre scritti di Schiaparelli su Marte e i "marziani" (Life on Planet Mars, Three Writings by Schiaparelli on Mars and the "Martians") – by Pasquale Tucci, Agnese Mandrino and Antonella Testa, Edited by Mimesis, Milan 2002, pages 28-29.

Geography. At that time, the Author was an emeritus member of the above-mentioned association. When the same service was proposed in Italy, it was promptly printed by the Italian magazine 'L'Astronomia'[2] and this act represented, in fact, the seal of the birth of 'Planetography'[3].

With regard to the above, I am pleased to present this very modest work as a contribution and tribute to the great Italian scientist Giovanni Virginio Schiaparelli, and all this has been made possible by the sensitivity of IF Press of Rome, to whom I extend my heartfelt thanks.

Gianni Viola

(*Roma/Italy, 2 September 2024*)

---

[2] Viola Gianni, 'Mars: physiognomy of a planet' – *L'Astronomie* – Science and Culture Monthly – no. 272 – March 2006 – 18 pages.

[3] See infra 'Scientific Glossary' – entry: "Planetography".

Preface and Dedication (By) Paolo C. Fienga

—Do what you can, using what you have at the time you do it—
(sentence attributed to Sir W. Churchill)

The volume created by my dear friend and colleague, Dr Gianni Viola, considering the complexity (and fascination) of the topic hereafter deeply discussed is—in the writer's opinion—simply perfect as regards its structure, articulation as well as its length.

We are in front of a work that can be adopted by experts and technicians of the field and, at the same time and with the same clarity, can also be read by amateur astronomers, astronomy enthusiasts and people simply fascinated by the work done by famous astronomers, their life and personal history included.

In a nutshell, we believe that this one is a very clear work as regards its goals and whose methodology should make it very easy to be understood.

So, how to end it and, above all, to whom it could be dedicate to (because a dedication is often, if not always, a "must")?

Well, since we all know that in the astronomical science nothing really "definitive" can still be written, i can say that Schiaparelli's work, just like the one carried out by many other astronomers who came before and after him, from the Mesopotamians up to now (even though only a few of them were similar to our protagonist), can correctly be considered as both a substantial and integral part of the word "history".

History, of course, has to be written a little bit at a time, perhaps one second after the other, certainly a discovery after the other and so forth. And the real historians tell us about the past. Always.

But who can fully tell and write the whole (personal) history and work of a scientist and then transmit the relevant outcomes to the future generations (especially when the personal facts of an individual mix with the history of science of an entire—or even more than one—age)?

Consider: if we think about writers, journalists and/or historians, we do know that, usually, these are people who were not physically there and, most likely, not directly involved in and by the queries they wrote about (whatever the subjects might have been).

And that's not all; do you want to know why? Because there are other people (mostly scientists and researchers) who spoke and wrote—in fact, I should say "speak and write" —about science and other scientists, even if, sometime, they clearly showed a certain lack of intellectual honesty and skill.

We are not naïve and we know that these people always existed and the "job" they perform is normally one of the consequences of their "privileged position".

Think now: who are these, let's say, "bad guys"?

Honestly, we decided to avoid naming them, for obvious reasons— However, we can define them like some "out of tune minstrels" who can only recall fragments of a distant past that is also difficult to define as a whole.

Due to this, many important questions remained and remain unanswered and the work performed by great minds sunk in a thick midst made of doubts and misinterpretations.

These "minstrels" intellectually out of tune are heavily influenced by the present (more and more "bulky", especially in these times) and they also are constantly victims of a very questionable future (such as the one hypothesised by all those who "think to know" and then decide who's right and who's wrong, extolling some and scuttling others).

As a matter of fact, these individuals grope in some sort of semi-darkness, but they keep trying to sell to others a lot of certainties that they have not.

Question: What if we are like them?

Answer: No, we aren't.

The reason is very simple. We are more privileged than the abovementioned minstrels.

We know that the truth is fleeting and it is always the product of a "Vision"; a vision which is very likely "incomplete". Because every vision lacks something and this is a fact: by definition.

However, "Partiality", as "incompleteness", is not and it cannot be a synonym of "Partisanship".

So, if you read carefully, you will see that was just this "partisanship" to relegate the great Schiaparelli among those who, in the opinion of many people (scientists and non-scientists), "have done much, BUT not enough".

In this "BUT not enough" lies the mistake or, maybe, the misunderstanding: Giovanni Virginio, according to his detractors, made a wrong interpretation of the available data.

Everybody knows that "making a wrong interpretation of the available data" is a natural risk for every scientist. But what about the "detractors"?

Were they in good faith? Were they the only ones who interpreted the available data in the right way? Or they may be acted in bad faith, just because they did not like what Schiaparelli had to say?

Of course, we cannot know what exactly Schiaparelli had in his mind when he observed Mars through the telescope and, at the same time, he drew it (left eye on the telescope, right one on a piece of paper: and that means *Will* and *Dedication*!).

We cannot know what he thought in his rational and scientific mind because that could be too close to what we now call "Informed Imagination/Speculation".

We don't know. But we can try to understand, putting ourselves in his shoes, in his time and in his resources and reasons—

We can try to understand why he wrote what he wrote and drew what he drew.

We can only try to imagine what his thoughts were after his eyes saw the "beautiful and yet volatile" Ares.

This is a very short description of what has been done in this treatise, brought forward with the greatest intellectual honesty, rationality and a pinch of speculation; so, writing, telling and explaining, when and where it was possible, some "simple facts", from ancient times to these days.

Possibly the good Giovanni Virginio had seen (and understood) the right things.

Just possibly.

Having doubts is licit because the work of the historian at every level is not simply to tell the truth, (that is the result of a philosophical, often random and essentially very personal vision) but a search for it that is only aimed to make other people capable to find it by indicating the way and showing some "facts".

And there is more: these "facts", once they are detected by acting—we repeat this concept—with intellectual honesty, methodological strictness and integrity, must be placed in their own specific and unique time and space to be properly interpreted.

It is very important to underline that an interpretation, here, in this book, has been given and the interpretation is the result of the personality of the Historian, the Researcher and, in certain cases, of the scientist as well.

This means, in other words, that it was (and it will always be) necessary to leave some space to doubts in order to be completely honest and reliable. The *doubt*, as described in Descartes or Hamlet—

Anyway, among the others, we have a very last doubt (as we said before): to whom we can dedicate this work?

It is difficult to answer because, in today's world, speaking about "doubts" seems to be an equivalent of "speaking in vain", because everyone (or at least most of us) is/are sure of everything. Well, "lucky them" I say!

But let's try to think about our present: we live in a so-called "virtual" age. That is a pseudo-scientific form of mystification of a very uncertain reality. Intellectually honest and sensitive people know this very well. And they are very often isolated or offended.

Great.

So, since we live in a time where nothing and, for absurd, everything is or seems to be absolutely sure, how can we find something or someone to dedicate such a fascinating work?

Well, the answer comes by itself: we dedicate this book with happiness and affection to all those who acted as Giovanni Virginio, using the available means till their maximum extent and then, at the end and in good faith, said "We need to wait for better times, more advanced and more reliable."

The *Technical and Scientific Evolution*, hoped for by the great Italian scientist finally arrived. But the so-called "understanding", in Schiaparelli's language, is still very far.

It is far in space, certainly way farther than Mars—And very far in time. Very far from our own daily life.

This enlightenment is, at the moment, apparently lost in the eyes of those Assyrian and Babylonian astronomers who were the first ones who, thousands of years ago, looked at the sky, eternal but variable and understood that the way to the stars, even though stunning and bright, is and will always be scattered with both human and divine harsh and treacherous pitfalls.

A wonderful and obviously endless way that, I'm afraid, none of us will ever fully see or understand.

Paolo C. Fienga
(*Lawyer and Dr in Astrophysics with a PHD in Planetary Sciences*)

PREFACE OF ANANDA SIRISENA (SPSR - Society for Planetary SETI Research)

In 1976, when Dr. Tobias Owen saw a "Face" staring up at him on one of the Mars Viking Orbiter photographs of a region known as Cydonia, there started a ball rolling which led to much speculation about the possibility of life on Mars, post Lowell's drawings of 'canali'.

Up until that time, the scientific consensus was that the 'canals on Mars' were nothing more than an optical illusion created by earthbound telescopes. The mainstream belief was that Mars had never harboured life, that conditions there were too severe for any life as we know it to have evolved as a species. Suddenly, a spanner had been thrown into the midst of this work of cherished belief. There may be no 'canals' on Mars but there was an unmistakable face pointing up into space, visible to our rudimentary spacecraft cameras of the time.

The Viking Orbiter 1 and 2 cameras were state-of-the-art for 1976; better resolution photos became available in the coming decades. At the time it was determined that the face was about a mile long and may have been carved out of rock or may have been a modification of an existing hillock or *massif*. The surge of books about a long gone civilisation on the surface of Mars, remnants of which were visible in the Viking Orbiter images, led to detailed image enhancement techniques and several TV shows about such a possibility. But the studies were localised, until planet wide coverage showed other intriguing features which might have bolstered the theory that such a civilisation, many millions of years ago, may have perished in a cataclysm. That event may have caused the planet Mars to lose its atmosphere, caused its water to become underground permafrost, reduced temperatures to what we would call 'very cold' and air pressure to be very low.

Gianni Viola presents a case here for a 'holistic' approach - a study of the whole surface of Mars with the entire stack of Viking 1 and 2 photos, archived by NASA. There is no doubt that there are pyramids on Mars: 3-sided, 4-sided and even one 5-sided mountain about a kilometre in height. Just what has been going on over millions or billions of years? Does our solar system have a history of life - humanoid life? Are there Martian connections with terrestrial pyramids and the Sphinx? Time will tell - for "boots on the ground" could provide details about the surface and eroded buildings. Let us listen carefully to Gianni Viola's case for large-scale artistry on the planet Mars.

There is evidence and then there is proof. In a court of law, layers of evidence are first presented, then a final proof is declared. However, science does not work that way. Progress in science is much slower (not so in technology, as there is something tangible); papers are written, peer reviewed and published. Those opposed to the ideas in the papers then write their rebuttals and get them published in the same journal – this goes on and on. Slow, painstaking.

The evidence for life on Mars is overwhelming. Photos of microfossils, crinoids, corals, lichen, seasonal greenery on rocks, occasional blue skies, methane, oxygen, water-ice, pyramidal hills, mounds the size of the Giza which form a precise mathematical pattern on the sands of Mars, all of which NASA quietly maintains in its archives, swept under a rug of silence. Space agencies from other nations will eventually break this stalemate and then there will be a deluge of information collected over many decades of remote sensing.

The ideas presented in Gianni Viola's book may be vindicated in years to come. In the meantime, there is much here to think about. The areology of Mars done in an entire-surface manner is to be complimented. It is fresh thinking. Pioneers are alone to begin with, then find many companions. Maybe you can be an esteemed companion by studying this volume carefully. To Mars, to the other planets, to the stars.

Ananda Sirisena
(*President SPSR - Society for Planetary SETI Research*)

Acronyms and abbreviations

3D: Three dimensions
ASU: Arizona State University (USA)
CAM: Satellite camera
CNR: National Research Council (Consiglio Nazionale delle Ricerche, Italy)
CPI= Center Point Intercept.
DAS: Data Automation Subsystem (Mariner 9)
Das Time: Mariner 9 image code
Doc.= documentes.
DF: Definite Resolution
EDU: Generic top-level domain, dedicated to higher education
ESA: European Space Agency (Paris-Framce).
ESRIN: European Institute for Space Research (Italy-Roma).
f: Focal Length of the lens
FLIP: Free Lance International Press (Italy, Roma)
Frame number: Numero del fotogramma
GR: Geometric Resolution
GSFC: Goddard Space Flight Centre (USA, NASA)
Heigth: Frame heigth
HST: Hubble Space Telescope
IAPS: Institute of Space Astrophysics and Planetology (Italy,Roma).
IAS: Institute of Space Astrophysics (Italy-Roma)
ID: Numero di identificazione
ID DAS TIME (see Das Reference Time-Scientific Glossary)
IFSI: Institute of Space Astrophysics and Planetology
INAF: National Institute of Astrophysics (Italy,Roma).
JPL: Jet Propulsion Laboratory (NASA)
Lat: Latitude
Long (or Lon) : Longitude
MC: Mars Chart (Mars territorial division)
MER: Mars Exploration Rover (NASA)
MGS: Mars Global Surveyor (NASA)
Modific.: Modification.
NASA: National Aeronautics and Space Administration
NE: Nord-East
Number Mars Map : Map Number at a scale of 1 :5.000.000 (Mariner 9)
NW: Nord-Est
OLAF: Literary works and figurative arts (Italy-Roma).
Planetogr.: Planetography.
R: Shooting distance
RR: Radiometrici Resolution
Repres.: repression
RPIF: Regional Planetary Image Facility (NASA)
SC: Sud-Centre
SE: Sud-East
SIAE: Italian Society of Author and Publishers (Società Italiana Autori ed Editori, Italy-Roma)
SP: Pixel Scale
SW: Sud-West
TAB: Photographic table
Tab1-52 (example) = Table 1 on page 52
TAPE IDE File: Tape Identification File (Mariner 9)
Text: reference text for an image
TM3D = Test Modelling 3D.
TR: Topographic Resolution
UCL: University College London
USGS: United States Geological Survey (USA)
Width: frame width

Part I
The History of Mars

1. The Search for Life[4]

Let us talk about life.

As a matter of fact, being able to give an explanation, as well as some evidence of the real existence of extraterrestrial life, would be a massive achievement.

But what is "life"?

With the word "life" we mean an organism that, once is adapted to its habitat, can reproduce itself by taking raw material (i.e. "*nourishment*") from the environment where it lives in order to be able, at some point, to create copies of (i.e.: *to reproduce*) itself.

An example of *intelligent life* (in a manner of speech) is mankind that has many questions about its essence and the consciousness (some time, it depends—) of its existence.

This *intellectual exercise* is called "Science" and it is typical of the so-called *conscience of species*. in other words, when looking for life, mankind looks instinctively at itself and what it thinks and believes to be.

How and where looking for "life" then?

The only way to answer this question is verifying, in analogical terms, similar scenarios in order to find similarities. So, since we live on a planet, it is likely that life, intelligent life, can be found on other planets as well and it is possible to verify this hypothesis with the tools we already have.

It is remarkable what the scholar Guido Ruggieri[5] wrote:

> *With Copernicus and for the very first time based on factual evidence, the unknown celestial bodies, even though still hidden and appearing just as little night lights, have been considered similar to Earth.*[6]

---

[4] Here we do not expect to know what life is. We refer to a shared opinion but not for this reason questionable. It is only right to highlight that in this framework there are several opinions and all of them deserve attention.

[5] Guido Ruggieri was one of the most famous and serious astronomer scholars. He was born in Faenza (Ravenna, Italy) in 1913 and died in Mestre (Venice, Italy) in 1976.

[6] Ruggieri, G. (1971) *La scoperta del Pianeta Marte* (*The Discovery of the planet Mars*), Arnoldo Mondadori Editore, Milan, pp. 41–42.

Already, in a far and distant past, great thinkers figured out that mankind was just one of the countless examples of life in the universe.

The philosopher Anaxagoras (Ἀναξαγόρας—499/98–428/27 BC), made it clear that the moon was not so different from Earth and in ancient Greece some philosophers gave proof of evidence about the existence of a plurality of worlds in the universe.

Among these great philosophers we can name Thales (Θαλῆς) of Miletus (Μίλητος—624–540 BC), Xenophanes (Ξενοφάνης) of Colophon (Κολοφών—570–475 BC), Heraclitus (Ηράκλειτος) of Ephesus (Έφεσο—544–484 BC), Democritus (Δημόκριτος) of Abdera (Ἄβδηρα—460–371 BC), Epicurus (Ἐπίκουρος) of Athens (Ἀθῆναι—341–271 BC) and Metrodorus (Μητρόδωρος) of Chios (Χίος), all of them who lived between the years 330 and 278 BC, who affirmed that *believing there is just a world that is inhabited is the equivalent of saying that in a huge wheat field there is just one ear full of grains.*

The philosopher Ōrigénēs (Ὠριγένης) of Alexandria (Αλεξάνδρεια—AD 183/85–252/53) asserted that Earth *is one of the minor worlds among the others in the immense celestial space.*

In more recent times, a scholar faced the problem of the plurality of worlds, the Cardinal Nicolò of Cusa, known as the *Cusano* (1401–1464), bishop of Bressanone and Copernicus' precursor (1473–1543).

Cusano asserted that there is no star, even though different from our world, we can deny there should live human creatures.

Even Galileo (1564–1642), at the beginning of the 17th century, wondered if the other planets could host (or hosted) Earth-like lifeforms and even more explicitly expressed himself Giordano Bruno in his work, *On the Infinite Universe and World.*

> *Universe is infinite and its centre is everywhere and nowhere at the same time. in the immeasurable and infinite space, there are countless stars surrounded by planets similar to those that spin around our sun. It is ridiculous thinking that on these planets there are no intelligent beings, even more intelligent than us.*

Descartes defined *bold but not absurd* the hypothesis of an extraterrestrial, intelligent life. From his side, Kant and Goethe shared the same thought and the astronomer Pierre La Place, recalling Kant's ideas, was able to go ahead the modern astrophysics by asserting that: *Man born for terrestrial temperature could not live on other worlds but should not be countless organisations as regards different temperature of planets and universes? If the only difference of elements and climate create so many different terrestrial lifeforms, there should be even more different (ones) on other planets and satellites!*

Fra' Angelo Secchi (1818–1878) director of the Vatican Observatory, when he worked for the Parisian Academy where those who believed in meteorites were bullied, bravely asserted that *Those who don't believe in the liveability of other stars are just like the inhabitants of the Celestial Empire who, since they were locked in their world by the Great Wall of China, thought and keep thinking that in our planet there are only Chinese people.*

The enthusiastic works by Camille Flammarion gave to humanity some brilliant intuitions and they still are incredibly valid. At the beginning of his book on Mars, he asserted: *We dare to hope that the day will come when unknown scientific tools will give us proof of the existence of inhabitants in other worlds.*

In a more scientific way spoke Kostantin Ziolkovskij, a Russian physicist and mathematician who wrote in 1896 in a more scientific style: *On several planets we can find all the evolutionary phases of life (—). There were men on Earth thousands of years ago: will they become extinct in a few millions years? The same process can occur on other worlds.*

So, which ones are the objects (i.e.: *Celestial Bodies*) useful to start (and at some point, to finish) the search for life?

Considering the Milky Way, such as the galaxy that includes our solar system, it has been estimated that it contains more than one billion planets which could host some kind of lifeforms. This news came from London and at that time was given by two British astronomers of the Open University. According to some results they obtained thanks to a virtual simulation, it was calculated (of course in merely speculative terms) how many solar systems of our galaxy could have the perfect characteristics and conditions to host intelligent lifeforms.

The press release, issued on 10 April 2002 by the researchers Barrie Jones, professor of Physics and Astronomy and Nick Sleep, biotechnologist, during the National Astronomy Meeting, asserts that:

> *Numbers go over last year figures spread by German meteorologists from Potsdam Institute. They suppose that the Milky Way contains at least 500.000 extra-solar planets that have a biosphere and they are nicknamed Gaia. So far, one hundred planets have been spotted in the stars' orbits. They are gas giants similar to Jupiter. However, in many astronomers' opinion, the cosmic space hides an enormous number of small planets of rocky nature that could be discovered with the help of an appropriate technology.*
>
> *Ten years later, the European group of astronomers and astrophysics, after observing 102 dwarf, red stars in the austral sky, fixed to 40% the probability that around them gravitates the so-called liveable Super-Earth and sustained that the Super-Hearths (planets that measure from one to ten times the dimensions of Earth), are more recurring in proximity of this kind of stars. Only in the Milky Way, the dwarf red stars are about 160 billion (80% of the whole galaxy) and some of them are only thirty light years away from our sun. This means that the number of liveable planets is enormous.*[7]

---

[7] Di Pasqua, E. (20 March 2012) "There are billions of livable planets in the Milky Way" on *La Repubblica* (Milan-Italy).

It is true that other planets, in addition to those present in the solar system, could or can present the conditions to permit the existence of life (not just its *development* because life could also have been *introduced*—) but it has to be added that there are also some planets of our solar system we know very little about; and as regards to those of other planetary systems, the only thing we can do is to count them—

So, which ones are the planets that, at the moment, can be studied *exhaustively* (just a manner of speech) and by following correct, scientific procedures?

The current level of the space exploration in the solar system considers Mars (recently, even Europa, a Jovian moon, as well as Titan and Enceladus, Saturnian moons) as the *only celestial bodies* where it is possible to carry on a systematic study aimed to the research of life.

To this purpose, the scientists Sagan and Slovskij, an American and a Russian respectively, at the time they declared that *Mars seems to give us, in the immediate future, the best opportunity to study extraterrestrial life.*[8]

According to the statements of researchers from *Project Sign*,[9] astronomers agree to a great extent to admit that only a *celestial body* located in the solar system, besides Earth, could be inhabited by *human* (but it would be better to say *humanoid*) beings and this planet is Mars. When liquid water was found on the Red Planet, Edward J. Weiler from NASA affirmed that among *all the planets of the universe, Mars is the first one in which scientists think to find life. This leads to the main question: Are we alone?*

Finally, a statement about Mission Mars Pathfinder of 1 December 1997, signed by Matt Golombeck, scientific responsible of the mission, informed us that *Mars is much more similar to Earth than what we expected.*

Robert Zubrin, Author of *The Case of Mars*[10] asserts:

> *The planet Mars will be the scenery of next century. It is the only world in the solar system where there is a high probability to find a past life and a present one.*

There is nothing new in this if we think that, in 1971, Guido Ruggieri,[11] quoting the work of Huygens and in contrast with Kircher, that considered Mars as the ancients saw it (*The planet of terror*), he wrote:

> *—in a very interesting work, Cosmotheoros, published posthumously in 1698 (—) there are no imaginary astronautical (the future sci-fi, Science Fiction, n.d.c.), but only considerations on the liveability of planets, masterfully developed and with an open mind. The Author finds that other worlds are equivalent to Earth because, like*

---

[8] Slovskij, J.S. and Sagan, C. (1980) *La vita intelligente nell'universo* (Intelligent life in the Universe), Milano: Feltrinelli, p. 188.

[9] The "Project Sign" was an official study of the USA government regarding self-propelled objects observed in the sky also named UFO or OVNI. The study was led by the American Air Force between 1947 and 1948.

[10] Zubrin, R. (2009) *The Case for Mars*, Aldebaran Editions, Curno (BG), p. VII.

[11] Guido Ruggieri, op. cit., p. 65.

*our "Blue Planet", they are enlightened and warmed by the sun and provided with water and air; they have to be inhabited by beings similar to us even though not necessarily identical—(**Table 1**).*

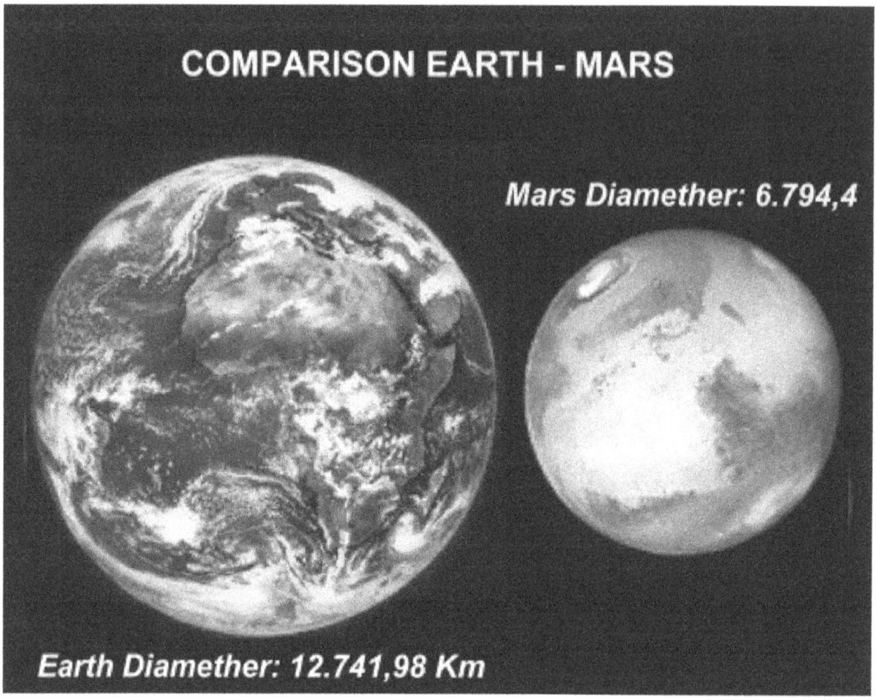

Tab. 1	Comparison of Earth – Mars.

2. Ancient Studies

Mars is one of the eight known planets of our solar system (excluding the so-called "Dwarf Planets", such as Pluto, according to the new nomenclature adopted by the IAU—International Astronomical Union) and it is the fourth starting from the sun so, in succession, we have Mercury, Venus, Earth and Mars.

Its diameter is about half in comparison with the Earth's one and it is the most similar to the Blue Planet.

This planet has been a riddle for science for ages and, in the last three centuries, it has been observed with the help of optical instruments of different kind, while before it had been observed only through the naked eye.

In ancient times it was known, along with Mercury, Venus, Jupiter and Saturn, while the other planets were still unknown.

Herschel discovered Uranus in the AD (*Anno Domini*) 1781, thanks to his calculations and visual observations; Le Verrier discovered Neptune in 1846, considering the *gravitational perturbations* of Uranus and Lowell, a famous scholar specialised in Mars canals, calculated the position of Pluto (discovered by C.W. Tombaugh in 1930) in the early years of the 20th century.

Centuries ago, in China, a *conjunction* of the moon with Mars, Jupiter, Saturn and Mercury was observed. It happened during the reign of the Emperor Hoang-Ti, in the year 2441 BC.

Chinese people called Mars *The Red Star* or *The Sparkling Light*. And we must not forget the name *Zhurong* (such as the *God of Fire* from ancient Chinese Mythology and now the name of the first Chinese lander on Mars, Utopia Planitia Region—already "visited" by a NASA Lander of the Viking Class).

In India, it was called *Angaraka* (i.e.: Burning Coal) and *Lohitanga* (i.e.: Red Body) and it was considered to be of evil influence.

In Arabia, it was considered as a *celestial warrior* always ready to hit humans with its foolproof spears.

Egyptians called it *Harmakhis* and in Persia it was *Pahlavani Siphir*: the *celestial warrior* (again!).

It was Nirgal for Chaldeans (a name borrowed by Sumerians and today attributed to a long Martian dried river—Nirgal Vallis) and it was considered the supreme hero, god of War and Death.

In the AD 1845, the archaeologist Henry Layard (1817–1894) discovered the legendary city of Nineveh, close to Mosul (Mesopotamia, now Iraq), a city forgotten for about 2400 Earth years under countless layers of soil that later became hills.

The most important discovery was its huge library of tablets: real books, most of them dedicated to celestial bodies. A whole book was about Mars' observations.

As the historians say, it was calculated that said book had been compiled around the year 1700 BC.

The *Book about Mars* was part of a greater work called *Observations of Bell* and it is one of most direct observations of Mars made in old (in fact, "ancient") times.

Greeks called Mars *Ares*; a name inspired by the flickering flame that the planet often emanates when close all the way to the horizon as the other stars (note: today we know that it is only an atmospheric distortion of the light).

In the fourth-century BC, Aristarchus of Samos hypothesised the presence of the sun in the centre of a system where the Earth was between Venus and Mars and the three planets orbiting around the sun.

In 357 BC (much time later found according to the computation of Kepler), Aristotle saw the moon passing over Mars but, in reality, he had just seen the moon overshadowing it.

Ares became Mars for the Romans and the month of March was dedicated to it.

March/Mars was the first month in the Roman calendar and it was reputed to be the father of Romulus, the mythical founder of Rome.

Mars was also the *National God* of the Romans and referring to this, Livius[12] wrote that

> *the Roman Empire was the most powerful ever after the celestial one (—) the Romans affirmed that its father was Mars.*

So it seems reasonable to think that, if the Romans imagined Mars to start an activity before the foundation of Rome, this could have meant that they transmitted a tradition where Rome had started its existence during a massive (and, of course, still and forever unknown) *event* linked to Mars.

Before the telescope was invented, the study of Mars was confined to mere philosophical speculations that ended when printing was discovered and the *Heliocentric Theory* announced, in the AD 1543, by the Polish abbot Nikolaj Kopernik or Nicolaus Copernicus (1473–1543).

---

[12] Titus Livius, *Ab Urbe condita*, Book I.

The German astronomer Johannes Kepler (1571–1630) accepted and extended the Copernican System, improving it through the study of Mars' movements that gave him the possibility to discover the motion of all the other known planets and that is the base of the so-called "Celestial Mechanics".

Copernicus had foreseen that the invention of the astronomical telescope, already envisaged in the 16th century, would have enabled to study the planets. Albeit nobody has ever written the history of this fundamental instrument, it is sure that it was conceived when the (basic) Laws of Refraction became known.

P. Borel, author of a work appeared in the AD 1660 (*De vero telescope inventore*), thought that the spyglass was invented by accident in 1590 by the two sons of an eyeglass maker from Middleburg (Germany), Zechariah Jansen.

Other sources mention instead Mr Jan Lepperskey another eyeglass maker.

The first man who investigated the sky by using a telescope was Galileo, who, in the Anno Domini 1610, at the age of forty-six, built a personal prototype that today can be found in the Academy of Florence.

He observed Mars and he noticed that the planet was not perfectly round.

He wrote the above-mentioned consideration in a letter dated 30 December 1610, that was sent to Benedetto Castelli, a Benedictine monk, Mathematics' professor in Pisa and one of his pupils.

Galileo started to give his opinion about Mars in his *Sidereus Nuncius*, published in Venice in March 1610. Quite obviously, at that time the book was deeply discussed.

The first targeted observations we know about were made by the Italian Francesco Fontana (1580–1656).

He declared himself the inventor of both the telescope and the microscope as (invented, respectively, in the AD 1608 and 1618). in 1632, he made the first drawing of Mars that was just a *shapeless* spot inside a circle.

In 1636, he eventually built a perfected model of telescope, discovering the rotation of the planet and published his works in 1655.

Two Jesuits also made interesting studies: Nicolò Zucchi in 1640 and Daniele Bartoli in 1644.

The astronomer Hevelius, who at the time had engraved the first lunar chart, realised he had to renounce to the representation of Mars because his telescope could not show significant/meaningful (for a man of that time) details.

Afterwards, the Dutch astronomer Christiaan Huygens (1609–1695) built a telescope by his own means. Thanks to that, he was able to distinguish actual luminous contrasts on the Red Planet. Huygens observed that some dark parts were not always in the same position, but they moved from west to east.

Observing this phenomenon, he inferred that the planet spun in a way similar to the Earth's one. As a consequence, he understood that Mars should have had days and nights.

In a note of his diary, dated 1 December 1659, we find the following statement: *Debet Martis conversio fieri spatio circiter diurno, sive 24 horarum nostrarum quemadmodum item Telluris* (the rotation of Mars seems to change as the one of the Earth in 24 hours).

He made what it is reputed to be the most ancient drawing of Mars' surface in 1672 and it is thought (not proven!) that his "map" (in a manner of speech) was the first one that clearly showed the South Pole of the Red Planet.

The astronomer Giandomenico Cassini (1625–1712) after being interested in comets for an exceptionally long time, studied Mars using a telescope of about 23 ft (approx. seven metres). Thanks to this instrument, he could see some successions of spots that created a sequence sufficiently typical to confirm the first hypotheses of Huygens on Mars' rotation time.

A long study led Cassini to think that Mars completed a full rotation in approx. 24 hours (and we now know that in reality the rotation lasts 23 hours and 39 minutes).

The real duration lasts only one minute more compared to what Cassini assumed. His results were reported in the *Journal des Savants* of 31 May 1666.

Towards the end of the18th century, it was observed that Mars' axis of rotation was remarkably similar to the Earth's one, so it would have been possible that there were different (four, it was said) seasons on Mars, just like on our planet.

From now on, it is common to speak about the possibility of life on Mars that was, is and will be a riddle for traditional scientists, even after the beginning of the space age (or even "space era", as you prefer).

Among the scholars interested in Mars and before Schiaparelli (Table 2), there was someone who had already understood how important was investigating the planets' surface: the researcher was Friedrich Wilhelm Herschel.

Of German origins, he moved to England when he was noticeably young. Organist and music teacher, he was also fond of astronomy.

Starting from the AD 1773, he built increasingly powerful telescopes (until a diameter of 1.20 mt) and he became a regularly active observer.

On 13 March 1781, he discovered a strange object in the constellation of Gemini. He thought it was a comet but in reality, he had discovered the gas-giant Uranus.

Herschel is important because after about a century of oblivion, following the observation of the Italian astronomer Francesco Bianchini in 1719, he was able to restart the observations of Mars that had been abandoned and opposed in the AD 1777, 1779, 1781 and 1783 and started a different approach defined as *Location Astronomy*.

In this context, Herschel proposed an alternative to the point of view of Newton and Laplace, highlighting the importance of the direct observation of *celestial bodies* led by reason in the attempt to represent them properly.

His works appeared on Philosophical Transactions (1781–1784).

Among the other sentences we have

> *On the configuration of the polar regions on Mars and its axis, the position of its poles and the spheroidal appearance with some explanations about its real diameter and atmosphere.*

Herschel treatises are the most important before the studies led by some researchers in the 19th century and they were considered significant milestones until the next discoveries in the AD 1879.

From a general point of view, the position of Herschel allocates astronomy the task to study and assign the reality of the sky through the observation instead of inferring the reality from the Law of the Universal Gravitation—in other words following the Principle of Newton-Laplace Theory.

As regards Schiaparelli, here is what he thought of the German scholar:

> *According to Herschel, the task of the astronomer is to rebuild the sky looking at the celestial vault instead of inferring it from the laws of the universal gravitation as it happened at the time of Newton-Laplace.*[13]

Incidentally, it is well known that, at the time, the only way to legitimate astronomy as *scientifically correct* was describing the planets' movement in the solar system, their natural satellites and the (periodic and non-periodic) comets, in relation to the gravitational action of the sun and the other planets.

Outside of this context, the astronomical subjects did not include the physical investigation of a planet's surface (and also its changes in time) and such studies were not considered important by the *Academic Authority*.

In this perspective, Herschel was the first astronomer to make observations in order to study the physical nature the solar system's planets using a rigorous method and, in the interpretation of the data found/available, he used the same means adopted to study the morphology of Earth. in other words, he used the "Earth Sciences" also known as *Geodesy*, *Geophysics* and *Meteorology*.

---

[13] Giovanni Virginio Schiaparelli, *La vita sul pianeta Marte* (Life on Mars), Mimesis Edizioni Milan 2002. Pasquale Tucci, *Fu solo illusione? I Canali di Marte tra il 1877 e il 1910* (Was it just an illusion? The canals of Mars between 1877 and 1910); p. 28.

Obviously, academic astronomers considered this method as inappropriate to their subject's traditional research.[14]

His works appeared on Philosophical Transactions (1781–1784) and among the titles we have:

> *On the remarkable appearances at the polar regions of the planet Mars, the inclination of its axis, the position of its poles and its spheroidal figure; with a few hints relating to its real diameter and atmosphere.*

Herschel treatises are the most important before every study led by the 19th century's researchers and at that time considered as great achievements; in fact, the discoveries of the AD 1879 would have only overcome them.

In line with Herschel's method, between 1785 and 1802 (the years of opposition) Johann Hieronymus Schröter made some observations from an observatory he had personally built in Lilienthal, a small town close to Bremen.

This scholar was the first to use the term *Areography*[15] (from Ares: Mars): the equivalent term for *Geography* (from Gea: Earth).

---

[14] When the Author proposed an article about Martian Planetography for the magazine "L'Astronomie" (official organ of the Association Astronomique de France in Paris), his proposal was rejected because (and it was correct after all) the article was not exactly about Astronomy but about Geography. At that time, the Author was an emeritus member of the above-mentioned association.

[15] From the Greek word Ares = Mars. It is the equivalent of geography, referring to the planet Mars.

3. The Beginning of the Scientific Study

As the scholar Guido Ruggieri suggested, *during the fervour of research, enthusiasm and denigrations, Mars was neglected.* After the planet was brought to light by Tycho Brahe's astronomical research (1546–1601) and by Johannes Kepler (1571–1630), Mars was put aside.

The reason of all this is quite simple: Mars did not offer anything to be observed studying Mars meant wasting time, in the tools of that time it appeared as a red spot apart from the periods of its oppositions.

Only at the beginning and even more at the end of the 19th century, scholars finally had optical instruments powerful enough to allow them to see something on the surface of the planet.[16]

The first relevant telescopic observations of Mars (1830–33) were held at the Astronomical Observatory of Tier Garten in Berlin, where Wilhem Beer, the brother of the composer Giacomo Meyerbeer worked.

These studies, led by Beer in collaboration with J. H. von Madler, were those that in 1840 helped to produce the first topographic map of the moon (1834) and, six years later (1840), the first topographic map of Mars.

However, this last work was definitely inferior to the one about the moon, but it represented the first attempt to put together, in some way, all the information available (at the time) on Mars' morphology.

Certainly, it was already possible to define the meridian of reference (called *Sinus Meridiani*) for the whole *Martian Cartography* (which is still valid).

Schiaparelli remembers the event with these words:[17]

---

[16] Along with the scientific study, some interesting data were found. They helped but not fulfilled the inadequacy of the tools in use. Jonathan Swift in the novel *Gulliver's Travels* published in 1726, told that astronomers of an imaginary "flying island" called Laputa, had discovered and fixed the orbital elements, the two moons of Mars, Phobos and Deimos and the data of Swift, resulted to be extremely similar to the real ones. From his side, the philosopher Voltaire (1750) in his short story *Micromégas* told the adventures of a giant coming from Syrio. He was travelling across the Solar System and when he had landed on Mars, he had noticed it had two moons.

[17] Giovanni Virginio Schiaparelli, *Il pianeta Marte* (The Planet Mars)—Book II—pp.3, 4.

As a faithful member of Aerography we need to consider the German scholar Maedler who in 1830, using an excellent telescope of Fraunhofer (famous optician of Munich who gave the primacy for the building of telescopes to Germany in 1820) he saw and described the spots on the planet better than any astronomer before. Maedler was the first to determine with well-organised measures, the position of a certain number of main points on the surface of Mars compared to the Equator and to the First Meridian that is numbered zero on the annexed chart.
(—) Putting in order the different topographic peculiarity compared to these points, he could make the first aerographic map that even though it is incomplete and limited to a few spots, it is anyway worth of being appreciated because it was made with diligence and as regards the description of Mars and it represents what Erathostenes chart was for Geography of Earth.

In this period, other scholars studied Mars: F. Kaiser in Holland, Sir Joseph Norman Lockyer in England, Dominique François Arengo in France and Father Angelo Secchi of the Vatican Specola in Italy. This last, in particular, was the first scholar to speak about the canals on Mars in 1858. Schiaparelli and Lowell resumed them in 1877.

In 1864, another year of opposition to Mars, the planet was observed by the Rev. W. R. Dawes (who worked in the Observatory of Hopefield in Haddenham—Buckinghamshire) and in 1867 the first map of Mars was published, thanks to the English astronomer Richard Anthony Proctor. The map contains some names of a few *peculiarities* detected on the surface of the planet. Proctor was the first one name placed on Mars (i.e. Proctor Crater, as it happened for the moon) after famous scholars of the planet: Kepler, Tycho Brahe, Madler, Cassini and Herschel.

Thanks to this, many astronomers were compelled to see with their own eyes these new discoveries using a telescope, often of an extremely low quality and definitely inadequate; others, instead, insisted that it was only an optical illusion.

The Mars case came out sensationally in the AD 1877.[18]

Mars was passing at the distance of sixty-four million kilometres from the Earth and at that time telescopes had already substantially improved. Astronomers started to check the sky in a systematic way in the direction of the Red Planet, but it was the Italian Giovanni Virginio Schiaparelli to produce the first universally accepted *toponymy of Mars* (**Table 2**).

To this purpose, the scholar Pasquale Tucci affirmed:

Herschel inspired a research program of "sky building" that could be developed only towards the end of the 19th century. The investigation on the physical structure of the Celestial Bodies was one of the directions on which the Herschel program could

[18] Thanks to the improvements of the telescope's observations, the two satellites of Mars were discovered by the scholar Asaph Hall and called Phobos and Deimos (1877).

develop and it can be proved that it influenced Schiaparelli's research about Mars and the Milky Way.[19]

It is important to remind that Schiaparelli was partially inspired during the first two decades of his activity by the Newton-Laplace method, thus gaining international fame.

None the less, even when he adopted the point of view of Herschel, he did not break the observational tradition of his century given that he continued with his study and he registered the typical data of the *Positional* (or *Location*) Astronomy going ahead after 1877 when his studies on Mars started properly.

Schiaparelli was not the only one.

As regards Mars, he expressed a certain interest for the physical structure of the Red Planet that led him to operate along with the spectroscopy of the stars and the sun as conducted by Angelo Secchi.

It is notorious that the two scholars appreciated each other and this is proved by their continuous correspondence.

However, the methodological hegemony of the Newton-Laplace Method notwithstanding, Schiaparelli was the first one to start the topographic study of a remote planet (Mars) and this represented a certain intellectual bravery we should appreciate.[20]

In this context, Schiaparelli asserted that:

> *It will not surprise that Galileo, whose telescopes never went beyond the amplification of 30 diameters, could have never discovered something on Mars.*[21]

Due to the several phenomena detected on Mars he strongly recommended:

> *what we need more now is the intelligent observation along with measures, even the most irrelevant, with a full description of them free from every possible theoretical concern.*[22]

---

[19] Schiaparelli, G. V. (2002) *La Vita sul Pianeta Marte,* Milano: Mimesis Editions, op. cit. p.30.

[20] F. W. Herschel's son, John F. W. Herschel, encouraged by his father, started his astronomy studies and in one of his works, he asserted the existence of an extra-terrestrial life, assuming that it was probable this was possible on the Moon and above all on Mars where "he found with perfect distinction the borders of what it could be continents and seas" (F.W. Herschel, *Outlines of Astronomy*, New Canaan, London, Readex Microprint, 1985: p. 132).

[21] Schiaparelli, (1893) La vita sul pianeta Marte, chapter II—Abstract from files No 5 and 6—Year II magazine *Natura ed Arte*, February 1893.

[22] G. V. Schiaparelli, *Osservazioni astronomiche e fisiche sull'asse di rotazione e sulla topografia e costituzione del pianeta Marte fatte nella reale Specola di Brera in Milano con l'equatoriale di Merz-Repsold durante l'opposizione del 1890,* "Atti della R. Accademia dei Lincei", series V, 1910 "Memorie della classe di scienze fisiche matematiche e naturali", in *Le Opere di G. V. Schiaparelli*, XI voll., Rome 1929–1943, Tome III p. 473.

Schiaparelli's scientific foresight was found in his clear and clean eye for every detail and for being free from every censorship aimed against his contemporaries' reactionary methodology.

The Red Planet was, by this time, more important of our Natural Satellite.

Regretting because of the bad results of the telescopic observations of the moon, Schiaparelli[23] wrote: *All our hopes are gradually posed on Mars, the only planet that can justify them at some point (—)* and against every odd, what he thought and wrote is still valid today.

On the contrary, the Milanese astronomer engaged in *believing* about *optical illusions*, started the new concept of *geometrical resolution* (something still valid and controversial at the same time) and said:

> *It is possible to detect on Mars a roundish object of 60 or 70 kilometres of diameter and as a fine line a strip 30 km-wide. The course of a river as Po would be easily distinguishable on the moon across most of its length but none of our terrestrial rivers could be visible for us if on Mars.*[24]

As the scholar Pasquale Tucci[25] affirmed:

> *At that time (1877, author's note), due to the lack of an acceptable photographic technique, the result of observing a planet was made public through some particular drawings of its handmade surface, a difficult and tiring procedure. An eye of the observer was glued to the telescope's eye cup (Schiaparelli observed with his left eye) and the other eye led the hand he drew with.*
>
> *The drawing of what had been observed had to be done in a truly brief time because the observation's conditions could rapidly change because of the terrestrial atmosphere's turbulence or because of changes on the planet's surface.*[26]

He could even calculate the measure of latitude's degree in the Martian Cartography: A geographic degree, which on Earth represents approx. 60 miles of 1852 metres each, on Mars, represents exactly 60 km.[27]

---

[23] Director of Brera observatory and considered by the modern science the greatest planetary astronomer of the time (1835–1910).

[24] Schiaparelli confirmed Secchi's fleeting observations: Mars was crisscrossed with a web of long and fine lines that cover almost all its surface. His photographs and drawings appeared on the world press arousing great impression in the scientific environment. in topographical terms, his observations (between 1877 and 1890) completed and developed those by Secchi and Proctor, maintaining 0 Meridian 0 established by Beer and Maedler and assigning to the new detected formations, a series of denominations borrowed from Latin and Greek classicism.

[25] Giovanni Virginio Schiaparelli, La vita sul pianeta Marte, Mimesis Edizioni Milano 2002; op. cit. p. 14.

[26] It is important to say that the first observations, done at the end of the 19th century, allowed the calculation of all orbital parameters of the Planet with an incredible accuracy.

[27] Schiaparelli, La vita sul pianeta Marte, 1893, chap. II—Abstract from files No 5 and 6—Year II of the magazine "Natura ed Arte, February 1893. Note 4.

He also calculated and considered an especially valuable information always neglected by every astronomer and planetographer, which is the circumference of the planet to the extent of about 21.600 km.[28]

Considering the contemporary calculations, the mistake was of about 1,3% compared to the real data, given that the latitude degree, calculated by Schiaparelli in 60 km, today it is believed to be 5924 km and the circumference of the planet, as calculated by Schiaparelli, is actually (approx.) 21.326 km.

Schiaparelli also elaborated the first Toponomy of Mars universally accepted where sixty-two important characteristics were reported.[29]

The clearest areas that Proctor called Lands or Continents were now named by Schiaparelli after real or mythical Earth's countries: Saudi Arabia, Syria, Arcadia, Utopia, et cet. The darkest areas had names linked to the sea such as Mare Tyrrhenum, Aurorae Sinus, Aonis Sinus, Margaritifer Sinus et cet. Schiaparelli's *deanglicised nomenclature* would have aroused a lively parochialism in many British astronomers.[30]

From this moment on, a series of works about Mars were published.

Only 2 years after Schiaparelli's first important observations (occurred in 1877), in Germany appeared the book *Mars, a second Earth* by the Climatologist Prof. J.H. Schmich from Köln.

In 1892, the famous French astronomer Flammarion, published the huge treatise known as *The planet Mars and its living conditions.*

Six years later, he made a Mars globe that at the time was available in libraries along with globes of the Earth.

In 1895, Percival Lowell published a book titled *Mars.*

In the same period, the writer Otto Dross published in Austria (AD 1901) a book *Mars, a world in a struggle for survival*, in line with a theory according to which Martians were engaged in a fight with nature and deter-

[28] Schiaparelli, La vita sul pianeta Marte, 1909, Abstract from files No 1—Year XIX of the magazine "Natura ed Arte".

[29] Schiaparelli's "deanglicized nomenclature" would have aroused a lively parochialism in many British astronomers. See the interesting discussion in K.M.D. Lane "Geographers of Mars: Cartographic Inscription and Exploration Narrative in Late Victorian Rappresentations of the Red Planet", Isis, 96, n. 4 (December 2005), pp. 477-506, especially p. 488. From: Sarah Stewart Johnson, Marte l'ultima frontiera, Sperling & Kupfer Milan 2021, p. 294 (original title: The Sirens of Mars, published in agreement with the author, c/o Baron International, Inc. Armonk, New York, U.S.A. 2021.

[30] See the interesting discussion in K.M.D. Lane "Geographers of Mars: Cartographic Inscription and Exploration Narrative in Late Victorian Representations of the Red Planet", Isis, 96, n. 4 (December 2005), pp. 477–506, especially p. 488. From: Sarah Stewart Johnson, *Marte l'ultima frontiera*, Sperling and Kupfer Milan 2021, p. 294 (original title: *The Sirens of Mars*, published in agreement with the author, c/o Baron International, Inc. Armonk, New York, U.S.A. 2021.

mined to save the minimum living conditions on the planet by using the little water that was still existing there.[31]

Tab. 2	Giovanni Virginio Schiaparelli – Photographic portrait of Paul D. Stewart (1870).

[31] In 1902, M. A. Mercier organised a meeting *about the planet Mars* that took place in Orléans' City Hall in France. During the meeting, it was proposed to find a way to communicate with Mars, but everything came to nought because it was not possible to find an adequate technique to achieve the purpose.

4. The Question of the Canals

Even though many people associate Schiaparelli to the discovery of the *canals*, he never claimed to be their discoverer.

In fact, even before Schiaparelli, Angelo Secchi, Beer and Madler had already drawn some (sort of) *canals* and also the Reverend Dawes already traced some lines (leading to Structures) that eventually could have been understood to be *canals*.

The same can be said for the maps proposed by Sir Norman Lockyer, De Larue of Warren and Proctor, all prior to the observations of 1877.

In relation to Schiaparelli's discoveries, the news seems to be not just the existence of structures on Mars interpreted as *canals* but above all the fact that the Italian astronomer observations had an organic dimension at that point.

In fact, he had succeeded to detect forty of these structures on the Planet's surface and they were clear *outlines*. They were described as highly accurate with very straight lines whose arches were traced with a compass and because of their disposition, *regular;* it was easy to imagine that an Extraterrestrial Civilisation was somehow acting on the planet.

The *canals* were clearly detected again in two more observations, which occurred on 6 March 1886 and on 31 January 1894.

Even though the *oppositions* were not very favourable, during the observation of the AD 1886, that was the less interesting, some observers announced that they saw some more *canals*. Perrotin and Tholon in Nice drew a partial map of Mars showing more than two dozens of canals and, in addition, the position of the canals in those occasions coincided with the positions detected during Schiaparelli's observations.

Other canals were observed in Belgium, England and the USA.

Many hypotheses arose and among them the hypothesis conceived by Percival Lowell (Tab. 15). He was an astronomer and a diplomat who had built his own observatory located in Flagstaff, Arizona (USA) in order to study Mars and in the AD 1900 the scholar was able to take pictures of strange formations and some spots later called *oasis*.

Lowell wanted to speculate that they could be *rivers or artificial canals*, but this theory was contrasted by many scholars like E.E. Barnard in the USA, Antoniadi in France and Vincenzo Cerulli in Italy.

To say the truth, Lowell at the beginning had reluctantly accepted the idea of the canals and of their artificial origin but, subsequently, he changed his mind and he expressed his adherence with these words: *Now, in the case of Mars, we have in front of us a world that is old, much older than the Earth. Its surface clearly proves this and it has an unmistakable sign of age. The time passing left its indelible sign: the continents are now flat and its oceans are now dried.*

Cooperation between Lowell and Schiaparelli started along with a dense correspondence. Lowell became famous in the Anglo-Saxon scope; Schiaparelli was famous in the world thanks to Lowell. This did not affect the fame of neither of them.

On his part, Schiaparelli praised Lowell for his study of Mars, especially when this last discovered some sort of *steam* on the planet[32] and added he was sorry that Lowell's pictures were still unpublished.

Lowell had Schiaparelli in great esteem and he considered him his teacher supporting him when the Brera's scholar showed a *drop* (in a manner of speech) of interest in his work.

In the introduction of one of his books, Lowell compared Schiaparelli to Columbus: *one discovered America, the other Mars*; Schiaparelli, a true gentleman, thanked Lowell for his stunning praise in the letter dated 1 February 1907.

It is remarkable that Lowell did not accept the collaboration of Antoniadi and he clearly stated this in a letter to Schiaparelli dated 16 March 1909.

He affirmed that only Schiaparelli could have accepted to examine his photographs, but he was perplexed because he had already figured out everyone would have been against him.

Percival Lowell studied Mars from his observatory in Flagstaff, indicated the canals with numbers and accepted the nomenclature given by Schiaparelli. The regularity of these *structures* verged on the absurd and it was impossible that such *structures* could have been natural!

This position was different from the one of Schiaparelli; even though the world had always ascribed to this last the role of the American one.

The scholar Desiderius Papp also dealt with the question of the canals establishing a relation between the *spots* and the *canals* and affirmed that: *Nature (—) has not a ruler to trace rivers as straights as the canals of Mars are (—).*

In addition to Papp, other scientists, among them the French Fournier, the British Slipher and the American Edward Pickering (Tab. 17) (this last in California, where he adopted the spectroscopy to proceed on an accurate

---

[32] in 1867, W. Huggiens had confirmed the presence of water steam in the atmosphere of the planet.

chemical and physical speculation about the atmosphere of Mars) they studied the canals taking pictures of the planet.

The proof that, in their time, was considered decisive as regards the existence of the canals, was given in 1939 by the previously mentioned Fournier. He took some clear pictures of huge grooves in the same positions detected by Schiaparelli and Lowell. in addition, other *official documentations* were produced through Slipher‘s surveys.

The last official observation of the Martian canals took place on 3 June 1956, thanks to Gerard Kuiper.

Eight years later, the *Mariner 4* would have broadcast to Earth the first close images (from about 6215 miles) of the Martian surface.

A new chapter about the surface of the Red Planet had just started to be written. Notwithstanding the negative comments that started to flourish at the beginning of the Space Age (1965), the clever Schiaparelli's (and other scholars) intuition about rivers, seas and lakes, either present or dried, on Mars, found new *impetus* with the space missions *Mariner 6* and *7* in 1969.

And it gets better! From the *Mariner 9* (1971–72), until the *Viking 1* and *2* (1976–1982) missions, the pictures demonstrated that the hypotheses proposed by the so-called *canalists* were correct.

From then on, however, the "Academic" Science, unfortunately known as the "Official Science", tried and still does to remove from our memory everything that those *pioneers* had managed to put forward.

The truth is that, in most cases, the canalists were right as some observed structures actually existed and it is important to remember that even Lowell had figured out that some *canals* could have been *rivers*, as eventually it has been discovered, thanks to the artificial satellites observations (started in 1971 and still in progress).

Anyway, about the question of the canals, there were—and there still are, in a way—many different hypotheses.

We believe that it is right to define Schiaparelli's position quoting parts of his works.

Furthermore, since he had foreseen the reactions of his contemporary colleagues, he also affirmed:

> *They are light effects in Mars' atmosphere or optical illusions, produced by steams or ice phenomena (typical) of (the existence of) a perpetual winter on the planet. They also can be double cracks on the planet's surface on which the image is doubled as the result of the smoke emitted on long lines or moved to the sides by the wind.*[33] *I will also add that the work of intelligent beings can explain the geometrical appearance*

[33] Giovanni Schiaparelli, *La vita sul pianeta Marte,* series of works published on the magazine "Natura ed Arte", Milan 1893–1909.

of germinations, but it is not necessarily related to this purpose. The geometry of nature is visible in other facts where the idea of an artificial work is excluded.[34]

In the same period, other scholars dealt with the same question.

In 1886, the astronomer Perrotin from Nice confirmed Schiaparelli's observations and he also was hardly criticised by other researchers.

So far so good, just a debate, some hypotheses, interesting ideas but at the end of the 19th century, some *crusaders* arrived even though, still in 1896–99, the astronomer Vincenzo Cerulli, from his observatory in Teramo (Abruzzo), studied Mars and inferred that its *canals* were the consequence of the low level of the tools used for that purpose and so, in this case, critics were not bad, and, in a different way, they repeated Schiaparelli's assertions.

Other researchers who were less accustomed to the scientific method (and among them we can find Newcomb and Barnard) had never seen Mars with a telescope; obviously, they knew the planet only through the popular spread, normally in opposition to Schiaparelli's rigorous study.

These people decided to go against Schiaparelli's ideas by using a very rude language and without possessing any scientific skill.

One of the slanders on Schiaparelli was that, at the time, he did not see more than the contemporary resources allowed him to see.

This means that those scholars did not understand the *phases of vision*, so they criticised Schiaparelli without actually knowing anything about his investigations.

And, eventually, the catastrophe arrived with Antoniadi, a French astronomer who in Meudon demonstrated that he did not understand neither the *phases of the vision,* nor anything about the power of a telescope.

As a consequence, he erased every detail from the deserts of Mars without taking into consideration, for instance, the *seasonal changes*, by saying that *they were false because Lowell had attributed those (details) to (the work of) Martians.*

This approach was very unfair and even though Lowell would have certainly not chosen such a compromising path, Schiaparelli did not share anything of all this.

To this purpose, Guido Ruggieri wrote about Antoniadi: some think in this way still now and he added: we quoted him only to forget about him forever.

We can and have to add that, unfortunately, this *way of thinking* is still present in most of today's scholars and represents the only obstacle to the correct method of conducting a scientific research in that specific (and "dangerous", in a certain way) field related to *extraterrestrial life*.

[34] Giovanni Schiaparelli, Il pianeta Marte, I book, p. 6.

5. The Rejection of the Scientific Method

As we already know, the study of the planets' characteristics had been supported by Copernicus' theory (16th century) when he extended—as regards *analogical terms*—the laws of physics (as experienced on Earth) and also the physical state of Earth, to the other planets of our solar system. On the experimental plane, this new doctrine was especially supported by the sky's observations, already used by Schiaparelli, near the end of the 20th century.

At the same time, but not necessarily in accord to what we mentioned above, a lot of observational techniques were developed to identify the planets' *right* conditions to host the life (in a "human" sense), such as the temperature, the pressure, the presence of water and, eventually, the existence of an atmosphere.

All these conditions are known as *The Categories of Life*.

These *conditions* can be useful for the progress of our knowledge about life on other planets but, very soon, their disproportioned use ended up in an almost insurmountable obstacle to improve the planets' direct observation as well as the correct comprehension of the images obtained by several *unmanned spacecrafts*.

This circumstance needs to be highlighted because it is something serious that could anyway be inferred by the historical reconstruction of the progresses of scientific research. It is possible to state that the science of the 20th century has reduced the pursuit of life in the solar system only to the search for *bacteria* or *fossils* on Mars, to the point that it is now the only kind of *life* to be looked for on all the planets!

All the above notwithstanding, the terrible opposition of many sceptics did not forbid the debate to continue in a scientific under-cultured environment and, as a proof, many important scholars participated in it from all over the world.[35]

For notable astronomers, the fact of being interested in participating to such a similar debate is not surprising, because Schiaparelli's study was based on a very rigorous scientific way.

---

[35] Steven J. Dick, *Vita nel Cosmo—Esistono gli extraterrestri?* (Life in the Cosmos—Do extra-terrestrials exist?)—Raffaello Cortina Editore, Milan 2002—p. 41.

What was abnormal (and it soon happened) was not to be interested in *a priori* (meaning without any preconceived idea), but only to oppose to the innovative ideas.

A very sterile approach.

In opposition to this antiscientific behaviour, Schiaparelli's position appeared very coherent as he affirmed:

> *notwithstanding the geometrical aspect of the system, we are keen on believing they are products of the planet's evolution in an equivalent manner as the channels existing on Earth.*

However, he left the door of the canals' artificial nature opened by asserting: *I am very careful not to contrast this hypothesis that is not at all impossible.*[36]

Schiaparelli was very attentive, but the situation was complicated by the general position of Lowell to the point of "pushing" Schiaparelli to affirm *if the environmental conditions had made impossible life on Mars, the theory of Lowell about the artificial canals would have been not plausible.*[37]

In a few words, the conditions about the existence of life on Mars were investigated only to contrast Lowell's dissertations.

Anyway, this would have not been a solid argument because the planet's *artificial structures* might have represented a past (and remote) heritage, even though the current environmental conditions of Mars may no longer be suitable to allow the presence of (intelligent) *lifeforms* (of course from a *terrestrial and conventional perspective*)!

The truth is that the adoption of *categories* never worked and works and it cannot produce any achievable and useful result. So therefore, it is important to reaffirm that nowadays, in the field of the now so-called *Exobiology*,[38] it does not exist any track of Schiaparelli's *critical thought*.

In fact, most of the scholars have always thought that Schiaparelli was convinced, beyond every scientific demonstration, that the canals were artificial.

As Guido Ruggieri said, the name of the Schiaparelli's detractors *were wiped out by time, while that of Brera's astronomer towers more and more*[39] and as a skilled researcher he did not exclude anything and in many of his works, called into question that the limitations of his considerations were just the limitations of the tools used.

On the contrary, his adversaries, who were very ignorant, were proud of the tools available to get to conclusions without any scientific background.

---

[36] Giovanni Virginio Schiaparelli—La vita sul pianeta Marte (The Life on Mars)—mentioned in Steven J. Dick, *Vita nel Cosmo* (Life in the Cosmos)—op. cit. p. 45.

[37] Idem, p. 49.

[38] It is the branch of the science that investigates the possibility of an extra-terrestrial life in the Universe. Astrobiology and bio-astronomy are synonyms.

[39] G. Ruggieri, op. cit. p. 121.

Briefly, if it is true that the canals were just "optical illusions" (the fact that Schiaparelli did not exclude, on the contrary he thought it was possible, demonstrating to be absolutely conscious), it was not true that from that simple finding could have been possible to infer that nothing artificial could exist on Mars!

After all, as the researcher Ilaria Arosio affirms,[40] in

no scientific publication Schiaparelli compromises about the origin of the canals, adding that in 1907 in a letter to Vincenzo Cerulli, Schiaparelli declared that the low resolution of the instruments created those optical illusions that the human eye interprets as canals and the same astronomer had clearly stated: We should trust a little bit more in the nature, thank to which sometimes unexpectedly rises a ray of sun to enlighten arguments that before were not accessible (—). We have hope then. And we need to study.[41]

On this occasion, a strange circumstance occurred. It seems not so interesting, but it is full of "historical" meanings in a written file of 1908. A certain Baroni quoted the AD 1971 as a *year* of one of the *oppositions* to Mars[42] and he wrote:

Last year, in these days, Mars was (—) in a very favourable position to be observed (—) being about sixty-two million kilometres while in more favourable circumstances can approach us until 55 million as it happened in 1924, 1971, etc.

The remarkable fact is that in one of these dates, when the technical evolution allowed it, (AD 1971), the first space mission gave crucial details about the true nature of Mars as well as about its surface[43] offering a valid proof as regards the intuitions of Schiaparelli (**Table 3**).

It was in place (and still it is) a persistent misunderstanding when it comes to connect a detail regarding Mars (yesterday the canals of Schiaparelli, today every possible artificial structure on Mars) to a possible "final solution" (or "conclusive"), as regards the presence of intelligent life on that planet.

It is easy to notice that it is a badly asked question, just instrumental and not at all scientific and this is so true because, as it happened in the past, Schiaparelli's canals were denied only to demonstrate that nothing artificial was on Mars, so now, in the same way, many more useless battles have been started, as if a certain element represents the key to solve the controversy about intelligent life on Mars.

---

[40] Astrophysics-INAF-Astronomical Observatory of Brera; from the website "Scienzainrete".

[41] Giovanni Virginio Schiaparelli, *Il pianeta Marte*, 1893.

[42] The oppositions are the various occasions in which two celestial bodies, the Earth and Mars, are at the minimum astronomical distances.

[43] The first discovery of an artificial structure on Mars from the B cam of the module Mariner 9 and on 23 December 1971 on Pavonis Mons.

As a result of these misunderstandings, both tragic and comical that originated in the first half of the 20th century, Antoniadi (Observatory of Meudon in France) and Maunder (Royal Observatory of Greenwich in England), even today we can see many other scientists and *institutional detractors* who were and still are only good at denying the existence of anything artificial!

However, excluding the *canals* does not mean excluding the (past or present) existence of some sort of intelligent life, as well as today to give a *proof* demonstrating that a structure on Mars is NOT artificial, in fact, does not demonstrate anything.

Antoniadi was considered by the contemporary astronomers and by the contemporary historians as the one who contributed more to the solution of Mars' canals.

Of course, all this does not correspond to the truth because, in reality, this astronomer did not solve anything; he did not understand that the structure of Mars was not uniform or deprived of structures as he had imagined after deleting every track of what was drawn by Schiaparelli, *because Lowell had attributed to the extra-terrestrials (Martians).*

Even Antoniadi, during his observations in 1909 with his telescope of 84 cm of Meudon, observed 36 *canals* in *stable conditions of good visibility* and he demonstrated that they corresponded to something "real" such as spread bands or dark areas edges, but not true *canals*.

Maunder also played his part, saying that in December 1909 he could be happy that scientists and the public *were no more obliged to feel obsessed to the idea of incredible engineers at work on Mars* and just to understand the level of their stupidity (today as like yesterday) he added that:

> *they could have slept peacefully in their beds without being afraid of a Martian invasion, even after the trend spread thanks to the vivid descriptions of H.G. Wells.*[44]

From his side, Antoniadi, according to *the mechanism of the blasphemy*, in one of his books, included the updated version of the Martian topography and he confirmed that only thanks to great telescopes it was eventually possible to understand the illusory nature of the "canals" even though they correspond to something real, such as stripes and small, scattered spots whose origins were unknown.

They imagined that the spots and the structures corresponded to something real, but they became optical illusions because everyone was afraid to say they were artificial!

It is assumed that after the brilliant Schiaparelli's intuitions, the shabby attacks from reactionaries whose head was Antoniadi, made possible what is considered the end of the scientific studies as regards the planet Mars, cir-

---

[44] Maunder Edward Walter—cit. in Steven J. Dick, *Vita nel Cosmo*—op. cit. p. 57.

cumstance that is known as *passage from the observational to the physical studies* and this happened "finally" during the 1924 opposition! (Table 3).

Mars was better seen and many enthusiastic scholars attracted by the question of the *canals* (that actually was a problem of the detractors rather than the supporters) said that we had passed *from the Aerography to the Aerophysics* and this notion was subsequently reaffirmed by the astronomer Gérard de Vaucouleurs, who in 1954 explained the new age in the field of the Martian studies.

Tab. 3	Italian commemorative stamp of Giovanni Virginio Schiaparelli on the 100th anniversary of his death – 1910 – 2010.

6. Observative and Physical Studies

After so many attacks by astronomers against the direct study of the planets' surfaces, the so desired "confirmation" arrived, but it would have just revealed an illusion as were the so mistreated canals!

Concisely, the physics study proved that Mars conditions were extreme on its surface and so there was no possibility of life on Mars *without specifying what form of life*.

In 1928, the *New York Magazine*, published an account of a meeting which subtitle was: *Some eminent astronomers explain why they are sure there is life on the fabulous planet Mars* underlining the differences and the contrasts at the time of Lowell to the point that in the article was argued the *traditional theory that Mars is a dead planet* and commented that the idea of life on Mars was accepted by scientists who a few years before were against this possibility.

On his side, W.P. Pickering (surprisingly!) declared to be "almost certain" that a form of intelligent life was present on Mars. On the contrary, G.C. Abbott, the director of the Smithsonian Astrophysics Laboratory, asserted that from his analyses, due to the actual physical conditions of Mars, only some kind of very primitive species of Flora could have grown up there. Since then on, this last assumption became an untouchable *dogma*!

Another group of scholars who represented the majority of the astronomers were there. We mention Henry Norris Russell from Princeton, Harlow Shapley from Harvard, William Coblenz from the National Bureau of Standards and Walter S. Adams from Mount Wilson, E.B. Frost from Yerkes and Earl C. Slipher from the Lowell Observatory.

They conveyed on the existence of vegetation and, in addition, they admitted the possibility of an animal form of life.

But these truly weak admissions notwithstanding, the attack against Mars' observations eventually arrived.

It started in 1930, when the first failures as regards the research and the presence of oxygen on Mars were registered.

Since then onward, the presence of life on the planet was considered *unreliable* (i.e. unacceptable, in fact) and since 1940 the probability got even lower, especially after the founding of a massive amount of carbon dioxide

(CO2) in the atmosphere of Mars and the acknowledgement of a possible *global warming* (two words that we now ear even too often as regards Earth).

From 1960 onward, the catastrophe was unavoidable. The surface temperature, (WRONGLY!) estimated to be close to 480°C, so therefore remarkably similar to the one of Venus, convinced many scientists and researchers to assume that every form of life on Mars was absolutely impossible.

At this point, it could be interesting to investigate about the level of knowledge of Mars at the time of the observations in 1924 and 1926.

It was a regression if we think about Schiaparelli's period and it was assumed that the temperature was under the point of water freezing, that the planet was surrounded by a thin atmospheric *stratum* of unknown nature and (this is the worst part!) that its surface was "variable and enigmatic". It was the moment to build "the mystery of Mars" still believed today!

The transition from the visual observations, such as the direct observations, of Mars to the use of Astrophysical Techniques—i.e., the chemical investigations of physical studies—started a new era (already ancient though) about the study of the Martian Biology. in this field, the search for intelligent life on Mars was neglected and gave way to the research of vegetation on the planet. This became the only possible investigation and because of this the contribution of many valuable scholars became extremely limited.

The consequences of this behaviour were the constant presence of physics studies (or even astrophysics) that reached its summit when giving way to a perfect and perverse correspondence between the *indicators* and the *real conditions* of Mars!

Briefly, this approach meant that there was no interest in looking for present life on Mars but for the past one considering some logical and biological premises previously established.

The physical studies were at that point a real regression in terms of studies about Mars' surface and the first negative consequence was the application of "categories" to the detriment of the visual observation!

The divulging article of 1925 titled "Climate conditions on Mars" is an example. in this article, Coblenz raised the possibility that

> *elevated temperatures on the surface observed in correspondence with the dark areas of Mars could be explained assuming that the alive vegetation could be overlapped on a stratum made of soil rich of organic material of vegetable origins that is a bad heat conductor.*[45]

[45] Coblenz William Weber—cit. in Steven J. Dick, *Vita nel Cosmo*, op. cit. p. 65.

If we think that it was accepted that there was no water on Mars,[46] as Coblenz affirmed

> *the process of rotting and decaying of vegetables should be very slow and this should provoke a slow accumulation of dried material from the previous season able to protect the living plants from the extreme winter temperature.*[47]

This was the consecration of Antoniadi destructive work!

However, it is remarkable that even though the *physics method* was in contrast with the *visual method*, the astrophysics supporters used to their advantage the visual observations and eventually they mocked those who made hypotheses about the possibility of an intelligent life (in addition to a vegetable one) on Mars!

In a few words, all data collected were used by the physicists' courses to reinforce the hypothesis of the vegetation that considered above all the elevated temperature values and also the clues regarding the existence of small quantities of oxygen and steam. But the reactionary attitude continued.

The search for the presence of oxygen on Mars revealed to be a perfect way to dampen the enthusiasm still on till the Twenties by proposing two very dubious data. The first was the level of oxygen sufficient to make possible the existence of intelligent human beings or at least a developed animal life, the other regarded the very childish usage of the oxygen level "found" to establish that at most it was only reasonable to speak about the existence of some kind of vegetation.

All this was the outcome of the research of oxygen on Mars made by the director of Mount Wilson Observatory, Walter A. Adams, who, at that time, cooperated with Theodore Dunham (1933).

Instruments involved were new photographic emulsions, a grating spectrograph and the telescope of 250 cm of Mount Wilson.

During their investigations, Adams and Dunham could not find traces of oxygen on Mars but notwithstanding this particular circumstance they felt the urgency to establish (of course!) a major limit for its presence that was less than 1/10 of 1% of the oxygen on Earth! in this way—as said before—it was possible to obtain discouraging data as regards the possibility of life on Mars, but it was not possible to exclude the existence of Martian vegetation.

In this context, it was compulsory to mention the red colour of Mars! Place it was still unknown (at that time but the doubt still survives) that the planet had to be observed before being studied, in Russell's opinion the oxygen could have been present on Mars in *free form* and later it got *fixed* and

---

[46] Nobody, among the supporters of physics have never noticed that the existence of vegetation, along with evidence of lakes, was proved in 1999 thank to Mars Global Surveyor module.

[47] Idem.

trapped in the rocks as iron compounds). in particular, on Mars it should have been occurred and *oxidation* in general!

Furthermore, just to attempt everything possible and—of course—at the maximum level of ignorance as regards the synthesis of colours perceived by observing a *celestial body* through a telescope, Russell added that this situation *would have explained the rusty colour of Mars*. Dunham conclusion was this: *It would be reckless to assert that steam and oxygen could be insufficient to allow any form of life that could adapt to the extreme conditions on Mars.*[48]

The perverse plan continued so that Russell wrote *the incredible lack of oxygen suggests that a form of animal life is improbable*[49].

However, in 1947, Gerard Kuiper (1905–1973), discovered carbon dioxide in Mars' atmosphere and since this gas has a significant role in the photosynthesis, that data stood as a truly relevant biological discovery.

[48] Dunhan Theodore—cit. in Steven J. Dick, *Vita nel Cosmo*—op. cit. p. 70.

[49] Russell Henry Norris—mentioned in Steven J. Dick, *Vita nel Cosmo*—op. cit. p. 71.

7. The Space Age

As the new, favourable oppositions of Mars were approaching in 1954 and 1956, the astronomer William Sinton from Harvard tried again to verify the existence of vegetation on the planet using physics tools and especially the spectroscopic method. The basic parameter was represented by the measurement of absorption bands of organic molecules in the infrared part of the *spectrum*.

The result was of about 3, 4 microns and beyond the wavelength interval previously measured by Kuiper, when it was comprised between 1 and 2.5 microns.

Sinton inferred that there was a high probability that only a spectrum of organic origins could explain data.[50]

Notwithstanding the scientific opening by Sinton and the low value on the general question on Mars, as the analysis of the original *structures* by visual means excluded their *artificial nature*, there was always someone who was strongly against the possibility of life on the planet.

For this reason, some thought that Sinton was just wasting his time.

In the Fifties, V.M. Slipher accurately led the research and it happened exactly what had happened a century before with Schiaparelli's observations.

He believed to have discovered oxygen and (*water vapour*) steam in Mars' atmosphere. He also noticed that his results had been contaminated by the terrestrial atmosphere especially by heavy water.

Among so many valid scholars, there were some of them who took their distance by the *scientific method* as the researcher Gérard de Vaucouleurs, who in his *physic of the planet Mars* (1954) barely mentioned the question of life because *we are convinced this sort of problem can be mostly beyond the positive knowledge's limitation and could be only a question of vague speculations where you are obliged to use as a guide general "principles" of metaphysic nature.*[51]

Some good news came from Kuiper:

---

[50] Sinton Williams—mentioned in Steven J. Dick, *Vita nel Cosmo*—op. cit. p. 73.

[51] Gerard de Vaucouleurs—mentioned in Steven J. Dick, *Vita nel Cosmo*—op. cit. p. 76.

If we could demonstrate that life exists only in two planets of the solar system, the only ones that can host it, we would be tempted to come eventually to the conclusion that life could spontaneously develop where conditions are favourable. As it is likely that there are many planetary systems, life (on the planet Earth) would not be an exception in the universe.[52]

At the beginning of the space age, the controversy about Mars' canals was sorted and it was possible to list several elements as regards the physical conditions of the so-called "Red Planet".

In this context, only some hard data about the possibility of vegetation on Mars should have permitted to demonstrate the validity of the scientific research regarding the planet and it would have had a vital role to transform Mars in a *key target* in the solar system's exploration program.

Obviously, this last *assumption* could have not the same value and importance of a possible intelligent civilisation. However, for academic researchers, the vegetation was still the best chance to cling to without the risk of being (too) close to Schiaparelli's hypotheses.

Let us put it in an uncomplicated way: in order to be appreciated as a scholar, it was compulsory at that time to dissent from Schiaparelli!

It was written that the space age made travels to Mars possible and doing so it changed the traditional approach to search for life in the solar system and it eventually brought scientists to convey that life on the planet's surface had disappeared forever.[53]

This was, of course, a wrong, hasty conclusion because it is like asserting that life could depend by the presence of the canals!

The results of explorations gave, instead—but only to those who could read between the lines—the opposite result: it confirmed Schiaparelli's brilliant intuitions even though it had never been declared officially.

Schiaparelli, in fact, did not exclude anything and his hypotheses were the most advanced available at his time.

If Schiaparelli would have been born 50 years later, he would have used the same methodology but with better instruments and more reliable data. On the contrary, epigones and disciples of his detractors did not make any effort to go farther notwithstanding they could.

---

52 Kuiper Gerard Peter—cit. in Steven J. Dick, *Vita nel Cosmo*—op. cit. p. 76. in this context, the discovery of Humans on Mars would appear in contrast with theories of evolution accepted in *toto* by academics because if humans would be born on Earth by evolution and if the same would be found on Mars the possible solutions would be two. First, Martian Humans would be part of the terrestrials who had been there in the remote past or Martian Humans are the Martian ones and they would belong to a cosmic and wider humanity present on both the Earth and Mars.

53 - Steven J. Dick, *Vita nel Cosmo*—op. cit. p. 77.

For this reason, the right, obliged choice to use space probes to explore Mars closer—as Schiaparelli had hoped on several occasions—did not exclude the debate the aim of which was to simply deny its scientific validity because the scepticism of Schiaparelli's time is still around and now possibly more active than ever!

In order to strengthen this trend, the space age did not mean to reject what already achieved or to diminish the importance of the studies from Earth. This was a big mistake because the new methodology nullifies in *toto* non-observational studies.

In other words, the direct observation of the planet made redundant or complementary the indirect physics studies (i.e., Spectrographic Analysis) in comparison with observational direct studies (i.e., Morphological Analysis).

Those who still wanted to go ahead in the traditional way were those who wanted to benefit from the public research fundings!

Even if not entirely, Science (better yet, scientists) recognised the importance of the space exploration. in this way, a few years before the beginning of the space age, the committee for the space science from the National Academy of Science of Iowa University, during the summer season in 1962, expressed the following opinion:

> *It is since Darwin's time and before him, Copernicus, that the science could not have the occasion of making a significant impact on humankind understanding. The scientific question of Exobiology is the most exciting, difficult and profound problem not only of this century but of the whole natural science's development in the last three hundred years. There is a possibility of a new perspective about the place of man in nature, to reach a new level regarding the meaning and the nature of life.*[54]

In the *media circus* of astronomy, it often happens that some inexperienced scholars without enough knowledge came out. An example is Abelson, such as the curator of the scientific magazine *Science*.

In 1965, he asserted that if we look for life on Mars, we run the risk of being reputed the silliest people of all times.[55]

On this purpose, it is important to say that sceptics are skilled in convincing others they are right to the point that, still, in 1965 even Sinton was convinced that the *absorption bands* were not due to the presence of vegetation on Mars but rather to the presence of *deuterated water* in the terrestrial atmosphere.

After the acquisition of Mariner 4 poor results, so having at hand only information of low value and technically useless, sceptics came out again advising against Mars' exploration.

However, the Committee for Space Science, in an appendix of the already mentioned biology and exploration of Mars, did not alter the recom-

---

[54] Steven J. Dick, *Vita nel Cosmo*—op. cit. p. 77.

[55] Abelson Harold—mentioned in Steven J. Dick, *Vita nel Cosmo*—op. cit. p. 84.

mendations of the account and NASA proceeded. Putting aside the comments about Mariner 4's probe and missions Mariner 6 and 7 **(Table 4)** we can now speak about Mariner 9.

We can read:

> *a milestone, being the first space probe to orbit around another planet and retransmitting to Earth high-resolution images (till 37 metres—author's note) renovating these questions but, at the same time, burying forever Percival Lowell's Mars.*[56]

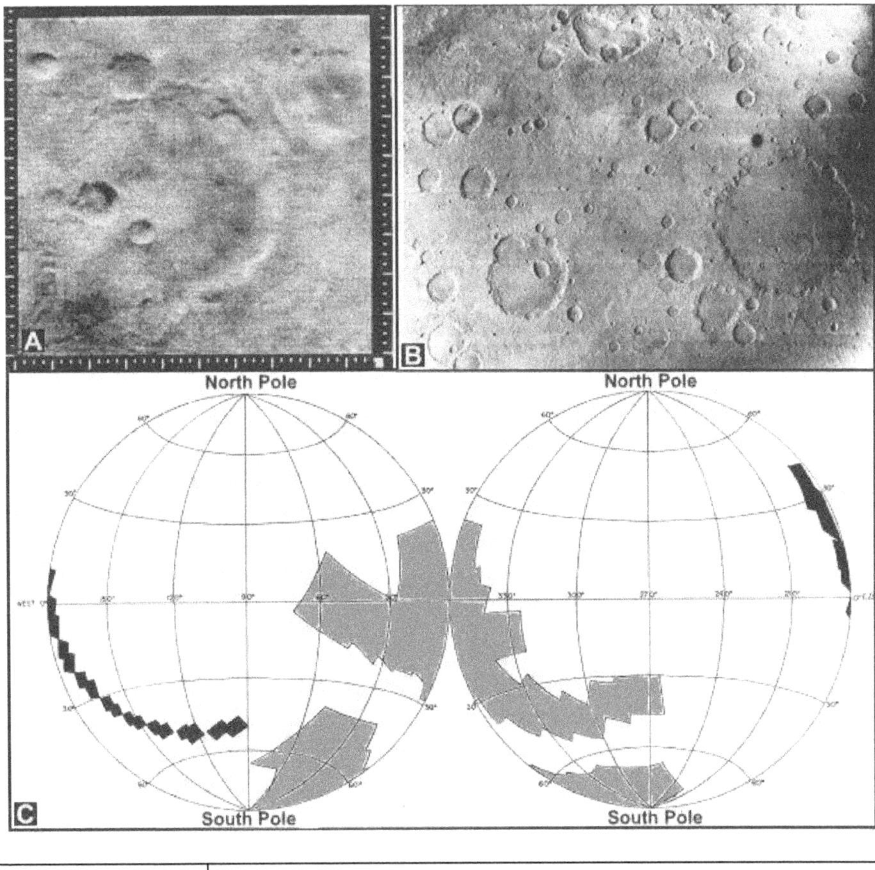

Tab. 4 Satellite shots: A) Mariner 4 (1965); B) Mariner 6 (1969); C) The parts in black indicate the surface portions surveyed by the Mariner 4 mission, the parts in grey those surveyed by the Mariner 6 and 7 missions.

8. The Hydrographic Structures

Three centuries after the first telescopic observations, mankind attempted to go to Mars. in the sixties, after the building of some probes able to make experiments, Mars was the main target along with, but never inferior, to the moon.

The first attempt to reach Mars occurred in the USSR in 1960 when two probes were sent there on 10 and 14 October from the Space Base of Baikonur (Байконур—Kazakhstan).

The launchings failed because the *rocket* carrying the probes could not even reach the terrestrial orbit.

Other failures followed on 24 October and 4 November, 1962, but finally the Mission *Mars 1*—still in the AD 1962—was (almost) successful.

The aim of *Mars 1* was to pass close to Mars in order to take hundreds of pictures of it and this happened.

In 1971, *Mars 2* and *Mars 3* (from USSR) were the first objects of the *contemporary* man to touch the Red Planet.[57]

The American mission that first reached Mars was the *Mariner 4* that was launched on 28 November 1964, from Cape Kennedy and reached Mars on the night of 15 July 1965, taking pictures from a distance between 17.000 and 12.000 km.

The only aim of Mariner 4 was taking those pictures. Shootings were done thanks to a camera of 305 mm of focal open at one-eighth and an exposition time of 1/50 of a second and all this coordinated with a broadcasting system.

The probe took twenty-one photographs and the fragment of a 22nd photo; among them 14 or 15 show the surface of the Martian hemisphere illuminated by the sun.

Unfortunately, no useful information could be obtained by the pictures of the night side.

They were images of an extremely low resolution including 1 or 2% of the whole surface of Mars in 26 minutes recording and storing them on a

---

[56] Steven J. Dick, *Vita nel Cosmo*—op. cit. p. 86.

[57] in 1965 the USSR sent two probes towards Mars, *Zond 2 and Zond 3*. The first one was a failure; the second failed the target but took beautiful pictures of the Moon instead.

magnetic tape. The probe detected three hundred craters and 10% of them were provided with a central peak.

It was the first attempt to take photos of Mars and the first to be successful asserts the writer Arthur E. Smith and then added: (Mariner 4) revolutionises the idea of the planetologists about the planet.

The photographs showed a landscape rich in craters and, in some way, pretty similar to the one of the moon.

Albeit it was a meaningless mission from the point of view of the pictures' quality, an array of sceptics scientists, whose main aim was to deny *a priori* even the evidence, considered these photographs related to the whole planet and so they concluded that the entire surface of Mars was similar to the moon and deprived of life or traces of it.

An article of the *New York Times*' first page said so: *The possibility that on Mars there is or there was life suffered a heavy or even a fatal blow* and somewhere else it was written that the hypothesis of a Civilisation's existence (current or ancient) on Mars was not possible.

However, these flights broke the veil of mystery that wrapped the Red Planet, the only relevant data that emerged appeared to be the demonstration of the inexistence of structures that at the time had been indicated as *canals.*

This result appeared to satisfy above all those scientists who since some time were waiting an element that allowed them to bluntly affirm that Mars, being an inhospitable planet, could not in any way to show traces of any past or present intelligent life.

Many scholars produced some hurried conclusions as regards that first space mission towards Mars; conclusions denied by the subsequent discoveries.

In these cases (and many others similar to them), instead of being cautious (a brilliant method used by Schiaparelli), scientists who had to analyse the images from the Mariner 4 mission reached negative conclusions.

Briefly:

1. Mars is totally covered by impact craters;
2. The dark regions do not show any structural difference when compared to the light regions:
3. It does not exist on the planet an orography similar to the terrestrial one and this means that orographic cycles (such as, for instance, the rise of mountains) never happened on Mars;
4. There is no trace of branched valleys and so no erosion.[58]

For those who saw the whole matter in a very superficial way, this could have been a data to be used against those who considered Mars a planet that host or have hosted intelligent life, but now we will see how different things are.

[58] G. Ruggieri, op. cit., p. 195.

After this stance, it took 5 more years to verify these assertions. The missions Mariner 6 and 7 detected on Mars many details similar to some terrestrial regions. This was the beginning of a new trend that saw the sceptics' theses as deprived of any scientific value.

On 24 and 27 March 1969, USA launched the *Mariner 6* and *Mariner 7* probes that started the research.

These two spacecrafts were heavier and more powerful than the Mariner 4 and they carried two cameras each one with wide-angle lens (camera A), equipped with a focal lens of 52 mm and one with telescopic optic (camera B), with a focal of 508 mm.

Mariner 6 reached the neighbourhood of Mars on 31 July 1969.

On this probe, two cameras put on a mobile tower started working while the probe approached the planet. Three series of images, seventy-two photos in total, were shot while approaching Mars, thirty-two while flying over it and others during the mission for a total of 428 images.

The optic resolution of these pictures was definitely superior to the other photos sent to Earth by the previous probes: in fact, 20% of the planet's surface was photographed with a surface resolution of 2 km and, in addition, a limited strip with a resolution of one hundred metres.

The resolution given by Mariner 4 had been multiplied 10 times and that of the terrestrial telescopes 1000 times! It was explored in detail about 10% of the planet.

Mariner 7 was close to Mars on the 5 August 1969. It photographed the same areas explored by the *Mariner 6* but with a better optical resolution thanks to the temperature of the camera that was 6 degrees higher. It took ninety-three photographs by a great distance that appeared spotted, thirty-three pictures when it was close to Mars—all of them in the southern hemisphere—and 17 in the northern hemisphere.

In addition, the probe took fifty-three photos of Mars approaching, therefore the images resulted to be 749 in all. The probe passed 3.425 kilometres from the planet.

The two missions Mariner 6 and 7 (and all the others so far), disproved all the catastrophic conclusions inviting the scientific community to commit itself to learn how to *read* and *correctly interpret* the satellite images.

The space mission that gave us a complete picture of Mars was the one of Mariner 9. It obtained excellent results, especially in relation to the photographs, even though the first images were blurred because of a sandstorm. Photographic tools were operated 14.004 times with a collection of 7329 images (five thousand real photographs and 2329 in the shape of a television shooting).

The probe photographed 85% of the Martian surface with a resolution of only a few hundred metres ("Camera A") and a small part of the surface with a resolution of 37 metres ("Camera B").

Cameras showed some craters that permitted to find the Austral Mountains already filmed by Mariner 7 and the attention of the scholars was attracted by the cone and shield volcanoes located in the Plain of Tharsis that were distinguishable from the others thanks to their gigantic dimensions (i.e.: between 7 and 25 km in height).

The probe *Mariner 9* gave incredible images included in thirty million bits of information and permitted to draw the first topographic detailed map of the planet and its satellites.[59]

The examination of the pictures taken by this probe revealed the presence on Mars of sinuous valleys that follow the sloping of the ground, they contain some tributaries and show the characteristics of riverbeds now dried, rich in alluvial deposits and containing drop-shaped islets oriented towards the sloping.

From these detections, it was very easy to formulate another hypothesis: if on Mars the canals were dug by flowing waters, these waters should have flown in an age when the Martian environment could not be so different from the current one and this environment should have been much more fit to allow the birth and the proliferation of Vital Forms based on those biochemical principals so familiar on our planet.

Even though the *official science* did not notice the question if the planet was dead or not, it was sorted after the transmission from *Mariner 9* in 1971–72, a number of photographs that permitted also to compile the first detailed cartography of Mars.

As regards the question of the canals, it was attempted to find in the pictures taken by the unmanned terrestrial satellites some elements that could be attributable to the alleged structures as they have been *seen* and described since last century.

The first reaction—that regards most of the scientists—is that in the images collected from the first *Mariner* it was impossible to see anything that could remember Lowell's canals.

However, not every scholar agreed with these conclusions. Clyde William Tombough declared that *those who achieved conclusions should need eye examination* because in photograph number 11 of Mariner 4, a crater and two faults seem to coincide with a truly short canal and Lowell's oasis. Moreover, in the photographs number 2 and 7, some dark spaces occupy oasis zones.

Already in 1964, the photographs taken from the American probe *Mariner 4* demonstrated that some alignments were not simple optical illusions as some detractors of the canals' theory asserted. in August 1967, an official American report regarding the results obtained by the previously mentioned mission, described long grooves comprised between 160 and 320 km.

---

[59] Thanks to the data from Mariner 9 was compiled the first general Atlas of Mars ("Atlas of Mars" cfr. Bibliography).

The question of the canals would have been finally clarified through the data acquired by mission *Mariner 9*.

After five years, the jigsaws composed with images from this mission were compared with the canalist cartography of E.C. Slipher of 1962, assumed to be the best and the apex of several attempts to map Mars through the use of the telescope. The difference between the two documents revealed to be very subtle.

In conclusion, *sometimes canalists were right.* in addition, as Ducrocq wrote, the canal they had called *Agathodaemon* is the colossal complex of canyons now known as *Valles Marineris Canyon System.*

The Canal *Styx* is identifiable with the mountains ranges known as *Phlegra Montes*; the Canal *Ceraunius* is the great *Ceraunius Fossae.*

Ducrocq points out that *Antoniadi had prepared a perfect map in which it was signalled a canal called Cerberus* (now known as *Cerberus Fossae*)*: a strip of grounds with a very weak reflexive power, possibly because of the accumulation of dust. in a different way, we can assert that, in some cases, the traces observed by canalists were founded on something solid.*

After the exploration of *Mariner 9*, examining the photos taken from about 2000 kilometres of altitude, the geologist Harold Masursky, declared: *if I could see something like that on Earth, I would say without any doubt that it derives from the action of waters* and the same position will maintain subsequently when he will become director of the Project *Viking* of 1975–1976 (**Table 5**).

Then, referring to the thousands of canals that wind on Mars, he asserted on another occasion: this interweaving of lines is, I believe, the final proof that massive erosion was created in old times by water.

The new images, revealing more details on Mars surface, created a new mystery, especially because they showed that the surface of the planet is actually scattered with canals of every shape and dimension. At that time, it was said that *canals of a different nature had been discovered!*

From mission Mariner 9, it emerged the existence of a series of canals of fluvial origin, thus confirming and clarifying the images transmitted by the previous missions *Mariner 6* and *7*.

David Pieri writes on The Planetary Report: Mariner 9 discovered these canals were real and natural. Thousands of them were of different shapes and sizes and they formed a system of deep, winding cuts with vast craters of dozens of kilometres persistent among them.[60]

A very essential point to keep in mind is the structure of these canals that reproduces the morphology produced by the terrestrial watercourses. Pieri added: *The dried riverbeds were one of those things that most astonished the*

---

[60] An area that aroused an immediate interest for its characteristics was the region of Mare Erythraeum that at the start was studied by Harold Masursky and D.T. Milton who finally accepted to be in front of a fluvial structure.

researchers of mission Mariner 9 and they are not normal rivers' beds. The Martian riverbeds are longer than the longest rivers on Earth.

In relation to the mission Mariner 9 and only after two years after the acquisition of the *essential data* coming from the cameras of the satellite, in 1973, prof. Henry Faul of Pennsylvania University presented the interesting possibility that an *ocean* could have existed on Mars but scholars Jeffrev S. Kargel and Robert G. Strom[61] commented like this: *due to the shortage of observations at the time, his article was never accepted for publishing.*

This is an assertion without any foundation because the probe *Mariner 9* was able to take pictures of a significant percentage of the Red Planet!

The same scholars, later, specified that *last decade other researchers, working on data of Viking missions* (of course based on data thought to be more significant of those by *Mariner 9* and this too is not totally acceptable—Author's note), *got back Faul's idea* and carried on confirming that *For instance, in 1989, Timothy J. Parker and his colleagues from the Jet Propulsion Laboratory of Pasadena in California hypothesised again the existence of a western ocean (as many characteristics of the northern plains seem to be the product of coastal erosion phenomena).*

In order to have more possibilities to publish, they camouflaged deliberately the revolutionary extent of their work with an awkward title such as "*Traditional morphology in the western part of the region called Deuteronilus Mensae of Mars: implications because of the modifications of the border between lowlands and highlands*". Partially motivated by this job, Baker (researcher for NASA), with the collaboration of many colleagues, denominated this hypothetical *Borealis Ocean.*

The same theses of Prof. Faul, but in bolder terms, had been formulated fourteen years before, above all in an environment where they had fully accepted, even at an official level from Prof. Josip Sklovskij, Soviet writer and academic (USSR Scientific Academy). Very active at the Astronomical Observatories of Moscow and Leningrad, the scholar, in 1959, proposed, through the pages of *Komsomol'skaja Pravda*, the thesis that millions of years ago on Mars should exist oceans and massive quantities of oxygen and, in the same article, he mentioned the possible *inhabitants of Mars* asserting that:

> *When oxygen and water were gradually needed, the inhabitants had been pushed to look at/for other places (—) and he added: two or three billions of years ago, the situation was very different: many scholars believe that a long time ago there was oxygen on Mars with vast marine extensions. It is probable that at that time there were intelligent beings who achieved a high grade of civilisation.*
>
> *I do not want to dwell on expressions of these civilisations nor about what could have happened after. What is sure, in a certain phase of their development, is that*

---

[61] Cfr. "Cambiamenti climatici globali su Marte" (*Global climate changes on Mars*), in "*Le Scienze*", n 341, January 1997.

the inhabitants of the planet abandoned its surface, and, on this purpose, it has to be observed that the cosmic flight is easier from Mars rather than from Earth because the Martian gravitational field of attraction is definitely inferior to the Earth's one.

The aspect of Mars' (dried) rivers is different from those dried rivers that we call uadi (that is rivers and streams of medium and large dimensions, dried or not, where flowed or flow—this depends on the observation period—quick watercourses with a non-perennial character and I defined them cyclic rivers and streams) that are usually found in the terrestrial deserts (i.e.: Sahara and others).

And all this not mentioning the titanic dimensions of the *Valles Marineris Canyon System,* in front of which every other feature of the Earth would shrink.[62]

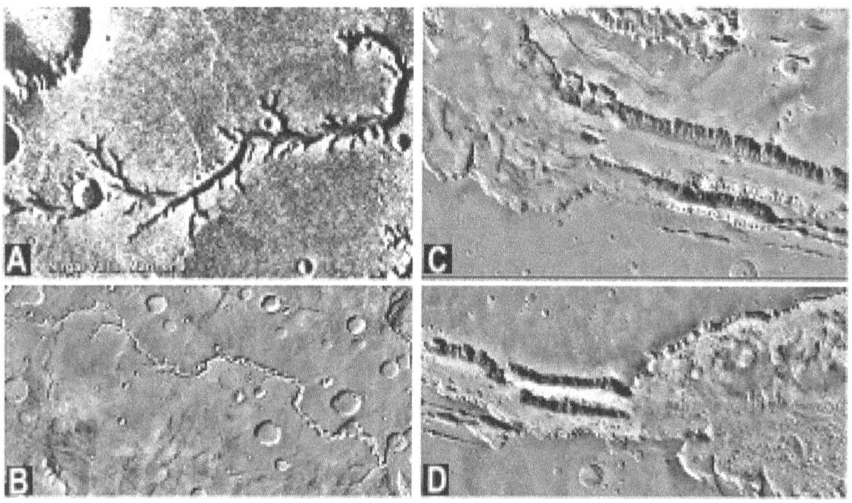

Tab. 5	Nirgal Vallis – A) satellite shot from Mariner 9; B) satellite shot from Viking 1- CD) Valles Marineris, portion of the Coprates Chasma (Viking 1).

[62] It is worth remembering what Gerard R. Kuiper and Barbara M. Middlehurst wrote in 'Planets and Satellites': 'At the same time, the development of planetary astronomy has also stimulated the enhancement of geophysics. The deepening of knowledge about one of the planets of the solar system has thrown new light on the problems common to all its

9. The “In Situ” Experimentation

In occasion of the mission called “Curiosity“ that reached Mars on 6 August 2012, John Grotzinger (head of the scientists of the California Institute of Technology) affirmed: *the probe detected a place that in a far past would have been liveable. This does not mean that in that place or in another place on planet Mars there could be life but that some physical crucial conditions would have permitted the appearance of it.*

Charles Bonden, head administrator of NASA, says:

> *The knowledge we hope to obtain from the observation of the crater Gale (descent place of the probe n.d.c.) it will give us a lot of information on tracks of life on Mars. Curiosity will bring advantages also on Earth and will inspire a new generation of scientists and explorers to prepare the way for a human mission to Mars in a future that is not far.*

It is easy to observe that in this case (as in many similar others), we are in front of sentences already said for more than half a century without any intention to change something.

The aim of the missions so-called *explorative* is different from what it should have been, because all the missions referred to Mars had always shown negative results, but all the collected elements had always suggested a positive solution.

It is evident that all those that lead investigations show an EXTREMELY negative attitude towards the idea of life on other planets.

As regards the official comments, it has been said that

> *sediments that form the elevation, for the scientists are a possible proof the crater Gale maybe was full of water and of its past existence there are the proofs that in the stones disposition that denote old streams of what some alluvial sediments are visible.*

As in many other occasions, this time too it has been said that *the probe Curiosity had allowed the possibility to find tracks of water courses now disappeared* and with the usual emphasis asserts that in *this sense is the best probe.*

---

other members'. Cited in V. N. Žarkov, Internal Structure of the Earth and Planets, Editori Riuniti, Edizioni Mir, Rome 1986, p. 267.

We wonder "in what sense" the "new" results had been partly detected by the probe Mars Pathfinder in 1997!

The researchers point out that water was present on Mars for thousands or even millions of years, even though they also think about the necessity to analyse all the available data in order to have a clearer idea of the age when water flowed on the Martian surface.

The scientists appointed to the analysis of the Mars Exploration Rover (MER) Curiosity's data announced that on the surface of Mars, today so barren, in ancient times water was flowing deep and with fast currents. It is believed that the presence of water gives hope to find tracks of ancient life, even microscopic *Surely, water flowing is a clue for microorganisms to have lived* while the emerging rock *could track some elements that are normally associated to an environment where life is possible* of course according to Grotzinger.

In the landing site, a strange stone has been observed in the shape of a pyramid large as a football field. Here it is what Edward Stoper referred one of the most important scientific experts in the mission:

> *Once again, the Red Planet surprises who studies it. The first rock analysed on Mars from the MER Curiosity has an unusual composition in comparison with stones so far known on the Red Planet.*

This assertion is quite important because referring to the planet Mars we have different forms of rocks (detected by the two Viking Descent Modules), so it is difficult to be surprised in front of such a phenomenon! This is the relevant declaration: *This kind of rocks is very similar in its chemical composition to the kind of stone that can be found in many volcanic regions but in this phase with only a Martian specimen of this kind it is difficult to understand if we are in front of a product created through the same processes known on Earth*, even though researchers always find something that prevents them to carry on and thus create a sort of demoralisation.

However, it is specified that scientists will analyse the chemical composition through the spectrometer mounted on the robot's arm that is part of a miniature chemical lab. These kind of devices can spray on soil and rocks to analyse even atmosphere and the mission of two years will focus on a mountain of layered rocks, Mount Sharp, which arises in the centre of the crater.

In particular, one of the ten instruments on board of the Rover Curiosity, is called SAM (Sample Analysis at Mars Instrument) and it is a device that looks for organic molecules, such as carbon compounds, which are associated to life as we know it on Earth and it comprehends three instruments: a mass spectrometer equipment, a chromatograph and a laser spec.

After extracting a sample from the soil of Mars, the researchers assert that the probe Curiosity could have discovered compounds without which life could not develop. This is the idea of Charles Elachi, director of JPL

(NASA). And he adds: *these are still preliminary data to verify* but he specifies that Rover could have discovered *organic molecules,* non-biological.

In order to make a comment on such a declaration, we have even a less balanced opinion: *If confirmed, this discovery will re-open the debate about a possible life on Mars* that means calling into question the brilliant results obtained in 1976 by the two Landers of the Viking mission! Grotzinger spoke about a sensational, even shocking, discovery and declares: *This discovery will be written in history books!*

The journalist there (and it does not matter who he/she is—) said, with a discouraged tone: *This is what they had said until data will be confirmed by new analyses. We hope it will happen very soon!* The announcement is fixed on 3 December in San Francisco, at the American Geophysical Union.

Surprisingly, three days before the fixed data, on 30 November 2012, at 16.01, NASA in its office in New York (the news was confirmed by ANSA) denies the discovery of organic molecules on Mars. The press release is the following:

Great delusion as regards the considerable expectations on the possibility of organic molecules on Mars. After the rumours and the declarations of the previous days on the possible discovery of organic molecules on the Red Planet, the NASA JPL reduces the importance of the results obtained by the lab robot Curiosity. One of the sentences textually referred by the agency note is the following: The voices and the hypotheses on a possible great discovery are not correct. in addition, we inform that life (literally speaking) was not found; we found, though, some evident tracks of molecules containing carbon that are the base element of life itself.

In his conference to the University "La Sapienza" in Rome, Elachi precises: Unfortunately, I had no occasion to speak to the researchers in these last days in which they are processing some information by Curiosity. He also added that in the last weeks, rumours abounded that there was a particularly important discovery but media exaggerated in the expectations. Preliminary data suggest that we have identified simple molecules, not biological (this opinion had been already expressed from the same person before the comment from 30.1.2012!).

From his side, the head of the scientists, Grotzinger, in some of his declarations, had in some way anticipated the *denial* that is a proof that the question on the overall plan took place according to a pre-ordered plan.

Something else even odder happened. "Curiosity"—said the director of JPL (denying what already declared by Grotzinger)—*is not equipped with instruments able to find biological tracks* that is molecules necessary to or produced by forms of life, *but it is able to recognise organic molecules—*

Basically, the probe had been built to recognise/identify only those *things* (let's say *elements*) that were planned to be found (just that!) and so it becomes obvious that the circumstances according to which nothing *im-*

portant was discovered did not depend on the absence of objective data (possibly detected from the Rover), but because the Rover could have not been able to *see and recognise* them in any case!

So confusing—But it is important to notice that the same motivations and justifications had been produced (even in similar occasions) to nullify and deviate the suspects (that are now proofs) that on Mars, the Descent Modules of the two Viking probes (operating since 1976) would have effectively verified the existence of life.

However, *nothing new under the Sun* as the poet Horatio would have said.

This false news was also backed-up by the media. On 30 November 2012, Raiuno announced the denial at 07.55.

The person who announced this was a researcher from the Planetary of Rome. *We still do not know anything* referring to that *they* know about Mars. The previous day on the Italian television channel known as "Raitre", during the broadcast of *Geoscienza*, a guest had shown a similar scenario. When he talked of Mars, we had the impression that he was facing a *paranormal problem* because he pronounced the word *mystery*—used by the sceptics as their last resort.

It is worth to remember at this point that, in this sector, Science had never been really insightful. We want to quote the incredible statement that appeared on 10 July 1997 (in occasion of the Mission Pathfinder) on a page of "Televideo RAI" according to which (Mars has) *a vastly different landscape from the one that is uniform found by the missions from the Seventies.* We wonder where the authors of these assertions found the material to write something like that.

In reality, the landscape shot by the probes of the Seventies, the Mariner 7 and 9 and the two Viking—in our case the two Descent Modules "Viking Lander 1" and "Viking Lander 2"—is far from being uniform and the topographic maps built on the base of those missions, are a solid proof of this.

In order to prove the presence of life on Mars, in 1976, the first biological experiments were performed using the probes Viking (**Table 6**) launched in August-September 1975 that reached Mars the following summer.

The two landers arrived on Mars' surface to run a series of experiments and analyse some soil's samples.

This experiment called "Liberation" gave radioactive carbon that could be released by the breath of living organisms. in every test, the samples produced measurable quantities of radioactive Carbon Oxide (CO): a robot arm collected soil samples putting them in a container along with a nutritive solution previously marked with radioactive carbon. If in the soil there was a living organism, it would have assimilated and processed the nutritive elements marked by radioactive carbon that would have been eventually released, giving us carbon in a gaseous state.

The experiment was checked through a detector of radiations installed on the lid of the container and linked to it through a pipe in which every gas could be eventually released and could flow.

The experiment gave correct results so that the researchers—Patricia Straat and Gilbert Levin—found proof of gas releasing.

Dr Klem, commenting on the reaction of Martian soil to the so-called *chicken broth* asserts that it is a biological phenomenon and this indicates that the microbial life is more developed there than on Earth.

However, notwithstanding the achievement of such an impressive result, other scientists suggested that the releasing could have been better explained as the product of chemical reactions of highly reactive components like *peroxides* and *superoxides*.

As the position of the science at that time was narrow-minded, the two researchers responsible for the experiments control could not prove that those gases were released by living organisms and they had to abandon the hypothesis they had supported before.

In order to not leave any stone unturned, other researchers investigated on the possibility of alternative solutions, but no solution was found that could avoid the involvement of living organisms.

Albeit several solutions were proposed, none of them was judged satisfactory. Levin's conclusions as regards the experiments[63] showed that:

> *They were all in favour of the presence of life. Since then, I have always tried to make those people understand the value of their discoveries, but it was in vain (—) All the microorganisms on Earth that metabolise organic substances release carbon dioxide. I thought they had' taken a sample from the Martian soil and they had strayed it with water and other nutrient substances containing radioactive carbon.*
> *In this case, the potential living would have nourished themselves with these substances so they had released carbon dioxide which molecules would have been composed also of introduced radioactive carbon, which is extremely easy to be detected. The test was tested on Earth to verify the validity and then performed on Mars using both the Viking probes in an automatic way. The experiment gave equivalent results: carbon dioxide was released as if it had been produced by a breathing, living organism.*

Michael Carr also dealt with this thing. He was a geologist at the control centre of NASA at the time of the Vikings.

When he checked the experiment data, he said:

> *Oh my God, I said that there is life on Mars. The answer was out of doubt. in the presence of nutrients, there was a massive presence of radioactive carbon dioxide that slowly diminished.*

---

[63] Panorama (Italian magazine—Milan 2001).

At this point, everything suggested that on Mars, there was life. But the situation worsened when another experiment on board of one of the two *Vikings*, whose aim was to find organic molecules on Mars' soil, could not find any. To this purpose Levin said:

> *It was not possible to find any organic molecule on Mars' soil only because the detecting system at that time was not so advanced to highlight Martian microorganisms.*

Other scientists dealt with the same problem such as Joseph Miller (professor at the Neurobiology Department of "Keck School of Medicine" at the University of South California—San Diego), fully supported by NASA. He was the second researcher after Levin to strongly support the hypothesis of life on Mars.[64]

What shocked him was a detail that 28 years ago, nobody would have noticed. in fact, during a new research from NASA regarding a future exploration of Mars, the scientist has recently analysed for the second time the graphics of Levin's experiment discovering that the chemical activity found on Mars' soil can correctly being explained only admitting the presence of living cells and, in Miller's opinion, all this is a proof that Mars can host life. He affirmed this on the 26th AGM of the American Society of Optical Engineering in 2011.

In particular, Miller asserted:

> *The sign of gas emission not only had a circadian rhythm but also a precise rhythm of 24,66 hours and this is particularly significant because it has the same duration of a Martian day (or "Sol"). I noticed that the release of carbon dioxide from the Martian soil was not constant, but it showed an increase by day and a decrease by night.*

The rhythms perfectly match with the Martian day and as Miller asserted, it is a *typical rhythm of a biological activity*.

Even in this case, chemists started to investigate to find an answer that did not involve the presence of living organisms, but Miller dismantled every possible proposed hypothesis pointing out: *Levin was right 25 years ago. Now, NASA should retry the experiment on board of a new probe to do the tests of the Viking once again, but they are reluctant.*

As regards the doubts of 25 years ago that were raised by chemists who supported the idea that the same signal could come from chemical reactions with some very responsive (but) nonorganic compounds on the soil, Miller asserts this scenario is impossible to figure out:

> *First of all, there are some researches showing that the superoxides exposed to a watering solution like the nutritive solution used by the Vikings, are rapidly destroyed. On the contrary, the circadian rhythms showed by the Martian soil persisted for nine weeks and there is not any explanation that a normal chemical reaction could be so strongly synchronised to so many minimal temperature fluctuations.*

---

[64] This same scholar prepared the sleep-wake schedule for the Shuttle's astronauts in the Eighties.

If we join the recent images of the probes that strongly indicate the presence of water on Mars even in the recent past, many conditions to allow the presence of life are there. I think that the Viking researchers in 1976 had good reasons to believe that had discovered life there.
Now, with this new discovery, I would say that life there is present with 90% probability. in addition, many biologists agree with me. On the base of the latest information I received from one of the biological experiments on Mars performed by the Vikings, I can state that the percentage that there could be the life on Mars is 90%.

Another researcher is Carol Stoker, biologist for the "Ames Research Centre" of NASA. She asserts that analysing the images received by probe Pathfinder that landed on Mars in 1997, there are some tracks of *chlorophyll* on the surface in at least two areas close to the probe. This time too it was opted to think about a *possible error*, but Stoker analysed those places in 15 wavelengths and in those two areas it is observable the wavelength obtained when the chlorophyll absorbs the sunlight.

In a few words, notwithstanding the overt validity of the biological experiments of the Vikings and all the positive news coming from Mars in the last 40 years, if an operator of a research centre or an astronomer or a teacher were asked to give their opinion about the presence of life on Mars (Intelligent life we mean!) they would say that it is not possible.

On the contrary, they would inform you that the probes proved the absence of life (while it is exactly the contrary) and because of such a statement, nobody ever produced any proof to support what was once affirmed.

Nobody could because in this sense there is no proof.

However, this *unpleasant* situation will end one day. In a near or a very distant future, we cannot say right now.

But we do know that it will.

In both life and Science and we know that, it is often just a matter of time. And since there is always time and a specific moment for everything, we just have to keep studying and waiting.

And, one day, the truth shall set us free.

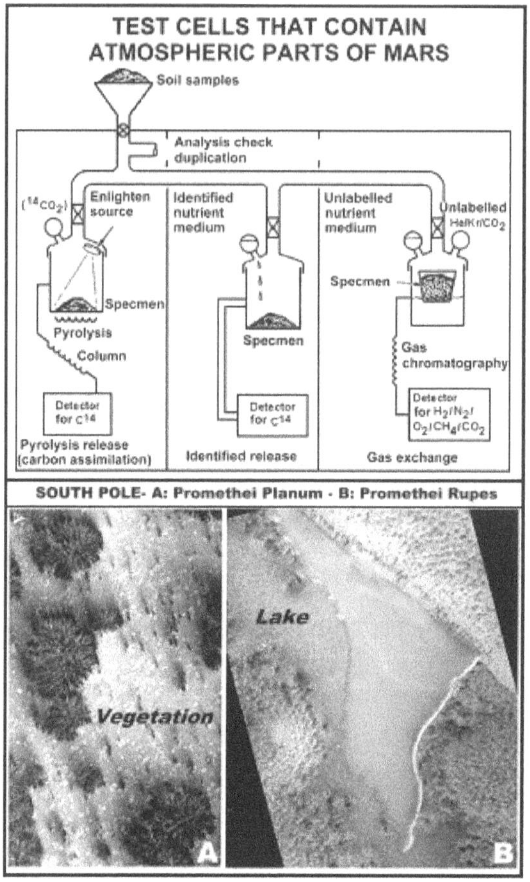

Tab. 6 Biological experiments on the surface of Mars (1976); A) Vegetation structure in the Promethei Planum region – Code M0804688 – 234.38°W – 82.02°S – Dimensions km 2.83x20.46 (minimum structure: m 286 x 332; maximum structure: m 1. 001 x 1,031); B) Lacustrine structure in the region of Promethei Rupes – Code M09-01354 – 292.28°W – 78.75°S; Dimensions km 1.41x7.25 (maximum length: m 1,976; maximum width: m 916.3; central width: m 693; central length: m 1,314.8).

10. The Scientific Intuition

Till Schiaparelli's time, astronomers had tried to get Mars closer to Earth but, of course, given the time, by staying on our planet.

This means that it was like holding a cable and lassoing the Red Planet to try a better observation. This was done with telescopes and this method could have obtained only approximate results because Mars is anyway a small planet and even the minimum distance from Earth is never inferior to fifty-five million kilometres.

All this remains a big obstacle for all those who want to use the optical instruments for a systematic study (that is *unpretentious observation*) of that tremulous red spot in the sky!

It was well known that the only way to investigate the cosmos was represented by instruments (and Schiaparelli was conscious of their almost absolute limitations, at least because they were used from the Earth's surface and so they were *victims* of the *anxieties—turbulences*—of our atmosphere).

He was saddened when he foresaw that all Mars' mysteries could be sorted only when it would have been possible to take measures in relation to the photographs of the Red Planet. It seems incredible, but Schiaparelli referred to future tight photographs, even though he was not able to refer to images that would have been shot by instruments way closer to the Planet. It is necessary to admit that the work done by the astronomers of the last century was a fantastic job, a massive commitment done using all means they could.

Those pioneers of the *Descriptive Astronomy* obtained controversial results and often very contradictory but of course, it was not their fault. Even at that time, studies were included in the denomination "Planetography".

Subsequently, those studies were abandoned and the remnant, now unrecognisable, was wrongly indicated as "planetology".

The first results referred to the investigations of Mars, were obtained in the last century thanks to great astronomers, pioneers both of modern science and of the Red Planet's study.

We should remember that, at that time, the Polar Caps of Mars had been already discovered, so the possible presence of ice (not necessarily composed by water), had been observed on what was after denominated "Mount Olympus", the tallest mountain of Mars and of the whole solar system *(that Schiaparelli baptised "Nix Olimpica" thinking the clouds were ice given that this extinct volcano rises to the height of about 26 km from the Datum or "Zero Altitude" of Mars (like any other celestial body that for Earth is given by the Sea Level) and about 22 km above the Plain of Tharsis, that is about 4 km higher than the Datum).*

These data caused heavy debates and confused reactions—especially of a conservative kind—where only a few scientists realised that the right thing to do was to wait for further checks.

The second operative moment was possible thanks to the observations done by instruments placed in Mars' orbit. The space flights took away the monopoly of Mars' observations from Classical Astronomy and the Descriptive Astronomy (done without the indispensable help of satellite photos). It cannot be forgotten that the space mission was made possible thanks to the elaboration of the massive amount of data collected by previous astronomers.

It seems obvious to imagine the existence of a methodological continuity between the studies made by researchers through the astronomical observations from the end of the 19th century (and the beginning of the 20th) and the studies that astronomers did (or would do) using sought after instruments on board of space shuttles in the Red Planet's orbit.

We need to say that it seems so because this *continuity*, in fact, did not happen. in a few words, the astronomers who dealt with Mars in the last century brought the research till the possible farther results.

This did not mean to abandon a scientific logic but (as it was later proved) an opening to a solution of problems always considering both data and the level of research available at the time.

Actually, nowadays, astronomers (those who still are linked to a methodology now useless due to the new observations' technologies) *simply renounced to accept the new reality arisen after the space era began.*

The researchers' attitude at Schiaparelli's time was to go against the hypothesis of the *artificial nature* of the observed canals. It did not start from scientific considerations (because science imposes to wait for better times to study a phenomenon with improved instruments); they *opposed* the hypothesis because, as it happens also today, they thought every idea that could (simply) suggest the existence of an intelligent life on Mars was unacceptable.

In this way, it appears that the "sceptic" position of Maunder, Antoniadi and Maggini (just to mention the most notable but not the most famous names) originated from the biased attitude that, until today, prevents other researchers to take into consideration hypotheses inadequate to the preconceived positions of absolute denial as regards the existence of intelligent life on Mars or elsewhere in the universe. In such a way and in a different field, an exception was the previously mentioned philosopher Ernest Naville[65]

[65] Jules-Ernest Naville, son of François-Louis, (Chancy, Geneva 1816—Vernier, Geneva 1909), evangelical, reformed pastor, professor of history of philosophy in Geneva. The idea that the hypothesis in the scientific research is central and omnipresent crossed "like a

(**Table 7**) a French-Swiss, not *by accident* a contemporary of Schiaparelli and today almost unknown.

Tab. 7	Jules Ernest Naville in 1868 (1816-1909).

He distinguished among *plausible hypotheses*, so entirely possible and *non-plausible* hypotheses.

---

simple glimmer" the spirit of Naville in 1844 while he was studying Bacon's philosophy. in 1874 he came back on the subject during a series of public talks where he spoke about philosophy and specific problems as regards the hypothesis' logic.

In 1880, he published in Paris, in the *Librairie Germer Baillière et C.*, "La logique de l'hypothèse", from which translation in Italian ("La logica dell'ipotesi", trad. by Anna Fabriziani, Rusconi Editor Milano 1989, with an introduction by Dario Antiseri) are extracted some references present in the present work. Among the philosophical Works we should also mention: *La définition de la philosophie (1894)*; *Les philosophies négatives (1900)*; *Les systèmes de philosophie ou les philosophies affirmatives (1909)*.

Of the two photos here reproduced (Tav. 2) the one put on the left was done in Geneva in the studio of the photographer Gaussen. An added manual note at the back indicates "Ernest Naville, 1868" when the philosopher was 52 years old. The image on the right is a portrait made by Elie Lecoultre in 1848 (when Navlle was 32 years old and the picture here reproduced was taken in the photographic studio Phototype Sdac and it was inserted in "Ernest. Naville, sa vie et sa pensée", written by Hélène Naville printed in 1913.

At the same way of Schiaparelli, Naville knew that the line between the two categories passed through *the perfectionism of the instruments* that he considered *one of the science's factors*[66] and he added that *the telescope was necessary to see the phases of Venus*—which was very important to verify Copernicus' theory[67]—concluding that *the impossibility of verifying in experimental science has then a provisional character related to the resources that science uses in a certain moment.*[68]

Facing the problem of the extraterrestrial life and considering the limitations of contemporary instruments of that time, Naville[69] demonstrated an open-minded attitude almost unknown to many of his contemporaries when questions like these were asked: *Are there other worlds beyond Earth inhabited by intelligent creatures?*

A very efficient analogy makes us think that in the multitude of planets that spin around uncountable suns, there are some that have the same conditions of life we have on Earth where life appeared and at the summit of life, the spiritual existence, but we have no means to make this hypothesis a thesis scientifically demonstrated.

Are there any intelligent creatures superior to man?

This should seem highly probable by analogy, starting from the consideration of the being's hierarchy. in any case, it is possible to admit it; this possibility is sufficient to exclude the affirmation that mankind is at the centre of universe. This is an affirmation that is deprived of prudence and humbleness, but it is impossible to see how it can be possible to find in the scientific field a probability that can become a fact.

(—) The hypotheses that cannot be verified have no right to access science, but this rule needs to be applied with prudence. It is a difficult limit, impossible to be traced in an absolute manner because the hypotheses that are not verifiable in a certain date could be so later (—).

Schiaparelli realised he needed extremely good optical instruments to study Mars.

It is not possible to accede to such an attractive study without an optical help proportioned to the difficulty of the thing[70] and the same concept resulted reiterated with passion in the first of his seven memories on Mars where he denounced the necessity of using means of a greater optical power such as *the gigantic telescopes of nations abroad* to obtain valid results.[71] (—)

---

[66] *Ernest Naville, La logica dell'ipotesi, translated by Anna Fabriziani, Rusconi Editore Milano 1989, with an extensive introduction by Dario Antiseri, p. 214.*

[67] Idem.

[68] Ibidem, p. 215.

[69] Ibidem, p. 194.

[70] Schiaparelli, La Vita sul pianeta Marte, 1893, cap II—Abstract from the files n. 5 and 6—Year II of the magazine *Natura e Arte,* February 1893.

[71] Raffaella Simili, *Giovanni Schiaparelli astronomo e uomo di scienza*, in: *"Giovanni Schiaparelli: storico della Astronomia e uomo di cultura"* (astronomer and man of science—

moreover in *America and in France it has been planned the construction of new telescopes of an incredible power never seen before that will cost millions, among all the time's signs* (note: a religious concept), *this will give us at least the right to hope for the better in the future.*[72]

This position is clearly expressed in the prevision of Schiaparelli on the future study of Mars as part of his last letter written to Lowell on 23 April 1910 where they knew that only more powerful instruments could have permitted a better knowledge of Mars.

> *The study of these enigmas has just started, there is still nothing sure about some principles on which a rational interpretation of them can be based on. Everything will depend on the future progress of the photographical representation of Mars. The question will do a great progress when it will be possible to take precise measures on photographs.*[73]

At the end of his life, with so much honesty and disappointment, emerged that he could not stand all the slanders he had to face.

In the letter of 23 April 1910, he remembered that, in the last times, scepticism enhanced about the configurations of Mars and he expressed the suffering of not being able to be sure of reality with his observations. *He thought that every confutation will finish when there will be available photographs of Mars so big to end all the uncertainties caused by the photographic granulation. (—) He ends proclaiming his unchanged interest to the great progress that the astronomy performs every day.*[74] **(Tables 8, 9).**

Currently and according to certain positions that are far away from the intuitions of Schiaparelli, the hypothesised existence of life on Mars should be related to forms extremely far in *time* (at least 3 billions of years ago) and in conceptual *space* (bacteria instead of intelligent beings, in line with the never abandoned physical study).

In the end, everything matches perfectly with the aim to "pull away" in both philosophical and physical terms, the idea that Mars could have hosted (and/or hosts) an intelligent life. The effort is to demonstrate this thesis and the language is on purpose (or constrictively) bombastic.

In this way, the scientists are close to demonstrate that Mars hosted life and still the results of this study could be a perfect point to start from to investigate the possibilities of present life on Mars, but without any certainty.

---

historian of Astronomy and scholar), edited by Antonio Panainoe and Guido Pellegrini, Editore Mimesis, Milan 1999—p.28.

[72]Schiaparelli, La vita sul pianeta Marte, 1893, Chapter II—Abstract from the files n. 5 and 6—Year II of the magazine *Natura e Arte*, February 1893.

[73]Schiaparelli, La vita sul pianeta Marte, 1909, Abstract from the file n. 1—Year XIX of the magazine *Natura e Arte*, December 1909.

[74] "*Il carteggio Lowell—Schiaparelli*" (The correspondence Lowell—Schiaparelli)—by Alessandro Manaro and Franca Chlistovsky—Acts of the XXI National Meeting of history and physics and astronomy. Department of Physics, University of Calabria, Arcavacata di Rende (CS)—6, 7 and 8 June 2001. p. 14.

In conclusion, the researchers "aligned" noticed the fossil lakes, dried rivers and the bottoms of oceans now without any water but they are unable to identify any "non-natural" structures already seen and decoded in relation to the findings done by several space missions (from 1971 to present). However, we know that today the resolution of the probes active on Mars is far superior to those used in the past. For this purpose, it is worth to quote prof. Jan-Peter Muller of the UCL (University College London), responsible for the 3D map of Mars' surface: *Now we can model the 3D shape of Mars to a less than 1 metre resolution like every good commercial satellite that orbits around Earth. in this way, we can verify our hypothesis in a more precise way than in the past.*

Why cannot scientists see anything in these readings?

The organic formula of Mars' reality, according to the scientific terms, implies that laws of physics are valid on Earth and elsewhere (in the universe) and this means that in presence of the same *technical* conditions, a satellite reading on Mars has the same objectivity and photographic success of a reading and a picture on Earth. On this purpose, Margherita Hack wrote:

> *Thinking that our planet is unique it is like to think that Earth is the centre of universe and that everything spins around it adding physics laws we know on earth are valid for the whole observable universe.*[75]

The researcher Viola Schiaffonati reminds us that starting from the telescopic observations of Galilei we "discovered" *the process that brings to the definitive unification of the terrestrial physics and the celestial one and to the overcoming of the Aristotelic vision*; all this contributes to prove that on Earth and other planets the same laws are applicable and it is a *nonsense* to support the Aristotelic division between the lunar world and the sublunary world adopted for such a long time.[76] The astronomer Erol Torun (recently passed away) asserts that

> *all the geophysics observations done on Mars till present, its gravity, meteorology, geomorphology, et cet., indicate that Mars is a place where Physics Laws and Geomorphology as we conceive them, find application with some variations due to gravity, atmospherical density and volume. It is nonsense to think there is only a small place on Mars' surface where the same principles are not observed.*[77]

---

[75] Pippo Battaglia and Walter Ferreri—*C'è vita nell'Universo?* (Is there life in the universe?)—Foreword by M. Hack—Ed. Lindau Turin 2008—p. 7.

[76] Viola Schiaffonati, "La rivoluzione scientifica del XVII *secolo*" in "Temi filosofici dell'informatica" a.a. 2010–11 (The scientific revolution of the eighteenth century—on "Philosophical themes of computer science")
http://home.dei.polimi.it/schiaffo/TFI/rivoluzione%20scientifica%20appunti.pdf.

[77] Erol Torun, "The Geomorphology and Geometry of the D&M Pyramid", unpublished, available in Compuserve ISSUES forum, section 10, file's name PYRAMI.RSH. Abstract by: Graham Hancock, Robert Bauval, John Grigsby, *L'Enigma di Marte* (The Enigma of Mars) Trad. by Lucia Corradini—TEA Tascabili degli Editori Associati—Milan 1998; p. 135.

This assertion supports our position. However, it has to be noticed that all literature in this field does not protect us from those who deal with Mars and in some cases do not consider all the data acquired from the explorative missions. This incongruity forbids, after more than half a century from the beginning of the space age, a correct and impartial scientific understanding of the "true" reality of the Red Planet.[78]

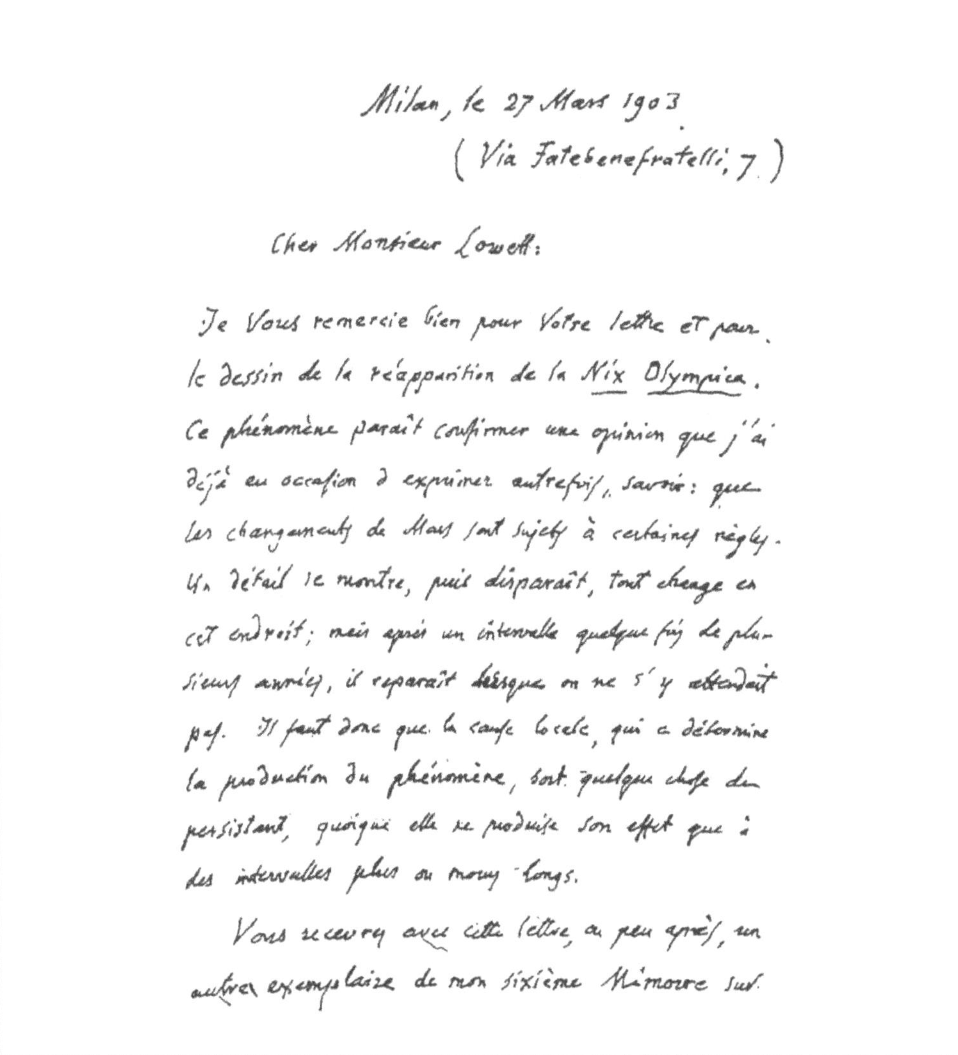

Milan, le 27 Mars 1903.
(Via Fatebenefratelli, 7)

Cher Monsieur Lowell:

Je Vous remercie bien pour Votre lettre et pour le dessin de la réapparition de la Nix Olympica. Ce phénomène paraît confirmer une opinion que j'ai déjà eu occasion d'exprimer autrefois, savoir: que les changements de Mars sont sujets à certaines règles. Un détail se montre, puis disparaît, tout change en cet endroit; mais après un intervalle quelque fois de plusieurs années, il reparaît lorsque on ne s'y attendait pas. Il faut donc que la cause locale, qui a déterminé la production du phénomène, soit quelque chose de persistant, quoique elle ne produise son effet que à des intervalles plus ou moins longs.

Vous recevrez avec cette lettre, ou peu après, un autre exemplaire de mon sixième Mémoire sur

Tab. 8	Letter written by Schiaparelli to Lowell (27 March 1903).

[78] It is worth to remember the words of Gerard R. Kuiper and Barbara M. Middlehurst in "Pianeti e satelliti" (Planets and satellites): "At the same time the development of the planetary astronomy stimulated the geophysics of astronomy. A further study of the knowledge on one of the planets of the solar system shed new light on the problems."

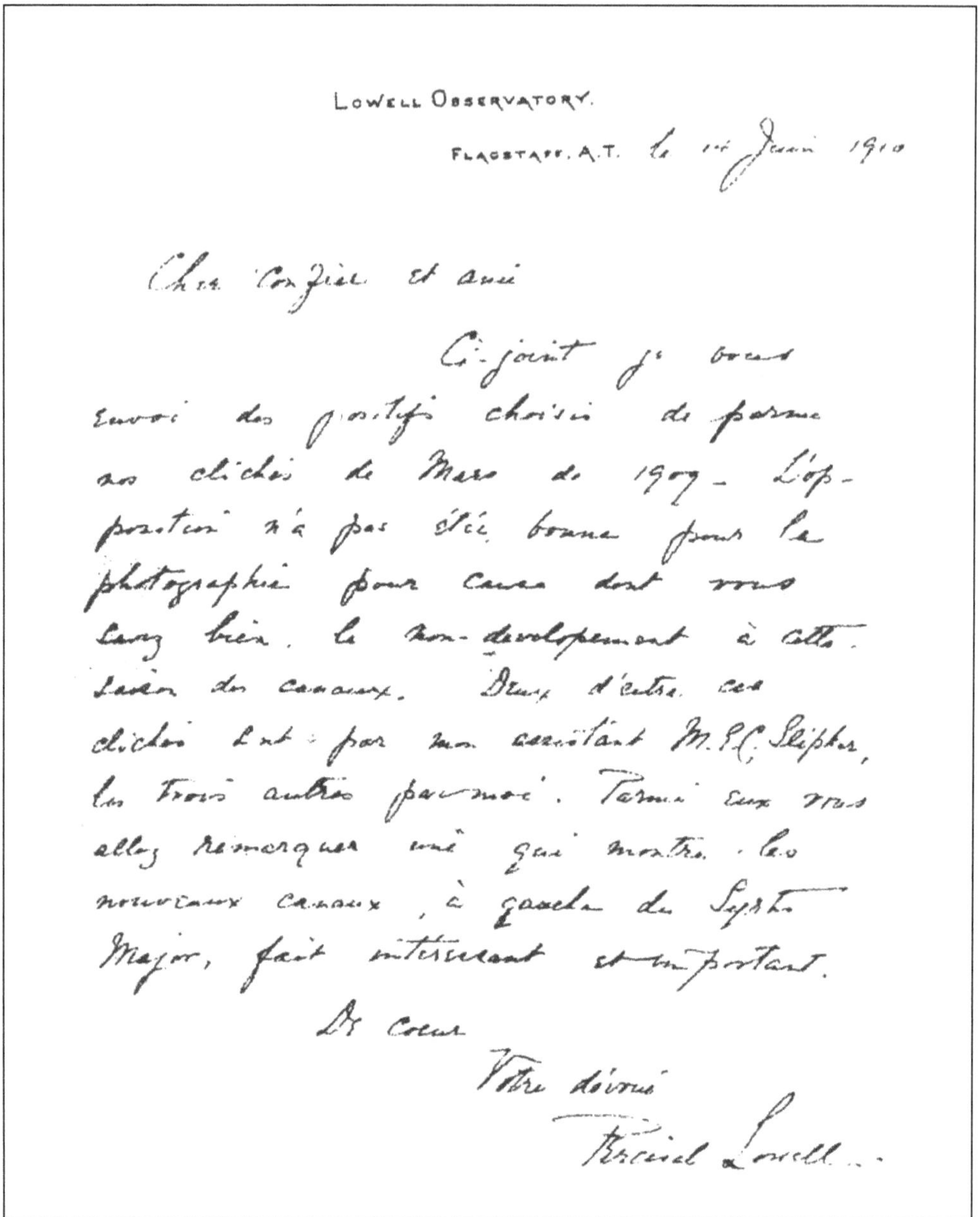
LOWELL OBSERVATORY.

FLAGSTAFF, A.T. le 11 Juin 1910

Cher Confrère et ami

Ci-joint je vous envoie des positifs choisis de parmi nos clichés de Mars de 1909. L'opposition n'a pas été bonne pour la photographie pour cause dont vous savez bien, le non-développement à cette saison des canaux. Deux d'entre ces clichés sont par mon assistant M. E.C. Slipher, les trois autres par moi. Parmi eux vous allez remarquer une qui montre les nouveaux canaux, à gauche du Syrtis Major, fait intéressant et important.

De Coeur

Votre dévoué

Percival Lowell

Tab. 9	Last letter written by Lowell to Schiaparelli (11 June 1910).

11. The Academic Contamination

It seems obvious that the Ptolemaic Theory (better yet: *approach*) created a division between sky and Earth considering the first unknowable and the second knowable.

The Copernican revolution, on the other side, unified the two worlds, but after about five centuries this unity is still imperfect because even though the Earth is no longer the centre of the universe, it is still considered an *unicum* because it is thought as *the only place where the life is possible*—at least as far as the solar system is concerned—and *the only place directly accessible not only to our instruments—for the scientific investigation—but also for ourselves, as "Physical Entities"*.

In this way, but from another point of view, Earth is anyway the *centre* of everything because its *unicity* depends on the exceptional fact that it can make life possible in the whole solar system.

Planetography is today the first, bold attempt to unify the study of *celestial bodies* by analysing a planet in accordance with modalities already adopted in the same perspective for the study of Earth and also realising the Copernican revolution in a field that is still excluded by its influence.

To this purpose, we want to quote the wonderful words by Guido Ruggieri when he wished the birth of a "new aerography"[79] and he added prophetically: *Every piece of information about Mars will be amplified or even revolutionised with consequences we cannot foresee.*[80]

However, all the work by Schiaparelli is a proof that his intuition was a scientific glimmer in the same way as the hasty and negative observations of his contemporaries—as well as those of the "scholars" of nowadays—demonstrate that the present science (as the past one) does not apply the scientific method because it does not take into consideration data coming from satellite detections.

---

[79] Guido Ruggieri, *La scoperta del pianeta Marte*, op. cit., p. 213.

[80] Idem.

Schiaparelli, who lived half a century before the beginning of the space age, had already realised the future achievements of science and he complained that he had not defined observation data better than those given by the telescopes available during his lifetime.

But the absurd is that today, the researchers who can benefit of thousands and thousands of images—with incredibly better-defined resolutions compared with the weak observations of the telescopes of one hundred years ago—are incapable of reading the satellite images!

For this reason, in front of these data, making comparisons and criticising Schiaparelli, is a big mistake and a disrespectful approach.

We like to remember the words of the Italian astronomer Mentore Maggini (who died in 1941) who prudently demonstrated that he had understood the whole matter when he asserted (in 1939) that *(—) the current image of Mars' spots represents just a "phase", "a stage" of its geographic unities' vision*[81] and in another part of his same work he added *This is what an optical hypothesis is: the nowadays telescope* (30 years after the negative assertions of Antoniadi and friends—author's note) *applied to the study of Mars is like a half blind eye.*[82]

Maggini said that the telescope was a *half blind eye*, Antoniadi and friends thought they had seen everything instead—

In 1909 Schiaparelli, was conscious he was observing Mars from a considerable distance and he wished for the day when the possibility to observe the planet from a shorter distance would have been possible.

Unfortunately, he passed away before this could come true—

At his time, in fact, his correct opinion was not fully understood and today the situation is still unchanged. The proof is that researchers, (with the coming of the space age) when they could observe Mars in the way Schiaparelli had wished for, instead of observing the planet as the astronomer had done so in a full *scientific serenity*, they refused to do and still refuse to take into consideration the data coming from satellite detections. This is what emerges from the current publications. We will consider one case randomly.

In a recent essay,[83] we read that astronomers William Campbell, Walter Maunder, Vincenzo Cerulli and Eugène Michel Antoniadi (all of them in a deadly coalition against Giovanni Schiaparelli—Author's note) *(demonstrated) that using telescopes as those used by Brera's astronomer*[84] *(Giovanni Schiaparelli Author's note) it was physically impossible to distinguish details on the surface of the planet (Mars).* This meant they had reinvented the wheel.

The Author of the quoted abstract is the researcher Laura Maraschi who adds: *It was 1909 and the ghost of a dried, desert planet had already been*

---

[81] Mentore Maggini, *Il Pianeta Marte*, op. cit., p. 173.

[82] Ibidem, p. 172.

[83] *Schiaparelli e Marte: Un sogno scientifico* (Schiaparelli and Mars: A scientific dream), Edited by Maraschi, Manfrini, Sandrelli, Edizioni Scientifiche Coelum, Milan 2003, p. 7.

[84] Schiaparelli is often called that way.

planned.[85] This comment was completely wrong and in contrast with the impossibility to distinguish the planet's details because of the technical insufficiency of the used instruments!

The question is: if it really were physically impossible to distinguish details on the planet's surface, in which way it would have been possible to materialise the ghost of a dried, desert planet?

We decided to omit the answer so to avoid offending the readers' intelligence. But there is more. Maraschi goes on asserting: Schiaparelli never was fully convinced. in December, the same year (1909 Author's note), a few months before his death, he concluded his article in this way, this too titled (as the precedent Author's note) "The planet Mars",[86] but (as Maraschi added Author's note) deliberately less "objective" of the 1893 article.[87]

In reality, Schiaparelli asserted that: Everything will depend on the progress that will be achieved in the next years as regards the photographic representation of Mars. A great step forward will be taken when we will obtain pictures that will allow us to take—on them—precise measurements.[88]

This is so true that Maraschi even affirmed that: The question had already gone ahead but the direction was not the one that Schiaparelli would have chosen.[89]

In addition to this, in order to make the readers aware of how much Schiaparelli was misunderstood, she added, *To demonstrate that a scientist is not a simple "scientific method" but also his own culture and his bias and that the evaluation of facts is far from being clear and evident.*[90]

It could appear superfluous that the misunderstanding of Schiaparelli's position was just this.

Actually, the scientist referred to *real and true progress* dreaming that using better optical instruments it would have been possible to achieve those results that in time were obtained only through the *satellite detection*, such as after the beginning of the Space Age, in the sixties! Of course, facts are quite different from the researcher's approach he refers to.

And all that was produced between 1892 (when the first work by Schiaparelli was published) and 1909, the year of Antoniadi's observations, was actually meaningless if we compare the elements with the *facts* that Maraschi speaks about.

This last, quoting Schiaparelli's idea/possibility to take *precise measurements* on Mars' photographs, spoke already about p*hotogrammetry*, while his

[85] *Schiaparelli e Marte: Un sogno scientifico*, op. cit., p. 7.
[86] Ibidem, p. 98.
[87] Ibidem, p. 7.
[88] Idem.
[89] Idem.
[90] Idem.

contemporaries still were stuck on the astronomical context where the *photogrammetry* did not and still does not occupy any place at all!

Schiaparelli had such an intuition (!) but he could not imagine that a space flight could replace the telescope.

Assuming that he could, we are trying to reassess his scientist's dignity and defending him from the disrespect of his contemporaries and detractors.

However, a progress actually had taken place, but the data to take into consideration to evaluate Schiaparelli's assertions could not be and are not the telescopic observations of his time but can only be the satellite detections operated more than fifty years after his death.

A real step forward was made not immediately after his death, but after 62 years (in 1971) with the first images of the mission Mariner 9.

Thanks to that mission, it was possible for the first time in the study's history of Mars, to take precise measurements on the planet's images given that the quality of the pictures achieved the maximum of thirty-seven metres of geometrical resolution in comparison with seventy kilometres of 1892 observed by Schiaparelli and 30 kilometres in 1909 reached by telescopes used by Antoniadi and colleagues.

In an article published in 1893 for *Natura e Arte*, Schiaparelli hypothesised that life, especially the intelligent one, was present in the whole universe. In particular, talking about the ancient history of the planet (what we call *Mythology*) he gave the reader the idea that intelligent life was present everywhere in the universe.[91]

The rightness of Schiaparelli's scientific position derives above all from the importance the scientist gave to the hypothesis (second phase of the experimental scientific method that follows the thesis and anticipates the check): in particular in the last letter written to Lowell on 23 April 1910 he affirms that

> *beyond the certainties given by rigorous mathematic and experimental demonstrations, the progress of knowledge comes also exploring the field of possibilities and probabilities. This happened for Copernicus theory that was confirmed only after 150 years.*[92]

The general setting of Schiaparelli's investigation gave as a possibility, the existence of intelligent civilisations in the universe and this was the trend of the most important scholars of the past and also the philosopher Jules Ernest Naville of Geneva, with whom he had no correspondence though.

Schiaparelli wrote that

> *No one had so much fame and provoked so much interest among men as that of inhabitants of other planets and no one more than that arouse so much speculation among the philosophers giving birth to strange thoughts in writers. Many (scien-*

[91] The correspondence Lowell—Schiaparelli—op. cit.—p. 6.

[92] Ibidem, p. 14.

tists/people) think that only astronomy and the optical instruments' improvement can give a perfect solution to the problems. However, these people exaggerated the power of our telescopes.[93] *More specifically, as regards Mars, Schiaparelli asserted that it is possible but so far nothing observed can allow us to assert they are real.*[94]

At that moment, the true aspiration of Schiaparelli was comprised between the complete realisation of the observational studies using adequate instruments that he inevitably defined *gigantic telescopes* which results are comparable to those of the big Hubble Space Telescope (HST) and to the artificial satellites:

Today, some of us have the hope that the main cosmological problem can be sorted by using the observations made of the surface with gigantic telescopes and the question is if on other celestial bodies there are intelligent beings or at least well-organised ones.[95]

The consideration towards the direct observational study of Mars emerges also from a comment regarding the study of Flammarion[96] where Schiaparelli asserts that

now he aims to a better result and anxiously he starts to spy on the possibility that a friendly voice could come from the cosmos and to have just a clue of this he is available to spend for a single telescope more money than in all the precedent centuries for science.[97]

In another work, he wrote:

He strongly wants to forbid poets and novelists to use the hypothesis of extraterrestrial life for their purpose. On the contrary, he wants to deal with this using the scientific instruments that today are possible to use to give consistency and empirical probability.[98]

---

[93] G. V. Schiaparelli, *Gli abitanti di altri mondi*, in *Le Opere di G. V. Schiaparelli*, (The inhabitants of other worlds—The works of G. V. Schiaparelli), XI vol., Rome 1929–1943, Tomo X, p. 122.

[94] Ibidem, p. 123.

[95] Schiaparelli, La vita sul pianeta Marte, 1893, Chapter I—Abstract from files number 5 e 6—Year II of the magazine *Natura ed Arte*, February 1893.

[96] As regards the relationship between Schiaparelli and Flammarion, it is important to highlight that in reality Flammarion amplified Schiaparelli's fame as we can read in a comment to the correspondence of the great astronomer: "Flammarion amplified the almost shy conclusions of Schiaparelli" (about the canals—Author's note). As he was a good communicator, he spread this kind of news about Mars everywhere and we also learn that Schiaparelli did not appreciate the interest of Flammarion and he did not comment on the praise he received: "(...) he did not write anything, any comment or interpretation because he did not appreciate what Flammarion had previously inferred by his thought". Also compare the correspondence Lowell—Schiaparelli—op. cit.—p. 6.

[97] Schiaparelli, La vita sul pianeta Marte, 1893, Chapter I—Abstract from files number 5 e 6—Year II of the magazine *Natura ed Arte*, February 1893.

[98] Cfr. *La Pluralité des Mondes Habités: Les mondes imaginaires et les Mondes rèels: Recits de L'Infini: Les Terres du Ciel: Contemplations Scientifiques.*

The wrong interpretations that today are present in the satellite images' captions are due to the ignorance of the scientific principles that we have mentioned so far.

In other words: instead of achieving a true (or, at least, a believable) interpretation of what an image really shows to the scholars, the *planetary scientists* (as well as the overvalued *satellite photos technicians*) gave and still keep giving us very questionable and unlikely *theories* about what we see, so that even some *artificial structure*—already detected—appears to us, in the end, as the final result (product) of (many) natural processes.

There are complicated interpretations of what is, instead, the result of a scrupulous geological analysis even though on Earth no scholar would even imagine applying geology to the interpretations of an artificial structure.

This is the result of the omission of the "Copernican Revolution" that since the 16th century applied the physics laws experienced on Earth to the rest of universe.[99]

An especially important data is that an image does not need an analytic examination if it would be recognised as the faithful replica of a real scene.

The proof that the detection is a faithful replica of the scene depends on the proof that there is a *favourable relation* between the dimension of the shot scene and the level of the obtained geometrical resolution. In this case, it cannot be said that the image is something different from what it shows to be and from the scene represented there, so the logical conclusion is that *what is showed is what exists in the reality*.

If we want to give a clinical explanation of this *cognitive deprivation* given as a result of one of our investigations[100] that detects the presence of a widespread *psychosis* among several analysed subjects characterised by the lost contact with reality, alteration of the interindividual ties and the usual disturbs' unawareness from the same subjects. This sort of *psychosis* is called *paranoia* and it is characterised by the slow and progressive building of a system that brings to the so-called *delusions of grandeur*.

From these two things derives the prevailing idea that invades and distorts the whole affective and relational system as well as the *ritual* where a series of actions is repeated always according to the same sequence (*like an obsessive/compulsive mental disorder works*).

This context coincides exactly with the reactions and the answers that several subjects have given and these reactions and answers do not depend

---

[99] In a nutshell, the qualitative analysis (or photo interpretation) is characterised by the following elements completely done by the expert; high level decision process; multi spec limited analysis; limitation in the distinction the grey shades (16) or colour tones; easy use of space information (shape, dimension, weaving); small-scale work in comparison with the dimensions of pixels; low accuracy in the areas calculation and finally it is subjective, practical and qualitative.

[100] ibidem p. 24.

only on the interviewed subjects but also on the context that had given them the same parameters of judgment and forbidden to all of them to acquire additional information if they wanted to achieve a professional career.

On the other hand, if this awareness would be brought to the knowledge of the interested subjects, these would know that the sandcastle built in years would be destroyed in a second, throwing away the "professional" commitment. Such a perspective is incredibly sad and it is avoided denying the real data and the invention of fake ones. in technical terms, this element is known as *cognitive dissonance.*

One of the *signs of times* is the reluctance of research towards the satellite images that should be the base for a decent (at least!) planetary research.

In the two images here attached, we plastically illustrated this tragic involution starting from a correct planetary research to come to a perspective in which the visual element is totally excluded from the list of the supportive elements. (**Tab. 9A**). After more than a century of scientific obscurantism, and the denial of Giovanni Schiaparelli's studies, based on the observational element – (today it corresponds to satellite surveying) our study brings planetary science back to its primitive bedrock that bases its method on the photographic element for the study and analysis of planets. Beyond this element there is only empty and sterile rhetoric.

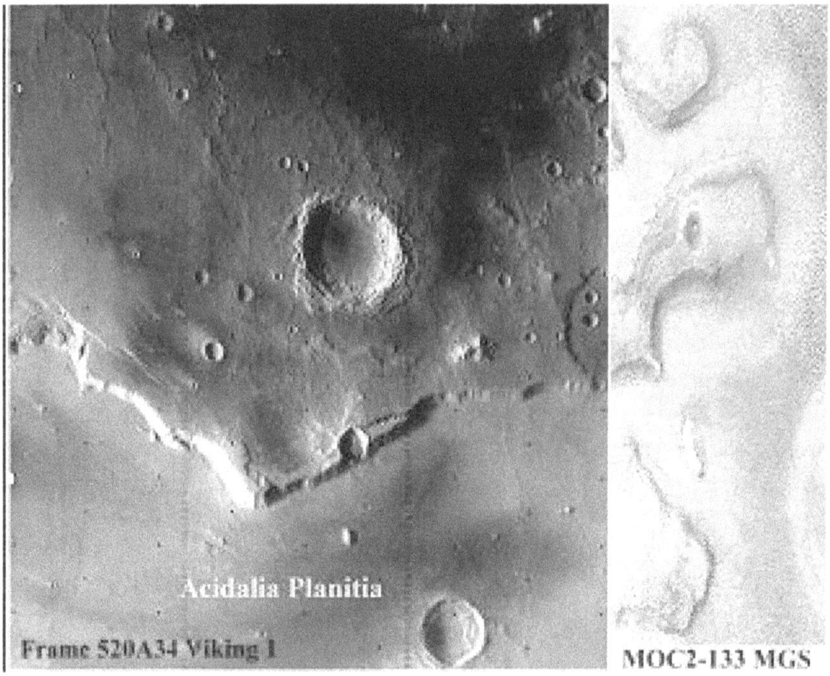

Tab. 9A	Left, authentic photo of the Viking 1 mission (520A34), right, artistic creation by Malin Space Science Systems, falsely attributed to the Mars Global Surveyor mission.

Part II
Hints of Applied Planetography

1. The Planetography

Schiaparelli was the first astronomer to notice the morphological variety of Mars. So, since the moment he had this perception, he also deducted that this factor implied, consequently, the use of a new studying methodology that was in assonance with those parameters previously adopted for our "home planet".

Schiaparelli imagined the creation of a new scientific discipline that he named "Planetography";[101] however, such a "new discipline", then as now, was considered like a "second class" Astronomy, considering that the "true" Astronomy was only the one that referred to the so-called "astrometry"!

The above-mentioned definition was not invented by us, but, instead, by the magazine known as *L'Astronomia*[102] (*The Astronomy*). On the other end and in our humble opinion, we must underline that such a new discipline was not and it is not "astronomy", but Geography applied to objects other than the Earth. That means "planetography": a science that has nothing to do with "astronomy", at least as intended in the classical meaning of the term.

This is even more true once you consider that, in the AD 2006 (when I still was "meritorious partner" of the "Société Astronomique de France", founded by Camille Flammarion!), I sent to the above-mentioned Société a report about Planetography applied to Mars, hoping that it might have been published on the official magazine "L'Astronomie". Unfortunately, the report was rejected and then sent back to me because (that was the specific motive)—*it was not a piece of information about astronomy (of course—) but rather geography—*

An accusation that, in such a case, also became a blatant defeat for them and an overwhelming methodological victory for me!

Here is what Raffaello Braga wrote about this subject on the Italian magazine *L'Astronomia*): "This attitude can even nowadays be found in many books about the history of astronomy, where only some fleeting hint

---

[101] The term 'Planetography' (together with the term Cosmography), was at one time used by Giovanni Schiaparelli in his work 'Life on the Planet Mars' published in 1893.

[102] "Mars at opposition: from the Renaissance to the present day", by Raffaello Braga, in "Marte all'opposizione", attached to "L'Astronomia" no. 242 May 2003, p. 59.

linked to the events of the telescopic (better yet: "orbital") observation and study of our solar system (beside the Galilean approach) is given".

So therefore, another great merit of Schiaparelli, beside the already well-known matter of the "canals", was also the one to raise the "observative planetology" (or "planetography") to the rank of modern science, worthy of the highest consideration.[103]

All that said, it now appears, despite the many reservations and the obvious embarrassments, it made its way the idea that, in a pure planetological (i.e.: "planetographic") field, the telescopic observations of any given celestial body represent an outdated and useless information to use.

And here is what Giancarlo Favero wrote on the already mentioned Italian magazine *L'Astronomia*[104]: "The just described surface formations (i.e.: *Albedo Features*) that have been observed for at least three centuries by the Mars enthusiasts, did not show, in the eyes of the probes and orbiters, any kind of correspondence with some truly peculiar geological process of the Martian surface. So, it may seem useless to observe them, but the fact remains that even the orbiters recorded them and they appear to be subject to modifications that deserve to be documented".

"Furthermore, this documentation must be done by using the same techniques used in the past: and that is the reason why the visual observations of the planet Mars keep, even now and probes, orbiters, Space Telescopes, CCD and webcams notwithstanding, all their scientific interest".

We remark that all the above-mentioned considerations represent an extreme attempt to save some data—the "Albedo Features") that are blatantly useless for the purposes of a technical study of a planet.

As we know very well, the Astronomy's development considered two different moments: the observation with the naked eye—*ocular astronomy*— and the instrumental or telescopic observation, such as the so-called *observative astronomy*.

The history of the *observative astronomy* started in the 17th century when the binocular started to be in use and, subsequently, the telescope.

Today, it is practiced in the *astronomical observatories*, by using optical instruments such as telescopes and radiotelescopes placed on the ground.

We have found an interesting definition of "astronomy":

> *Astronomy is a discipline that fascinates many people but at the same time it is scaring because of its difficulties even though it is studied at amateur level (—) engage in observing the sky is not that simple but it is not difficult to reach a proficient level of satisfaction.*[105]

---

[103] Idem.

[104] "Mars at the opposition", by Giancarlo Favero, in *L'Astronomia*—Marte all'opposizione—n. 242 Maggio 2003; p. 24.

[105] Emilio Ricci, *Telescope no problem*, Biblioteca Antares, Milan 1992, p. 9.

How can this person understand complex concepts as "angular resolution"and "magnitude limit"?[106]

Such a definition allows us to highlight that all this does not relate to our study neither to the analysis of the planets' surface as this is just the "Geography of the Planets" and it is not necessary to be a mathematician to observe and understand a planet's landscape!

The difficulty of the astronomy is not anymore used to hide a real incapacity of nowadays science to study planets in valid scientific terms, which is with the same tools and the application of the same principles used to observe our Earth.

According to another definition, the astronomy's field of expertise is restricted to the study of celestial phenomena (and), the telescope is the main instrument of astronomy (—) and the telescopes (are) the instruments to study astronomy,[107] but what most interests our study are the satellite detections that use the optical instruments mounted on board of space probes (i.e.: cameras to take pictures and shooting) and they are guided by operators located on the ground.

These two methodologies (*telescopic observations* and *satellite detection*) even though they have different perspectives, however—and this is the real problem—they are part of situations that ignore these differences. It is like a person who is at the same time short-sighted and farsighted and uses the same pair of glasses to observe close and far objects.

To this purpose, we read:

As a short-sighted finds helpful a pair of glasses to recover his normal visual acuity, in the same way the telescope helps the astronomers to look farther, beyond the natural possibility of their eyes.[108]

In this context, it is possible to detect that, since the beginning of the satellite space exploration's development (such as 1965 if we refer to Mars), there is a lack of distinction between the *telescopic observation* and the *satellite detection*.

This involved a parallel issue, which is the distinction between the study of the *substance* and the study of the *shape* of a planet: two things that are still indistinct.

This concept is found in the work The Key of Cosmos by Conrad A. Bohm[109] where it is said that: After a long thinking about many searches whose aim is to unveil the substance of celestial bodies, searches that never

[106] Ibidem, p. 7.

[107] Ibidem, p. 9.

[108] Ibidem, p. 7.

[109] Conrad, B. A. (1989) *Le Chiavi del Cosmo* (The Keys of the Cosmos), Franco Muzzio Editore, Padua, p. 205.

achieved their aim, we now highlight those that examined the shape. in order to detect the morphology of the celestial bodies, the observation, more than the pure calculation, was also, in this case, the main method to follow.

At this point, it could be right to ask ourselves if the close observation of a planet can, in some way, be related to astronomy in accordance with the terms that were given when the space age had not begun yet.

Our opinion is that astronomy has nothing to do with the close observation of a planet bearing in mind that it studies the *celestial bodies*, considers the position they occupy in the space, gives them a name (or an identifying code) but it does not require (and it is not possible to do it, at least up to now) the study of the related surfaces.

In this context, the astronomer could say something about the movement of the planet but it would be quite difficult to speak in relation with the nature of what it is observed on its surface (this situation turns out from the press investigations led by the author from 1995 to the present), unless he had the intelligence—and, above all, the humbleness—to dismiss the role of *simple* astronomer and proposed himself as the *observer* of a landscaped reality.

All that said, throughout the *satellite detection*, he *moved in* good time the terms of the scientific research, focusing primarily on the latter, in lieu of the *telescopic observations*.

However, such a "revolution" was never really operated and/or accepted to an academic level, where also the *satellite observations* are still perceived according to *wrong* astronomical parameters.

Since today it is possible to watch and study a planet in the same way, we can watch and study our own Earth, the *descriptive astronomy's* development, today called *planetography*, should constitute a progressive abandonment of the classical telescopic [110]observations and a *complementary* application of them in addition to a very clever usage of the data coming from the *satellite observations*.

Just consider the astronomic language applied to the topography of the planets where a *celestial body* was described with terms borrowed from as-

[110] All this, does not exclude the fact that space missions have been made possible thank to the processing of the enormous mass of data previously collected by astronomers, a heritage that is still used and permanently updated for use by space exploration projects. It also concedes that without telescopic observations, there could be no overall view of the (visible) Universe but it is equally true that the methodology proper to astronomy must be abandoned when it comes close-up observation of satellite images. Astronomy also makes use of astronautics, not to assume data that would alter the basic approach of astronomy itself, but simply to reconfirm the same parameters by using information from methodologically different sources. It is enough to consider the absurdity of the astronomical language used to describe the topography of a planet, something that has never been done in relation to the Earth that is indeed a celestial body too.

tronomy, while it will be more logic and correct to adopt a *geographical* (and *topographical* of course) language.

This vice is one of the traits that indicate how, in the academic field, the space revolution is not completely accepted yet.

The proof is that the institutional research was not prepared for the "space" appointment, biased at the point that in the examination of the satellite images of the first celestial bodies observed in the solar system, the investigation was not *favourable* to the presence of life on Mars and nevertheless to the geological past and present dynamism.

As regards the planet Mars, such conclusions, in the following years, have always been denied.

Albeit the material coming from Mars contributed to the heritage about the knowledge of the planet, thus making possible a global consideration of the Martian reality, it continued to endure without any reason the "telescopic" loan where there should have been considered the results of the satellite detection from photogrammetric perspectives.

Some people thought that it was not possible to interpret Mars in the same way adopted for Earth, even though the *satellite images* had been shot with the same criteria of the terrestrial ones. This is because it was not applied the first principle of the Copernican Revolution according to which the terrestrial and celestial reality (we would say *space*) are to be considered identical and analysable with the same methodological and technological method used for the terrestrial images.

Such a *remora* is still present in the institutional research and this blocked in some way the scientific research towards Mars so that some operators' falsified data, including images that demonstrated the preconceived denial's contradiction of the presence of intelligent life discovered thanks to the artificial *structures'* detection.

In fact, the satellite's detection is the only normal way to look at a planet considering it a *celestial body* similar to Earth and not a "mysterious" object.

In addition to the above, the satellite's detection offers the possibility to study the *celestial bodies'* surface with the same precision with which it is possible to observe the terrestrial surface by using the same instruments of observation.

A fundamental data is that Schiaparelli's approach had set the foundation of the future planetography.[111] This word was already used at that time and it is more correct than the current *planetology* that is not fully wrong, but it becomes inappropriate when it goes beyond its specific field that is the physical study of a planet.

---

[111] It appeared on specialised magazines in 2005 and 2006 (*L'Astronomia*, November 2005–March 2006).

A clear reference to *planetography* is in a definition given by Schiaparelli who outlined a topographic representation of the Planet—a schematic character's figure and *its aim is not giving a picture that imitates the aspect of the planet as in a painting, as in geographic maps but facilitating the descriptive exposition*[112] so not an *artistic representation* but a *geometric description.*

In a memory dated 1878, Schiaparelli described the results of his observations on Mars from August 1877 to March 1878 where he clearly exposed the aim and the method of his study:

> *My fundamental purpose is to proceed with a description of Mars but not through records or portraits of it based on measurements done by eye but with the aid of geometrical principles and methods.*

In 1878, he added that Mars was not

> *a desert of dried stones, it lives and its life appears on the surface thank to a very complex variety of phenomena and part of them develops on a large scale to be observable by the inhabitants of Earth.*[113]

This last assertion recalled the concept of *geometrical resolution* that at the time started to appear in the new-born planetography.

Another cue can be inferred by the comment made by Schiaparelli on the work by Maedler:

> *Putting in order the different topographic peculiarities, he succeeded in building his first aerographic map. Albeit it is incomplete and shows only some important spots, it is a proof of his care and commitment and it represents as regards the description of Mars what two thousand ago represented Eratosthenes' map for terrestrial geography.*[114]

Regarding the main aim of the observations of Mars, it is possible to infer Schiaparelli's position thanks to a document found among the correspondence with Lowell:

> *it was the research of life and the research of analogies with Earth done by applying the Copernican principle! (—) in addition to the main problem that at the time increased every kind of study of the planet that was the research of life on it.*

Eventually, an interesting observation: As today we are led (unfortunately because we are not focusing on our solar system. Author's note) beyond the solar system, at that time the analogies with Earth favoured the hypothesis of Mars being inhabited.[115]

---

[112] *Il Pianeta Marte*, *Natura ed Arte*, 1909, see *Le Opere di G. V. Schiaparelli*, XI vol., Rome 1929–1943, vol.2.

[113] *La vita sul pianeta Marte*, *Natura ed Arte*, 1895, in *Le Opere di G. V. Schiaparelli*, XI vol., Rome 1929–1943, vol. II, p. 88.

[114] Schiaparelli, La vita sul pianeta Marte, 1893, chap. II—Abstract from files number 5 and 6—Year II of the magazine "Natura ed Arte", February 1893.

[115] Il carteggio Lowell—Schiaparelli—op. cit. p. 1.

In his investigations, Schiaparelli mentioned some precise measurements by determining the axis of rotation and reworking a new map very which was different from Proctor's one and he did all this using the same procedures of the geographer who describes a terrestrial area.

So, how are those subjects that investigate the planets' surface designated today?

Astronomy's sector that studies planets is indicated with the term p*lanetology*, that it's about the morphology, in theory and the chemistry in practice, the meteorology and physics of the planets.

It is inspired by the terrestrial geography and it was born by accident thanks to the first description made by Galileo about the moon, Mars and Saturn.[116] Subsequently, it was enriched thanks to the contribution of Schiaparelli's scientific graphic method that went beyond the initial "Aerography" (as it was indicated at that time the observative study of Mars).

A clear functional distinction between the *telescopic observations* and the *satellite detections* appeared for the first time only in 2005 thanks to a contribution written by the author[117] (**Table 10**) and published on the magazine *L'Astronomia* where it was written about structures detected by telescopes from great distances and with geometrical resolutions always very low (till tens or hundreds kilometres) denominated *Albedo*[118] as well as structures detected by satellite (*topographical*) that are close detections with geometrical resolutions that can be medium-low (from some kilometres till 500 metres) till reaching high levels (under 100 metres) and very high so a few centimetres or even less, as in the case of the pictures taken by descent modules.

---

[116] Introduzione alla Planetologia (Introduction to Planetology), Nemesis Project Research.

[117] See. "Contraddizioni areologiche", *L'Astronomia*, November 2005.

[118] Albedo: a fraction of incident light reflected by a body.

AREOLOGICAL CONTRADICTIONS
From telescopic observation to satellite detection

Planetography: a new science and a new language for studying the planets.

The beginning of the space age, in this case, the close observation of the planets and Mars in particular, effectively shifted the terms of scientific research by putting satellite photos in the foreground instead of telescopic observations.
In addition, it should not be forgotten that space missions were made possible precisely by processing the enormous mass of data previously collected by astronomers.

Tab. 10	Viola Gianni, 'Contraddizioni areologiche', L'Astronomia, no. 268, November 2005. This is the first founding report of Modern Planetography.

In such a distinction, it is helpful the following assertion in which Geography possesses a precise peculiarity and an epistemological autonomy that gives it the possibility to offer valid instruments to understand the terrestrial reality.[119] However, it is just the "lacking" role of geography that de-

[119] Mario A. Gomarasca, *Elementi di Geomatica*, Associazione Italiana di Telerilevamento, Milan 2004.

termines the absence of an adequate terminology related to the geographic study of the planets.

It is evident that we are facing a very confusing situation where terms used do not correspond semantically to the proper meaning of the words. This is also due to the fact that, in time, words have acquired different meanings.

For instance, *astrology* means something different from its etymology and the word *astronomy*, which etymological meaning is the enumeration of stars. Actually, it has a wider meaning, so it is *the science of the laws that rule the movements of celestial bodies, their dimensions and distance.*[120]

This calls for a general reform of the current terminology so that every definition should have its *scientifically correct position.* in addition, today many terms in use before are now forgotten and this is a great damage to the scientific research (**Table 11**).

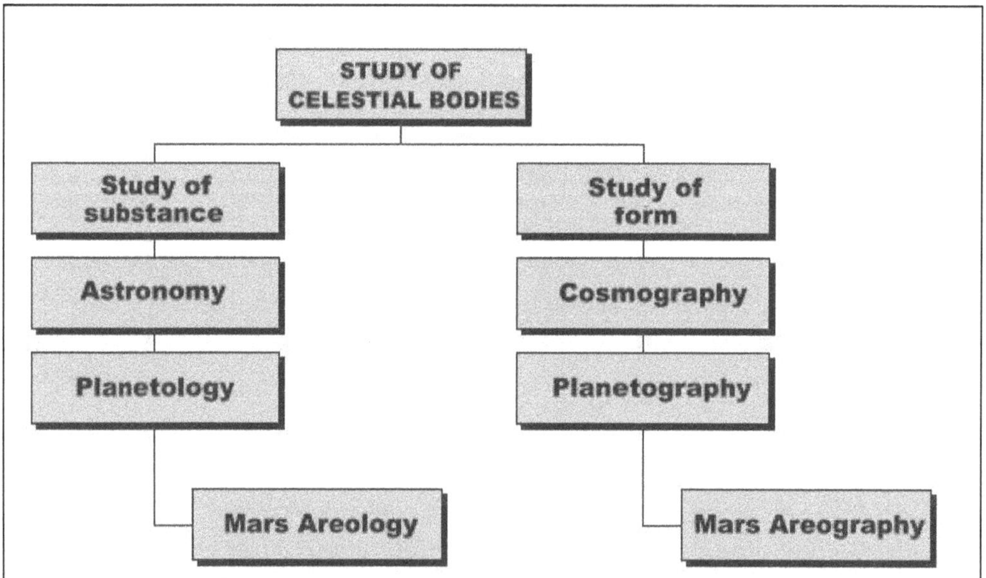

Tab. 11	New terminological formulation of the study of Planets (Gianni Viola).

The word "Cosmography" (from Greek Kosmographia—κοσμογραφία) is That part of the mathematical science that deals with our planetary system considering it in itself and in relation to the whole universe's order.[121] Another word in use was "planetography". "cosmography", along with the term planetography, were used by Giovanni Schiaparelli in

[120] Vocabolario Etimologico della Lingua Italiana (Etymological Vocabulary of the Italian Language) by Ottorino Pianigiani, on internet.

[121] Idem.

"La Vita sul pianeta Marte" (life on planet Mars) published in 1893 and it is for this reason that we decided to use them now in this book.

As regards the term "Planetography", it resulted used also in previous times. For instance, we can find it in the book "Gramatica delle Scienze Filosofiche" (Venice, 1795) by Benjamin Martin,[122] a scholar that wanted to systematise the knowledge of the time and it used the term "planetography" because *the science that studies the planets* and nevertheless as *theory of planets.* **(Table 12)**.

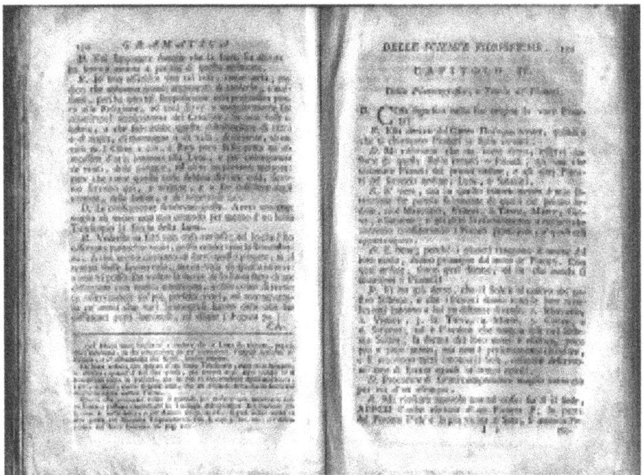

DELLE SCIENZE FILOSOFICHE. 131

CAPITOLO IV.

Della Planetografia, o Teoria de' Pianeti.

D. Coſa ſignifica nella ſua origine la voce Pianeta?

R. Ella deriva dal Greco Πλάνομαι *errare*, quindi è che ſi chiamano Pianeti le ſtelle erranti.

D. Mi riſoviene che mi avete detto, eſſervi due ſorte di queſte ſtelle erranti o Pianeti; gli uni che chiamate Pianeti del primo ordine, e gli altri Piane-

Tab. 12	Beniamino Martin, Gramatica delle Scienze Filosofiche o breve analisi della Filosofia Moderna Appoggiata alle Sperienze – Bassano 1795; p. 131: citation of the term 'Planetography'.

[122] Gramatica delle Scienze Filosofiche o breve analisi della Filosofica Moderna Appoggiata alle Sperienze (Bassano 1795), p. 131.

Today, referring to the planet Mars, it is used the expression "Aerography" that suggests the existence of a general equivalent term, i.e. "planetography", but this does not result because the term "planetography" (already in use) indicates something that deals only with the census of the celestial bodies and not with the study of the planets' surface (as it will be a right thing instead). It is an improper use of the term with a meaning that is unparalleled.

This semantic accuracy is a parameter of linguistic quality and it is defined as a correspondence between reality and the qualitative attribute associated with the object.

In March 2006, the first planetographic study of Mars appeared (on behalf of the writer), outlining an outline of Martian geography according to non-telescopic parameters[123] (**Table 13**).

<table>
<tr><td colspan="2">

Mars: physiognomy of a planet

The current morphology of the red planet bears witness to times when its surface was hydrogeologically vibrant, to the point that it can be likened to today's Earth surface. Seas, lakes and rivers shaped the Martian moorlands for a long time, and an approach that is not so much astronomical as geological (geographical) allows us to relive that distant period of activity and to understand why the exploration of the small planet is becoming increasingly pressing today.
</td></tr>
<tr><td>Tab. 13</td><td>Viola Gianni, 'Mars: physiognomy of a planet', L'Astronomia, no. 272, March 2006. This is the second founding report of Modern Planetography.</td></tr>
</table>

[123] "Marte: fisionomia di un pianeta" of Gianni Viola, on *L'Astronomia n. 272, Marzo 2006*

2. The Vision Phase

It is a *non-scientific concept* that some structures observed from a certain distance would show something different if observed, in a later stage, at an inferior distance.

Furthermore, as regards some structures detected with a high geometrical definition, *a series of subsequent shootings with enhanced geometrical resolutions, would have annulled* (i.e.: modified or cancelled) the images shot before (these last taken with a lower geometrical resolution).

As it will be explained, these considerations depend on an illegitimate mixture of a series of "astronomical" parameters and another of "photogrammetric" parameters. For this reason, this affirmation contradicts the experimental science and the logic on which science is based.

Briefly, if we observe a planet as, for instance, Mars, with a telescope from about 200 million km and until 55 million km (the minimum distance between Mars and Earth occurs about every 15–17 years), we can only detect some *macrostructures* BUT not in details.

This situation could improve when we get "closer" to that planet more than the telescope could allow us "moving" on Earth toward Mars with the help of automatic probes. These last, when close to the planet and in its orbit, with the help of cameras mounted on board, will carry out the so-called *satellite detection* that is *the automatic space exploration* in absence of a human crew (so-called "unmanned spacecrafts").

At this point, the non-scientific literature appeals to the concept of *optical illusion* along with the so-called *pareidolia* (from the Greek εἴδωλον, image with the prefix παρά, similar).

This is a subconscious illusion, the instinctive tendency of our brain to link to familiar objects a series of messed images. It is particularly recurring in human figures and faces.

It indicates an incorrect perception of a *stimulus* as an object, pattern or meaning known to the observer, such as seeing shapes in clouds, seeing faces in inanimate objects or abstract patterns or the faces of Jesus or the Virgin Mary in spots formed on walls or the face on the surface of the moon. Another example is recognising images of various natures during the observation of constellations.

The word *pareidolia* has been abused and its semantic value has not been considered. It should be noticed that all the previously mentioned cases do not include close visions similar to those from a satellite. Notwithstanding this, still a pseudo-science (but many believe it is actually *science*) obsessively insists that it is possible to apply these considerations to some close satellite images (high and very high resolution) ignoring the fact that pareidolia could subsist only if there are the conditions for the phenomenon to occur and, in scientific terms, *pareidolia* can occur in a *non-favourable* relation among the base elements of the visual perception.

In this context, the first base element to take into consideration is the *shooting distance*, which is the minimum distance between the cameras mounted on the edge of the satellite and the structure detected on the surface.

If this element would not be functionally linked with the dimension of the photographed structure (the second base element to take into consideration), giving origin to an image with a favourable "distance-dimension", could verify the presence of pareidolia.

This is common for the lunar surface's naked eye observation and even for the telescopic observations of the solar system's celestial bodies.

In the case of the same *celestial bodies'* satellite detection, it subsists only for shootings beyond 15.000 km and thanks to cameras with inadequate focal length, so with a NON-favourable relation between the structure's dimension and the shooting distance.

On the contrary, under such a distance and relating the focal length to adequate dimensional parameters, it is *never possible to speak of pareidolia* so perpetuating in using this term means ignoring the most important principles of the vision's stages and of the satellite Aero-photogrammetry.

The level of 15.000 km is deduced from satellite detections carried out with reference to Earth. This because the same parameters related to satellite detections operated with reference to planets have been adopted with reference to Earth.

This allows us to compare and relate two realities analysing the results of both contexts.[124] (Table 14).

---

[124] Of course, to observe bigger structures the shooting distances should be increased. in this way 15,000 km as detection threshold is the minimum distance level that, in any case, could be increased in relation to bigger structures.

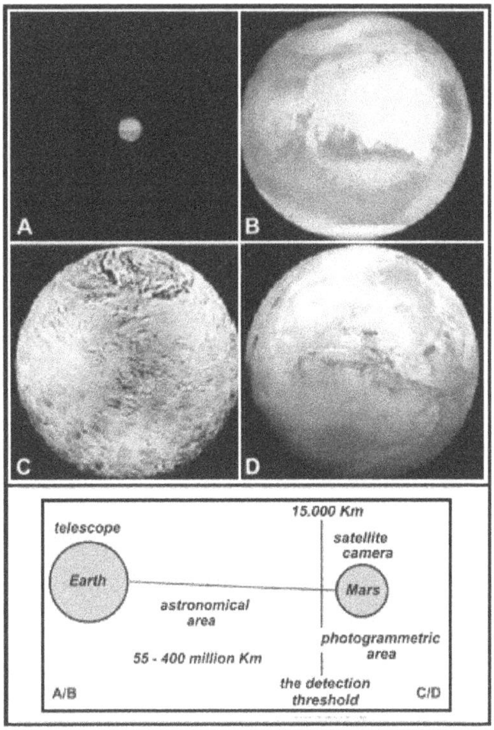

Tab. 14	Detection threshold; A) Mars observed by the terrestrial telescope; B) Mars observed by the Hubble Space Telescope (1990). C) Mars in a mosaic of images from Mariner 9 (1972); D) Mars in a mosaic of images from Viking 1 (1980). Below we have the explanatory diagram of the astronomical area (A/B) and the photogrammetric area (C/D).

In other words, when on Earth the satellite detections obtained a clear vision of the artificial structures (such as NOT created by the action of *natural agents*) universally *known* (such as the Roman Coliseum or the Pyramids of Giza) and when at the same time calculated and evaluated conditions of similar structures or even conceptually analogous (at least from an architectural point of view) have been detected on Mars (and "somewhere else too—"), we should agree that the *artificialist* interpretation is acceptable, real and objective.

The point where the telescopic observation's area of competence ends and the satellite detection starts, is called *detection threshold*. This ideal line of demarcation divides the astronomical parameters typical of the telescopic observation from the photogrammetric parameters which are typical of the satellite observation.

This change includes the assumption of different parameters of judgment because the detection *threshold* is conditioned by the *shooting distance* and the length of the focal lens.

In the close-up view, the detailed image allows to obtain, by using the right instruments and applying the appropriate rules, some precise measurements, such as the so-called *photogrammetry*. When it is possible to *focus* on the chosen image, we are in the *detection threshold*.[125] (**Table 15**).

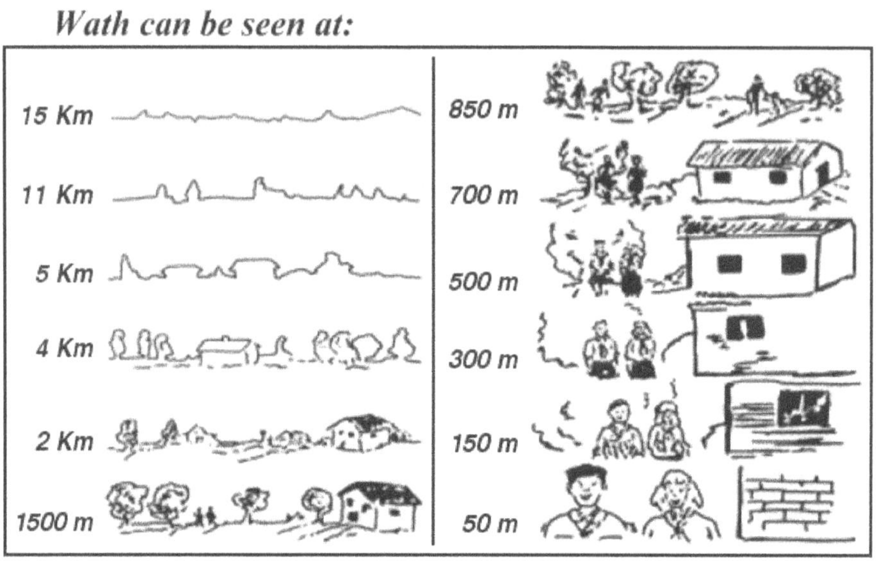

Tab. 15	Visibility table.

If in the astronomy field, one of the basic rules says that it is not possible to reconfirm later what has been previously observed; on the other hand, in photogrammetry, it is exactly the opposite: what was observed before can be observed again and better, later.

[125] Minimum distance to observe a certain structure according to the most relevant photogrammetric terms. in the specific case, it is the shooting distance between the satellite and a portion of the planet's surface. Surface Surveys between 0.5 and 2 km in size (Lateral or Longitudinal) are perfectly resolved, using a modest telephoto lens, in shots taken from an altitude of over 380 km up to about 430/450 km. Approaching 280/300 km altitude, it is necessary to shoot them, in order for them to be "understood" (i.e. captured in their authentic nature – in other words, without enlarging them too much) by the observer, almost in Direct Focus, i.e. with a maximum magnification of 4/6 X (basically an amateur binoculars). This is in the case of a shot taken in full daylight or even at dusk, but with a very low Emission Angle (i.e. the orbiter must be almost on the vertical of the surveyed reliefs, to avoid optical distortions) and an extremely clear or thin Atmosphere. Clearly, the more the Emission Angle rises, the less reliable is the interpretation of the Survey, and the denser or more turbulent the Atmosphere becomes, the more it will be necessary to enhance the cameras and possibly resort to filters (UV, or IR or NEAR-IR). (Note by Prof. Paolo C. Fienga).

Being beyond the *detection threshold* from that point on, going closer with an observation instrument or an approaching unmanned satellite, we will be able to better redefine what we have observed before.

In a home example, if we look at a person from a distance of 1 km, we will not see anything because at that distance the person would be not visible to our eyes but getting closer, we could see the silhouette and then, by getting remarkably close, we will be able to see even details.

Using the same example for Mars, we observe a view of the apparent horizon shot in 1976 from the lander of Viking 1 in the Region of Chryse Planitia, from 10.000 metres where everything is not distinguishable.

Only two vague protuberances were visible and these can be seen in detail in the selected part of the precedent table, verifying that the two already mentioned protuberances were and are the summits of pyramidal structures (**Table 16**).

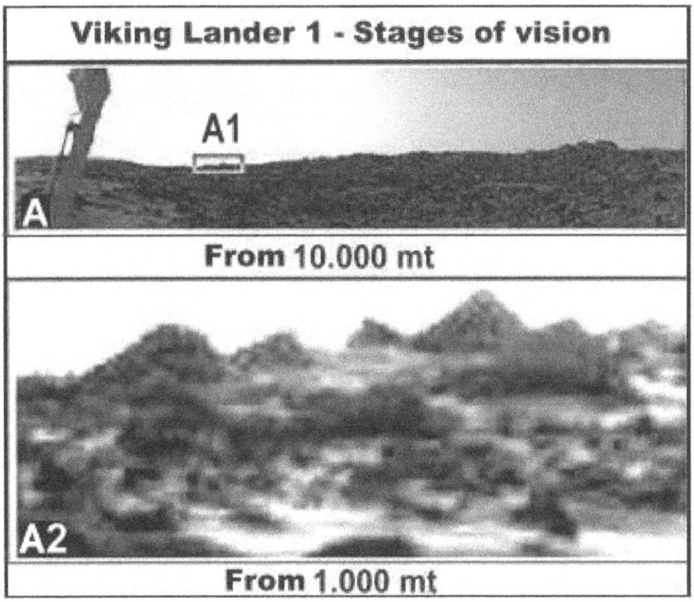

Tab. 16	Visibility progress of the Viking Lander 1 probe's camera (1976); A) Vision at 10,000 metres with highlighted portion; A1) Analysed portion; A2) Magnification of the A1 portion, as a simulated shot at a distance of 1,000 metres.

In this context, where images that present modifications of the structures shot before or elements of distortion subsist, they should necessarily be considered as products of manipulation and lies.

However, it is possible to notice that in the daily institutional research and consequently as regards the news coming from space, the astronomical and photogrammetric contexts got confused and mixed-up so they cannot be considered scientifically acceptable.

On Earth, we have some examples that cannot be proposed. There are uncountable natural shapes that remind us of human faces, but they are always in profile.

For this reason, these "illusory shapes" disappear when the same structures are observed by different points of view or in a different light condition. in a few words, in those cases, they are *conjectural transient* shapes that do not present shapes and characteristics of *objective* structures.

It is not possible to speak of *pareidolia* without knowing of how and when this can happen. However, it is true that every factor needs some necessary conditions for the *pareidolia* to subsist and when these factors are not present, it is only a word just to find excuses.

It is true that every image obtained via photograph presents (or it can present) a certain *illusion* (optical illusion) but not all images represent the inexistent! It is different to say, *this photo is an optical illusion* rather than *This photo presents*—shows—*an optical illusion*. There are, in fact, two distinct types of optical illusion; the *absolute* one that subsists because the detected structure does not exist at all and the relative one that happens when the structure exists, but it appears distorted by the effect of a particular geometry (of lighting and/or shooting) determined by the *emission*, *phase* and *solar incidence angles*.

In practice, the optical illusion is *absolute* when it shows a reality that simply does not exist, while it is *relative* when it shows an objective reality BUT some essential traits are (appears, to the observer) modified.

> *In the first case, it is meant that a given image represents an inexistent reality so what is shown does not correspond to nothing real because the optical illusion produces a photographic image that does not correspond to an objective structure. in the second case, the photographic image shows a real structure even though with some modified traits.*
>
> *This is due to the satellite shooting's technical conditions because of some effects created by the solar incidence angle, which is the height of the Sun on the local horizon at the exact time when the satellite detection occurred (when the picture was taken).*
>
> *The optical illusion therefore subsists but it does not relate to the existence of the mentioned structure but to the way the structure itself has been portrayed.*

And all this is due to several factors.

First, the relation that exists between the position of the Probe in relation to the shot dial (surface/object) and its perpendicular to the ground (emission angle).

Second, the Solar Incidence Angle or S.I.A. (for instance, because the sun is too low on the local horizon—i.e.: *high S.I.A. means low sun on the lo-*

cal horizon and vice-versa—and it causes long shadows and dark areas where the area/structure appears modified) and the phase angle (that is the angle described from three fixed points: the sun, the shot object—or a portion of it—and the position of the shooting probe).

Obviously, in a subsequent image, we could easily get the proof about the validity (for strictly interpretative purposes) of the previous shooting and, above all, the explanation of the present (actual) configuration of the photographed *structures*, as they appear to us.

At this point, it is possible to adopt a general argument under which:

1. In the images showing some, *strictu sensu*, *artistic structures*, the objective quality of the vision will diminish when the solar incidence and emission angles increase.
2. Vice-versa the quality (ALWAYS referred to the vision) will increase—obviously maintaining the constant angular parameters—in the images that could suggest and/or to point out the presence of always, *strictu sensu, geometrical structures.*

In simpler words: in the images with *artistic structures*, the quality diminishes when the solar incidence angle increases while, vice-versa, the quality increases in the images with *geometrical structures.*

To this purpose, we mention the great astronomer Mentore Maggini who, in 1939 (such as 30 years before the Space Age), wrote that *the image of Mars' spots represents a "stadium", a "stage" of its topographical units' vision.*[126] (Tables **17, 18, 19**). If we consider this scholar's work, the other assertions become ridiculous.

PROGRESS OF VISIBILITY OF PLANET MARS (1609-1982)		
Years	**Used tools**	**Resolution** *(max in metres)*
1609	Galileo's Telescope	6.000.000
1892	Schiaparelli's Telescope	70.000
1909	Astronomical observatories telescopes	40.000
1952	Astronomical observatories telescopes	6.000
1965	Mariner 4 Camera	1.250
1969	Mariner 6-7 Cameras	1.000
1971-72	Mariner 9/A Camera	430
1971-72	Mariner 9/B Camera	37
1976-80	Viking Orbiter 1&2 Cameras	6
1976-82	Viking Lander 1&2 Cameras	0,1

Tab. 17 Historical linear evolution of Planet Mars visibility progress (1609-1982).

[126] Maggini, M. (1939) *Il Pianeta Marte*, Milan: Ulrico Hoepli Editore, p. 173.

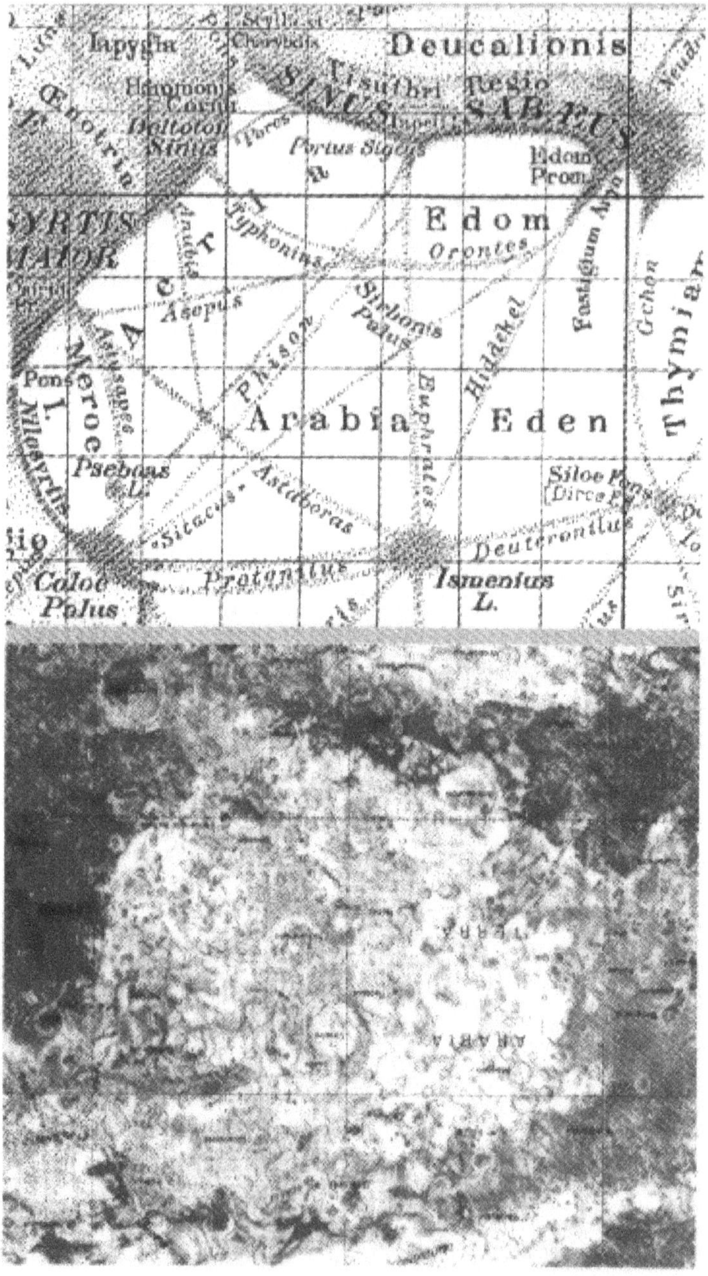

Tab. 18	Portion of the surface of Mars, observed and drawn by Schiaparelli, compared with the same portion, photographed by the Viking 1: MC.5 Arabia Earth probe.

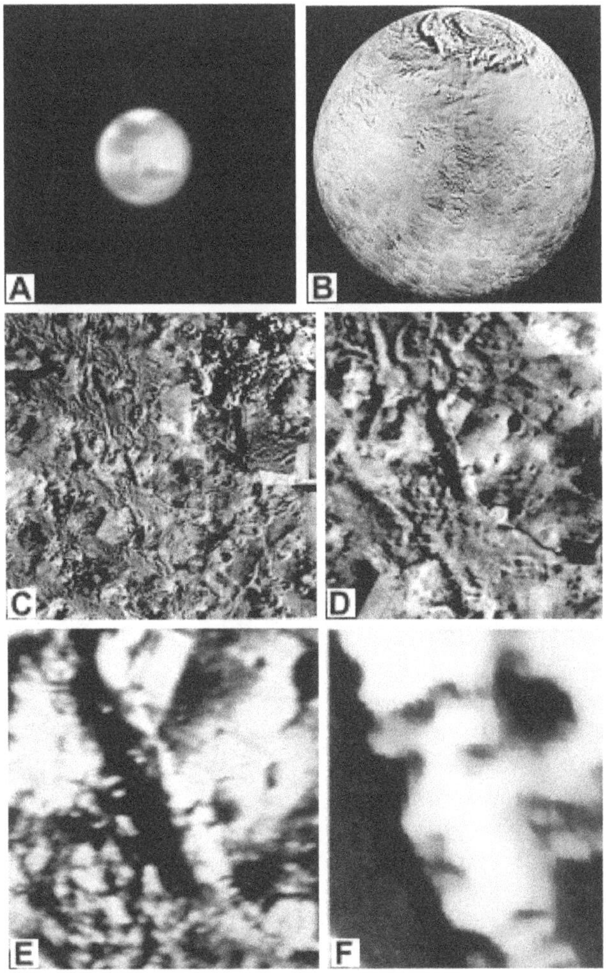

Tab. 19

Graphical evolution of the progress of visibility of the planet Mars; A) Telescopic observation from Earth; B) Mosaic of close-up satellite surveys (Mariner 9); C) ditto – enlarged portion of surface; D) Highlighting of an anthropomorphic structure; E) ditto; F) Magnification of the structure surveyed.

3. Study of Images

In order to perform scientific observations and to describe a given phenomenon and to consider all this a universal truth, it is necessary to follow a series of operational, progressive phases. These steps, applied to satellite detections, originate an algorithm called "planetographic scientific method" **(Table 20).** The first phase is the observation that starts from experience, as "personal knowledge" not necessarily conscious, and it can be voluntary or provoked. The second phase is the collection of information, which consists of the basic technical data of a satellite image. The third phase is the choice of laboratory physical quantities. This operation is the search for the properties possessed by a given element, and It distinguishes between subjective and objective properties that take the name of "physical quantities". The fourth phase is the description (or measurement) of the phenomenon, through the assumption of physical quantities, used to translate data observation in numbers (dimensions, distances, times, etc.). The fifth phase is the experiment, originated from the observation of a phenomenon that implies the solution of a problem. The experiment is the answer to the initial question about the nature of the observed phenomenon. In a few words, the experiment is set up as a planned and controlled observation, repeatable and complex, and characterized by the monitoring of variables and a manipulation and registration of dependent variables as well as an analytical, exact, and operational description. The use of the experiment in a systematic form, sets up the so-called "experimentation".

EXPERIMENTAL SCIENTIFIC METHOD PLANETOGRAPHIC SECTOR				
OBSERVATION				
FIRS PHASE	SECOND PHASE	THIRD PHASE	FOURTH PHASE	FIFTH PHASE
Experience Observation of the phenomenon via satellite survey.	**Gathering of the Information** General data acquisition.	**Choice of the laboratory physical quantities** Assumption of the reference parameters. Resolutions: Radiometric, Geometric, Topographic and Defined.	**Description of the data recording** Measurement carried out by applying the assumed parameters.	**Experiment** Repetition of the satellite survey.
HYPOTHESIS' FORMULATION *(Photointerpretation: modification processes)*				
EXPERIMENTATION *(Experimental verification of the hypothesis by repeating the experiment – Reconfirmation and image restoration processes – Application of the verification test)*				
FORMULATION OF A THEORY *(Assumption of General Scientific Principles: Scientific Law)*				

Tab. 20 Experimental scientific method – Planetographic sector

Since "planetography" stands as the first attempt to study a planet according to methods already adopted, from the same perspective, for the study of planet Earth, it analyses a planet's surface, observing its "landscape", which becomes a sensitive geographical landscape through the presentation of objective elements noticeable to direct observation, considered in itself and in its various mutual spatial relationships. On Earth, the landscape can be "natural", if you consider the elements pre-existing any human intervention and do not depend on his action, while we can call "artificial" the transformations produced/caused by human beings (also called: "*anthropic modifications*"). The landscape undergoes changes through time and the causes of such a transformation can be natural or artificial.

Natural causes can result in both abrupt and violent changes, as well as example the great natural disasters such as floods, earthquakes, volcanic eruptions, and tornadoes, but also slow and almost undetectable changes as the flow of water whose motion causes the erosion of rocks and the transport of sand and debris that accumulates downstream.

The artificial causes are to be ascribed to human intervention, so they are also called "human causes" and, considering any non-natural changes observable on other celestial bodies, the more general term of "intelligent causes" is designated. The consequence is the so-called "*modified landscape*", so that the analysis of the landscape distinguishes two study areas that are the analysis of the natural landscape and the analysis of the modified landscape.

If we apply the scientific method to planetography, the study of a phenomenon or a surface, carried out with remote sensing techniques, provides for two distinct phases that are data acquisition from the ground, by aircraft or satellite and the analysis of the captured images. This last includes the analysis for the extraction of information regarding the represented objects consisting of an "interpretative" phase with a qualitative and a quantitative analysis, and an "elaborative" phase through experimental verification **(Table 21).**

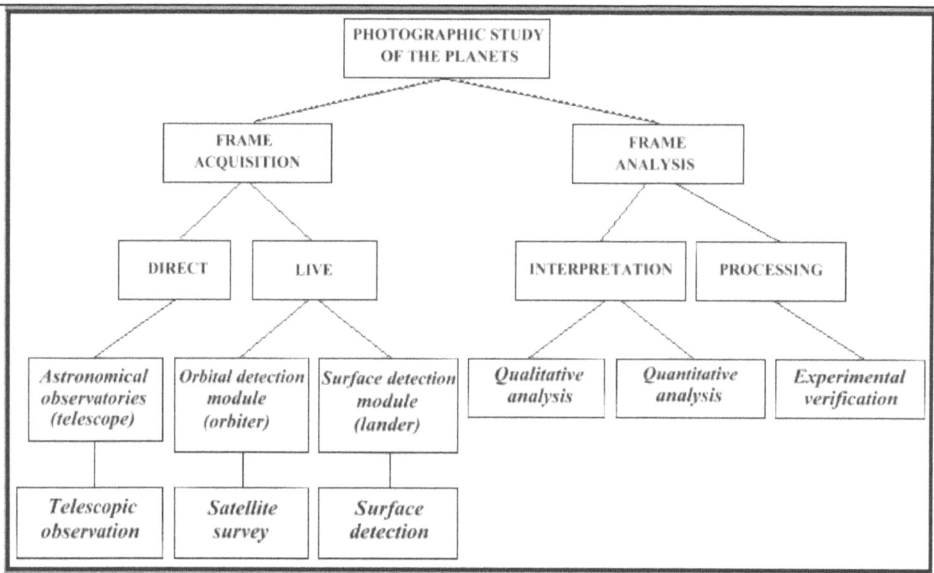

Tab. 21	Photographic study of planets.

These two different forms of interpretation – Qualitative and Quantitative – have substantial differences. The first is linked to the interpretative capacity of logical mechanisms of the human mind (visual interpretation or photointerpretation), with the consequent use of the acquired data, the second is related to the decisive contribution of the calculation systems, particularly the automatic interpretation using a machine (digital processing) or semi- automatic (if checks are carried out by man in one or more stages of processing) and also includes the classification of images.

In this regard, it should be noted that life must be sought in all possible places and, after having found it, we can say that in this place it exists. It can also be said that if the Indicators are essential in the research carried out (as far as this is possible), beyond the Solar System, given that we would be in the presence of a very few elements to be analyzed and it is true that they would be - as mentioned above – always complementary in the research conducted with reference to our neighbors. This represents a central and neuralgic point in the conduct of the Planetographic Research.

The qualitative content relates to the type of cover: dry seabed, eroded plains, mountainous formations, etc., while the quantitative content relates to geometric information, compared to the position and contours of the different qualities identified. Image processing techniques include a series of operations for a better qualitative use of them and to facilitate the extraction of information of objects represented in the qualitative analysis.

They consist of certain operations performed on an image. Finally, the experimental verification uses those data deduced from the two interpretative analyses and checks, in detailed terms, the validity of these analyses establishes their belonging to the field of scientifically verifiable elements, i.e., the result of tricks and various types of manipulations of a digital nature.

A fixed point in the morphological analysis of planets concerning the search for life outside the Earth concerns the use of so-called 'Life Presence Indicators'. Indicators are those elements through which, directly or indirectly, the presence of life on a given celestial body can be assumed or inferred. Indicators are, for example, the presence of water, as well as certain given chemical components. The use of them as an alternative to the algorithm proper, of the scientific method has to be avoided as they can only play a complementary role, together with all the other elements of analysis.

In this regard, it must be made clear that life must be sought in all possible places and, having found it, we can say that it exists there. It can also be said that if the Indicators are indispensable in the search carried out (as far as this is possible), beyond the Solar System, given that we would be in the presence of very few elements to analyse, it is true that they would – as aforesaid – always be complementary in the search carried out with reference to our neighbours. This represents a central and pivotal point in planetographic research.

4. Qualitative Analysis

The photointerpretation can be defined as the systematic collection of georeferenced information played on the base of targets and predetermined criteria by using images taken from a distance, like aerophotography or satellite imaging. The photointerpretation is then the technique that allows us to determine and classify specific phenomena by the reading of aerial photos or remote sensing images (in our case detected by artificial satellites).

The images that are the target of the photo interpretation are both *analogue* so produced by the recording of the visible light (in a few words wavelength of the electromagnetic spectrum perceived by the human eye) and images derived by information regarding the bands not visible in the spectrum.

These are made by a vast number of unifying elements called *picture elements*: such as the smallest elements in a photograph.

The *analogue* photo is a photographic format that saves all data in a continuative way so that any "visible" information is present in the frame without interruption.

The *digital* format, instead, saves every portion of information in a separate way so that by enlarging the image it is possible to distinguish many squares (*pixels*) of different tones and brightness.

The planetary images include both structures detected by artificial satellites orbiting (i.e.: orbiters) around the planet (*orbital structures*) and structures detected by the descent modules *(landers)* and this is what we define *terrestrial* or *superficial structures.* in the first case, the structures are of different dimensions that vary from a few tens of metres until hundreds of kilometres.

In the second case, we have structures that vary from a few centimetres to tens of metres.

The aerophotographic investigation assumes a refinement path to be developed in many years of observations and the possibility and the regularity to read and interpreting a substantial number of photograms but also maps made of photograms later elaborated and corrected.

The process of analysis and visual interpretation of images, highly dependent on the possibilities of human mind, works by generalisation and it is possible to go from the detail to the single observations of the abstract model. This is an analogue type of process of generalisation.

The basic preparation of a photo-interpreter (technical or a thinker of images), assumes a strong experience and the ability to recognise and define geometrically the several elements that compose the territory through the images they have.

Once they acquire the technique, it becomes a practice and so simple for a used eye that it is possible to progressively express deeper judgments on the quality of orbital photographs referring to the photographs and shootings from ground stations.

In order to be able to identify an object on an image, it is necessary to refer to previous experiences with images.

The procedure to analyse the photo interpretation include a first phase to investigate the territory that has to be interpreted thanks to information contained in the images especially using enlargements that allow to have a general vision of the area to be investigated.

In particular, the image can be read using the identification and the measurements of what is present in the picture and it can be interpreted thanks to the research of the meaning and the formulation of hypotheses as regards the context (the territory) and the functions developed by the scrutinised structure.

The photo reading (or the reading of the photograms or frames tout-court) is the knowledge of the object or of the work visible on the photo. It includes the analysis and the evaluation of its dimensions and its correct classification. It is like a road: we need to value the length and the width.

In the qualitative analysis, the data is fundamental for information and in general represents the measure of the external world or of a specific characteristic and only an expert system (human or not) can convert a data in information so reading it following some rules will result useful. in this sense, the acquisition of information from a subject goes through a cognitive process based on data.

The data that is like letters scattered without a particular order that becomes *information* when they become words. A rough satellite image forms the data and its elaboration generates information (themed images, topographic maps).

The *interpretation of data* aims to understand the real potentiality of the observed area putting into relation the objects with the surrounding environment. in a few words, it is the definition of time-space relationship between the single analysed objects to identify phenomena and situations that are not visible directly on the photographs.

As in this case the interpretation of data presupposes the use of socio-historical elements (inexistent in the specific case of the planets' study beyond Earth), it is conceivable that the interpretation uses some situations previously considered and similarly detected on the terrestrial surface.

An especially important form of analysis of the detected structures regards the observation of the homogeneous identified area. Compared to the detected forms, the irregular ones are often associated to the so-called *general morphology* (natural structures) while the surfaces with a geometrical and normally regular shape are of an artificial nature.

After the first synoptic survey of the territory, some homogeneous areas with peculiar shapes and dimensions are outlined as a starting point to identify the classification categories of the present structures. It is important to achieve familiarity with the investigated territory especially from the morphological and landscaped point of view to go into details.

The photo interpretation of the architectonical structures shall use the *architectonical photogrammetry* that is the application of photogrammetry techniques of notable buildings in the detected landscape.

It is important and recommended to use not only the images on a screen but also printed photographs to have a better interpretation of the images.

In order to analyse a structure, we must distinguish the object from its shadow. This is particularly important because the length of the shadows depend on the solar rays' inclination. in some cases, the presence of the shadows can interfere with the detection or, vice-versa, helping with the reading of photograms that is the photo interpretation. The line of the objects needs to be always neat and clear and never blurred.

When an image is analysed, the first thing to do is to quantify the possible geometrical characteristics using as a foundation the photograms scale, the measurements of the distance, the shape, the distribution of the particles and the structure characteristics.

The characterisation of an object happens through the definition of the shape and dimensions and the knowledge of the relations of its points' positions in relation to the external environment (that is the context where the structure is inserted).

A general approach to the morphological study of the detected structures consists in the transformation of an image into another where the information results more comprehensible.

All this can be done with the help of the schematic depictions' construction, which the aim is selecting and highlighting some elements in the image. in addition, the *geometrical* and non-*geometrical structures* of possible artificial nature should be detected.

In this procedure, the emphasis techniques are particularly important and they are applied to help with the images' interpretation. They are useful to give information that could be not evident in the original images. For this reason, discarding non-relevant characteristics and highlighting parameters and models of interest is a must to do.

Generally speaking, it is important to theorise the existence of *works inserted in a landscape that is completely different as regards morphology and geometrical shape prevalent or dominant.* This is the procedure that normally allows determining that some objects detected on a planet's surface *cannot be natural.* in this context, the anthropic element is the detail on a map that represents objects of the landscape created by the human hand.

All this is needed to understand the modified landscapes and also to identify anthropic sites also denominated urban *landscape* that includes settlement structures and artistic buildings. During the data elaboration phase, it is fundamental to isolate the areas of interest cutting out those parts of the image where the structures of artistic interest[127] have been detected.

In the specific field of the research for intelligent life on planets, this technique tries to detect—in relation to a natural landscape—some artificial morphologic forms such as geometrical structures and also non-natural forms even though they cannot be geometrically defined.

In general, it is possible to affirm that in the *anthropic derivation structures,* straight or broken lines prevail, sometimes with curves and regular trend while in natural structures, it is possible to recognise irregular shapes with corrugated, jagged and sometime indefinite margins.

Briefly, it is needed to detect signs or tracks of intelligent life's presence distinguishing the typology of constructions of obvious or probable artificial nature that is the possible engineering work of the biopic[128] activity.

The different detected structures can be classified using the methodology related to terrestrial geologic investigations. One of these classifications distinguishes the geomorphologic structures, the hydrographic and topographical structures. These last comprehend the *architectural structures* and the *road infrastructures.*

As an example, we enclose some elements of analysis accompanied by the relevant hypotheses of interpretation. This is a quality selection made within several hundred similar structures (**Tables 22, 23, 24, 25, 26, 27, 28, 29, 30**).

---

[127] In the terrestrial satellite research, the difficulty in recognizing the "shapes" (the detection of big urban centres is a reality limited to particular geographic areas), made the aerial photograph unsurpassed as regards the acknowledgement and the study of settlement areas.

[128] Activities linked to the presence of intelligent life.

ELEMENTS OF QUALITATIVE ANALYSIS			
Mars Map (tab number)	**Region / code (geometric resolution)**	**Coordinates (measurements)**	**Typology of structures surveyed**
MC.3.1352 SE **(23)**	Tempe Terra Viking 1 519A33 (186 m)	68.0-76.0°W 30.0-34.0°N (Km 100x200)	Architecural structure with interweaving of overhead tunnels
MC.7.1398 SC **(24)**	Galaxias Chaos Viking 1 86A11-13-15 (40 m)	208.0-211.50°W 35.17-35.23°N (Km 30x50)	Buildings connected by road structures
MC.7.1398 SC **(25)**	Galaxias Chaos Viking 1 86A33-35-37 (37 m)	211.39-212.87°W 33.93-33.97°N (Km 30x60)	Architecural structure with interweaving of overhead tunnels
MC.7.1521 NW **(26)**	Utopia Planitia Viking 2 Lander 22A252 (13.3 m)	225.73°W–47.96°N (Side m 33.0) (Height m 25.0)	Pyramid structure with a quadrangular base
MC.28.1451 NE **(27)**	Reull Vallis Viking 1 97A62 (215 m)	254.5°W – 42.5°S (Km 14x34)	Male human face, very detailed, placed in profile
MC.30.1625 H **(28)**	Promethei Terra Crater Hutton Viking 2 472B25 (384 m)	254.0°W-70.5°S (Km 10x32)	Arch structure with polished walls
MC.30.1625 H **(29)**	Promethei Rupes Viking 2 383B69 (172 m)	229.0°W-77.3°S (Km 10x70)	Arched structure with worn walls
MC.30.1625 H **(30)**	Promethei Terra Crater Ry Viking 2 510B52 (124 m)	239.2°W-66.5°S (Km 4x12)	Structure of unknown nature and metallic appearance

Tab. 22 Elements of qualitative analysis; structures surveyed and analysed (hypothesis).

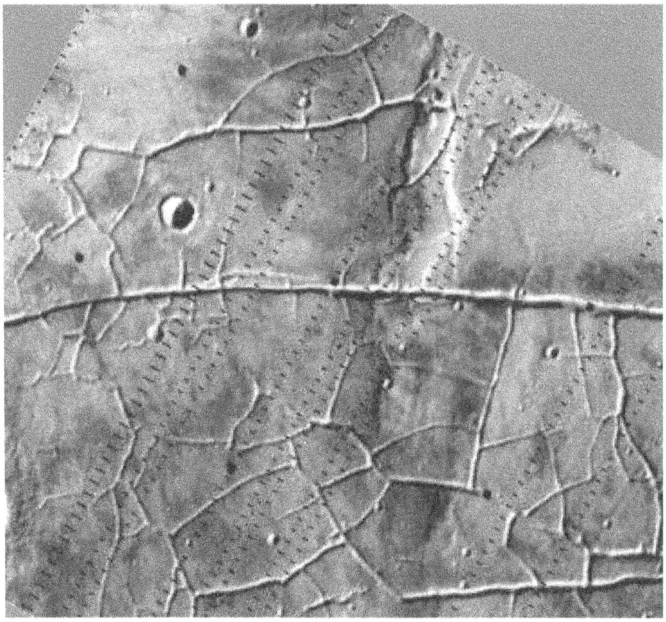

Tab. 23	MC.3.1352.SE-Tempe Terra – Architectural structure with a network of elevated galleries.

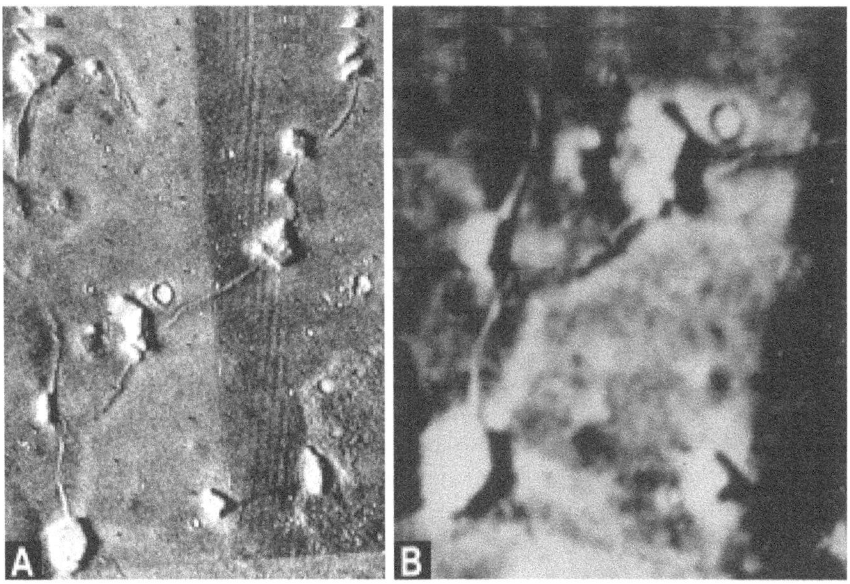

Tab. 24	MC.7.1398.SC- Galaxias Chaos – Buildings connected by road structures.

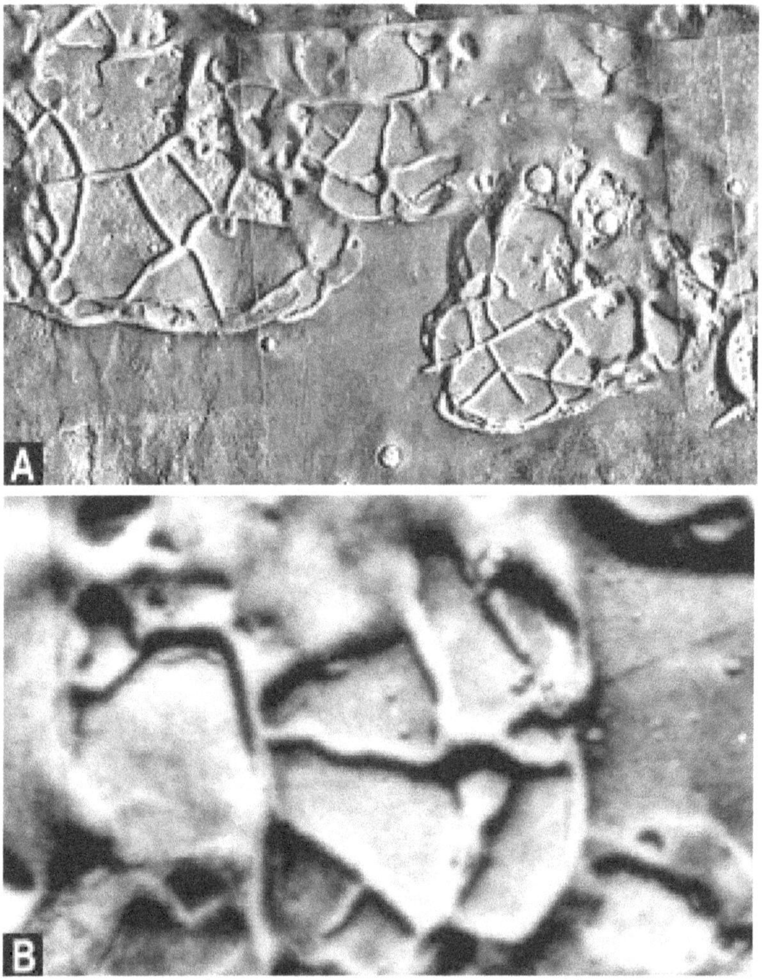

Tab. 25	MC.7.1398.SC – Galaxias Chaos – A) architectural structure with interweaving of elevated tunnels; B) enlarged portion.

Tab. 26	MC.7.1521.NW – Utopia Planitia: A) Pyramidal structure with quadrangular base; B) enlarged structure.

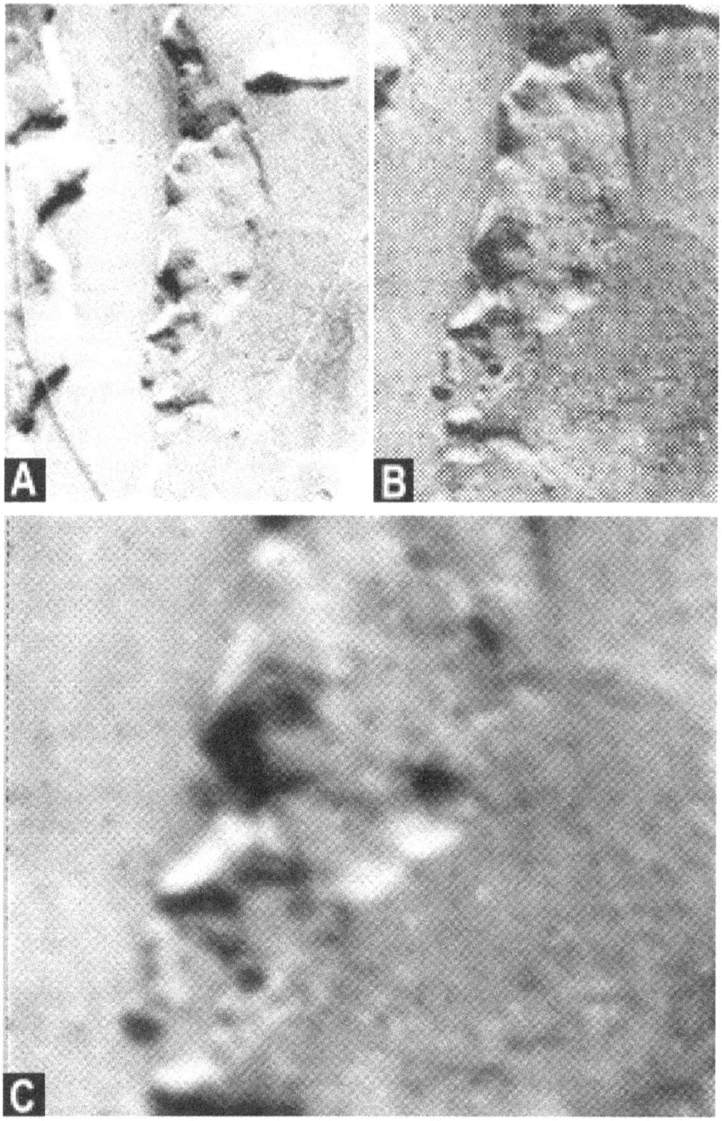

Tab. 27	MC.28.1451.NE – Reull Vallis – A,B) Male human face, very detailed, placed in profile; C) Enlarged structure.

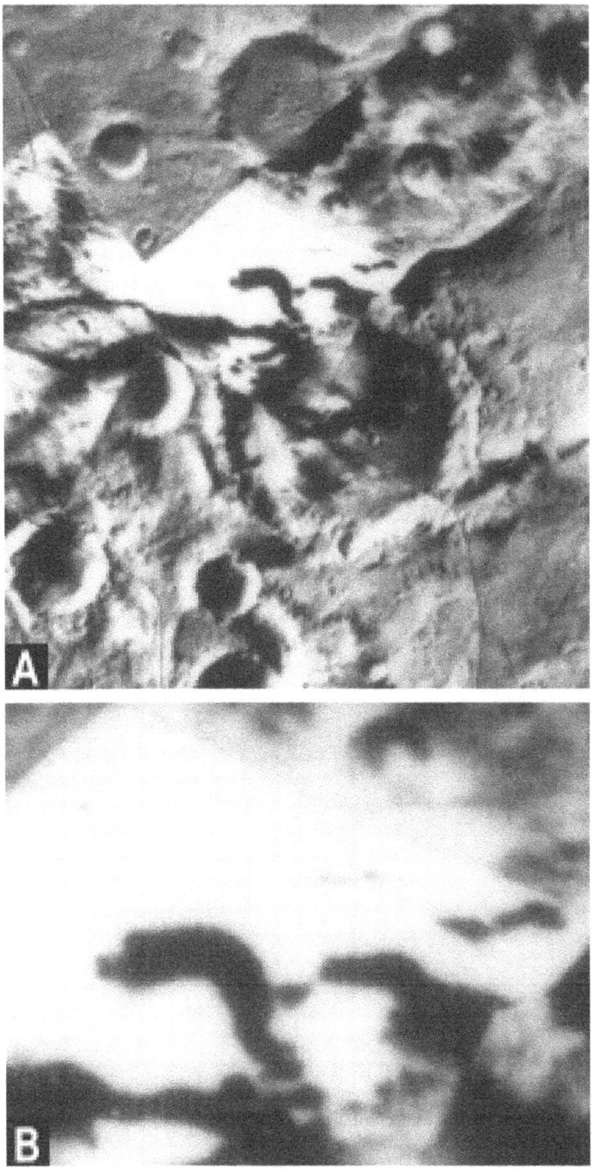

Tab. 28	MC.30.1625.H- Promethei Terra – Hutton Crater A) Arched structure with polished walls; B) Enlarged structure.

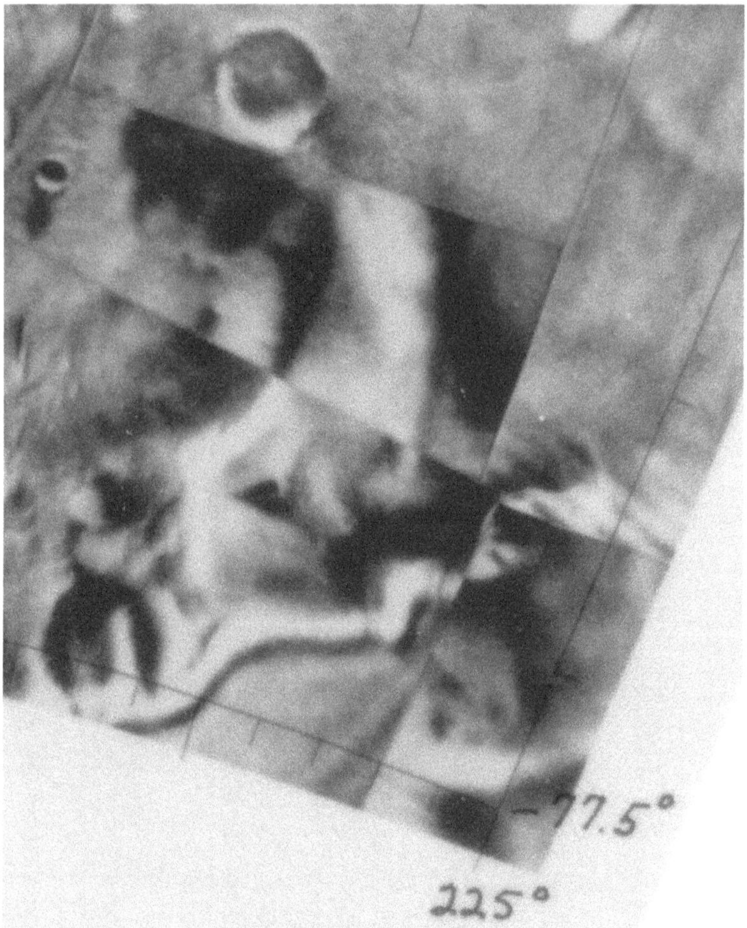

Tab. 29	MC.30.1625.H – Promethei Terra; Arched structure with smoothed walls.

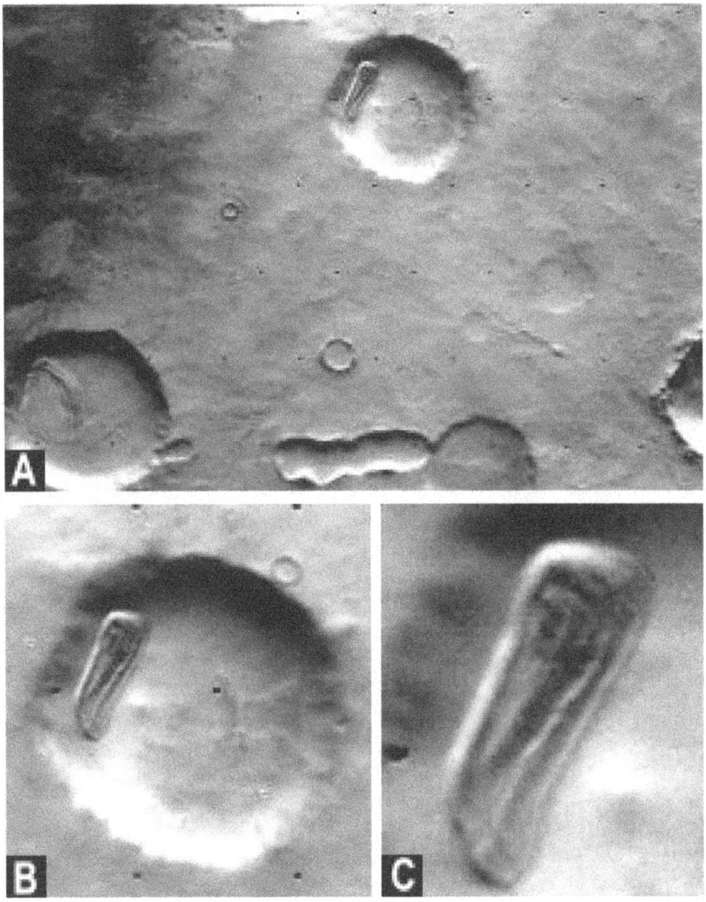

Tab. 30	MC.30.1625.H – Promethei Terra – Crater Ry; A) Structure of unknown nature, of metallic appearance; B,C) Progressive enlargements of the indicated part.

5. Quantitative Analysis

The visual interpretation of satellite images which, on occasion, could be subject to potential errors on the part of the interpreting analyst, is functionally replaced with automatic interpretation, technically decipherable as an abstract, formally defined, quantifiable, and repeatable process.

This is the Quantitative Analysis of images, conducted through the use of technical instruments (computers), which allows a comprehensive and organic review of the elements present in the various images.

Among the most important measurement values in the Quantitative Analysis there are the Photographic Resolutions (Table 31), such as fundamental evaluation parameters without which any type of satellite image analysis would be impossible.

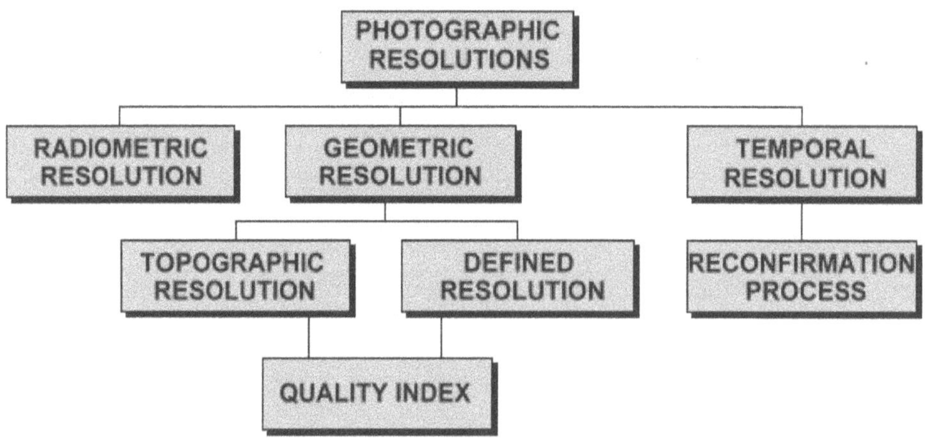

Tab. 31 Photographic resolutions.

The term “Resolution” refers to the separating power of a measuring or observation instrument, where the detected scene can be measured and divided into its various composing elements.

The basic Photographic Resolutions are the following ones: ***Radiometric***, ***Geometric***, ***Topographic***, and ***Defined***.

Radiometric Resolution (RR).

Every digital photograph is composed of a grid of small squares (or "*matrix*"), each of which is an "*image element*," abbreviated as bytes or pixels. It represents a portion of the elementary surface of which the image itself is composed and is the result of the combination of 8 elementary bits and is identified by an integer digital number (DN - Digital Number), between 0 and 255, which varies depending on the number of bits used in the acquisition. The value of each square corresponds to the intensity of the radiation (radiance, or brightness or energy level) reflected by the observed object, and measures the brightness value, where the image is displayed on a monitor using a scale of 256 gray levels.

The "Radiometric Resolution" refers to the levels of recognizability and representativeness of objects, especially with reference to the contrast of their respective tonal values, and is an essential trait in defining detectable detail, and therefore in the visual perceptibility of parts of an image. It represents the sensitivity of a given sensor's detector in perceiving and encoding into a signal the differences in radiant flux reflected or emitted by the analyzed surfaces and the number of levels into which the original signal can be broken up (Table 32).[129]

[129] Remote Sensing Course – Lesson 3 – Filming Tools – Resolutions – Satellites for Remote Sensing. (from: 508_2008_73_3588.pdf).

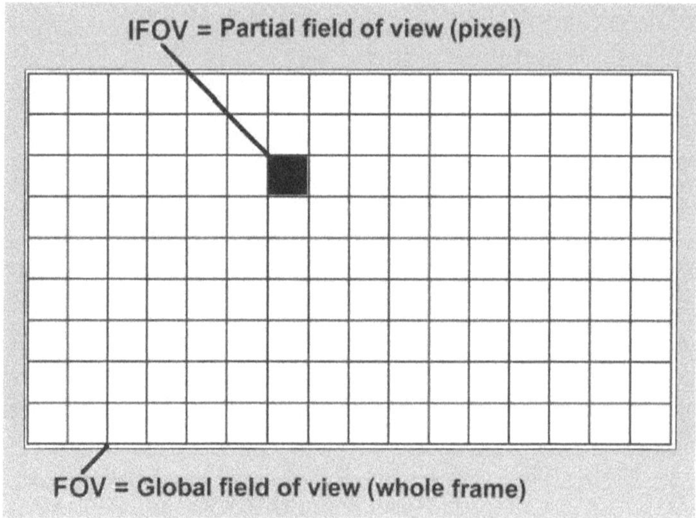

Tab. 32	Photographic grid (IFOV-FOV).

Geometric or Spatial Resolution (GR).

The Geometric Resolution refers to the ability of objects of a certain size to be more or less visible (in our specific case) to satellite cameras and measures the minimum distance at which two adjacent objects appear individually separated in the original image.

In technical terms, it is the resolving power of the image point, that is, the pixel representing the smallest unit defined by the observation geometry of the instrument's sensor, located on board the satellite.

Representing the pixel as a small square, the length of its side must be taken into account, which represents the Geometric Resolution and is indicated by length measurements; therefore, a Geometric Resolution of 50 meters means that the minimum unit is 50 meters on a side and that two objects less than 50 meters apart are not resolved, meaning they cannot be seen individually separated as two different objects.

Thus, in an image with a resolution of 800 meters, objects the size of a truck, for example, would be indistinguishable, meaning that small objects are invisible to satellite cameras.

We can fully explain the concept of Geometric Resolution with an example from optical physics. In vision (the eye or the photographic lens), at the nodal point, where the object is converted into an image, these two elements subtend an equal angle called the visual angle, which represents the angular measurement of the size of the retinal (or photographic) image. [130]

Let's take the case of the same object (or two of equal size), observed from different distances: it will form photographic images of varied sizes depending on the distance from the camera lens from which it is positioned. The size of the photographic image is inversely proportional to the distance from the lens, and the resolution (or resolving power) of the optical instrument used is inversely proportional to the diameter of its lens, essentially the focal length.[131]

In the attached table[132] (Table 33): we have the diagram of the human eye, with the object O'B', which, despite having linear dimensions equal to

---

[130] Resolution can be described in various ways. However, a completely satisfactory definition does not still exist since factors such as the interpreter's experience can determine the actual resolution of an image. A list of factors that control the resolution includes contrast level, resolution feed system, atmosphere, image motion, and geometric display. Mutch et al., The Geology of Mars, Op. cit., p. 327.

[131] The Geometric resolution is technically influenced by the aperture of the light beam entering the lens. The focal length corresponds to the maximum aperture of the diaphragm, therefore to the brightness and measures the power of the camera's lens; in practice, it is the distance traveled by the light beam from the lens to the focal plane and is measured in millimeters.

[132] See Gian Paolo Paliaga, The Vices of Refraction, Edizioni Minerva Medica, Turin 2008, p. 45.

OB, subtends a wider visual angle (a'>a) and forms a larger photographic image because it is closer to the lens. This makes us fully understand the importance of the shooting distance in constructing the value of the Geometric Resolution.[133]

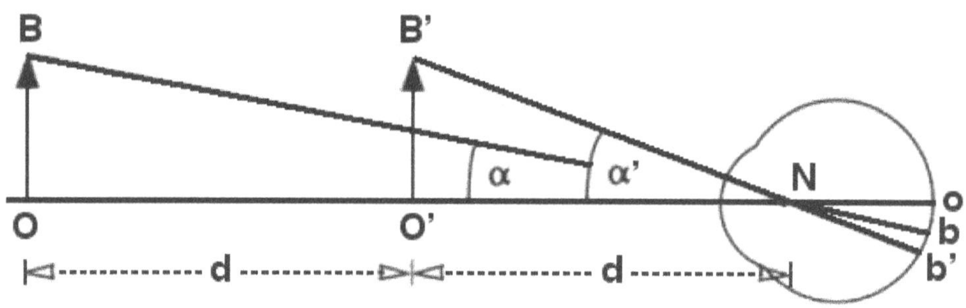

Tab. 33	Geometric resolution (optical physics).

[133] The Italian Space Agency's website states that "The first important thing to know about satellite images is their resolution." From: ESA – Eduspace – July 29, 2017.

The GR value is relative to the ground pixels visible in the original image, and each pixel represents a portion of terrain whose size is a function of the orbital height, since a smaller distance corresponds to a higher resolution and vice versa.[134]

For the planet Mars, the Geometric Resolution data was presented analytically only starting with the Viking mission (1976-83), while for the previous mission, Mariner 9[135] of 1971-72, the same data was not inserted into the frames (labels) of the frames and was obtained through the application of a formula resulting from the combination of three elements: the shooting distance (R),[136] the focal length of the lens/lenses (f), and the pixel scale (PS) (Table 34).

---

[134] Land Resolution or, more correctly, "*Ground Resolution.*" The original definition is as follows: it is the minimum distance on the ground between two closely located objects at which they are distinguishable as separate objects. The methodologies for determining the capability of individual sensors to achieve a specified ground resolution, including the minimum altitude from which such a resolution can be achieved, were developed by the OSCC during the Treaty's period of provisional application. The Ground Resolution is the Minimum Distance between two objects (bodies) relatively (somewhat) close, both on the ground, at which they appear (to the eyes of an Observer — i.e., to the Orbiter's cameras) as two separate objects. The methodologies adopted for determining the capability of individual sensors (optical cameras) to achieve a specific Ground Resolution—including the minimum altitude from which such a resolution can be achieved—were developed by the OSCC (*), during the period of provisional application of the treaty it signed. (*): Open Skies Consultative Commission (Note by Dr. Paolo C. Fienga).

[135] In the Mariner 9 Mission, we have two sets of data, since that mission used two different types of cameras. The first (a WAC telephoto lens) called "Camera A," with a focal length of 52.267 mm and a pixel scale of 75.0 pixels/mm, the second, called "Camera B" (also called "high resolution"), was a NAC with a wide angle and a focal length of 500.636 mm, as well as a pixel scale of 75.0 pixels/mm: these represent the constant data referred to the two cameras, to which a third (variable) data relating to the shooting distance must be added.

[136] The shooting distance (R) is a variable data in relation to the orbital motion of the satellite detection station; it measures the distance between the camera and the detected surface; the focal length (f) determines the resolving power, i.e. the magnification capacity of the optical system used; the pixel scale (PS) provides the size of the pixel on the image, present within 1 square mm, i.e. the resolution with which the image was scanned and determines the size of the smallest part detected on the ground (pixel).

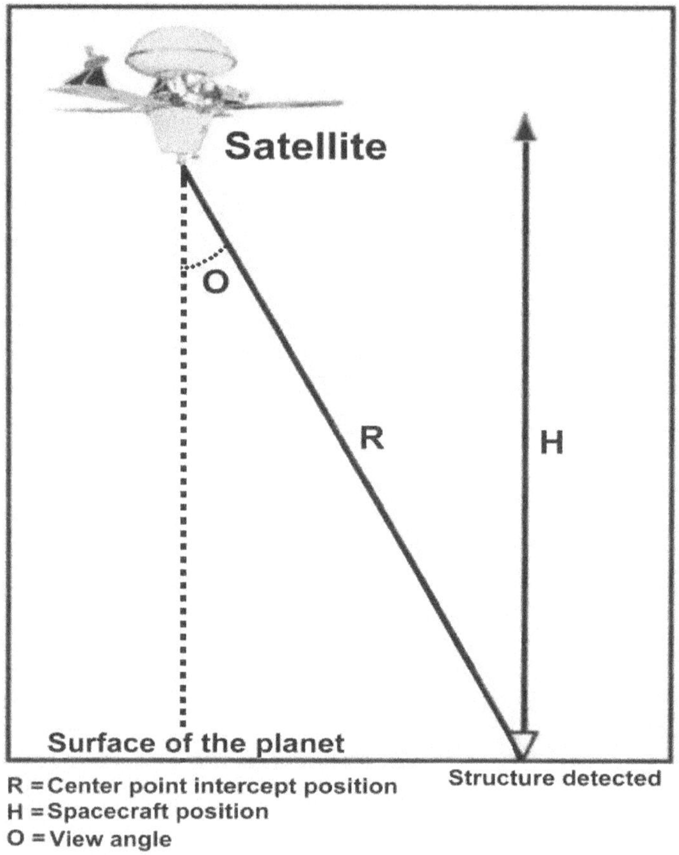

Tab. 34	Satellite tracking.

In the planetographic application, the relationship between the three data is expressed by the following proportion: PS: f = GR: R, where PS is the pixel scale, f is the focal length, GR is the Geometric Resolution and R is the shooting distance. Setting GR as the unknown to be found (fourth proportional), we have: GR = R: f x PS. Therefore: given the focal length (f) and the shooting distance (R), it is possible to calculate the Geometric Resolution of the object, or the Geometric Resolution (GR) near the projection of the principal point "on the ground" of any aerial image, given the Resolution with which the image was scanned (PS) (Tables 35, 36).

LEVELS OF GEOMETRIC RESOLUTION		
Measure detectable / pixel	Name	Acronym
Up to 10 meters	*Very High Resolution*	*AA*
From 10 to 100 meters	*High Resolution*	*A*
From 100 to 500 meters	*Medium Resolution*	*M*
From 500 meters to 1 km	*Low Resolution*	*B*
From 1 km upwards	*Very High Resolution*	*BB*

Tab. 35 Geometric resolution levels.

COMPARISON OF RESOLUTION LEVELS AND CENTER POINT INTERCEPT					
Intercepting range (kilometres)	GEOMETRIC RESOLUTION (metres)				
	Mariner 9 Camera A (1971-1972)	Mars Odyssey (2004)	Mars Express (2006)	Mariner 9 Camera B (1971-1972)	Viking Ortbiter 1&2 (1976-1980)
15.000	3.826	660	655	400	378
10.000	2.551	440	436	266	252
5.000	1.275	220	218	133	126
3.000	765	132	131	80	75
2.000	510	88	87	53	50
1.500	382	66	65	40	38
1.000	255	44	43	26	25

Tab. 36 Comparison of resolution levels and shooting distances.

Topographic Resolution (TR).

Based on the value of the Geometric Resolution, such an image is said to be "High," "Medium," or "Low" Resolution (Table 35).

Geometric Resolutions are high if they have low numerical values, while they are low if they have high numerical values. Thus, 500 meters represents a medium-low level, 50 meters a high level, 10 meters a remarkably high level, etc.

What value do these parameters have, and could they provide us, in a technical sense, with an idea of the quality of the image itself?

In this context, it should be noted that Geometric Resolution is not an absolute value, i.e., "*detached*" from the entire context, but depends on at least three factors:

1) the technical characteristics of the cameras mounted on the observation satellite, and in particular the focal length of the lens (essentially, the magnification capacity of the lenses used);

2) the spatial position of the observation satellite, specifically the distance between the camera lens and the imaged Structure;

3) the dimensions of the Structures detected on the ground.

Factors 1 and 2 are automatically included in the calculation of Geometric Resolution, but it appears that the Factor 3 did not receive the right importance. This means that the definitions of High or Low Resolution take on a relative value, since – for example – to better detect a Structure many Km across requires a lower resolution than to detect a Structure just a few hundred meters across!

Again, to clarify the concept, we will use optical physics. Referring to the example of the human eye, as it has been mentioned above, let's now consider two objects (or Structures) of different sizes, placed and observed from equal distances: the width of the visual angle is directly proportional to the size of the objects; that is, the larger is the object, the larger shall be the photographic image.

In the first attached diagram[137] [9] (Table 37), the visual angle of object AA' has larger dimensions than object BB' and subtends a wider visual angle (a>b), consequently the retinal image a'a is larger than the image b'b. In the second attached diagram[138] [10] (Table 38), the same "*alpha*" angle is subtended by objects of different sizes, when their distance from the lens is directly proportional to their dimensions.

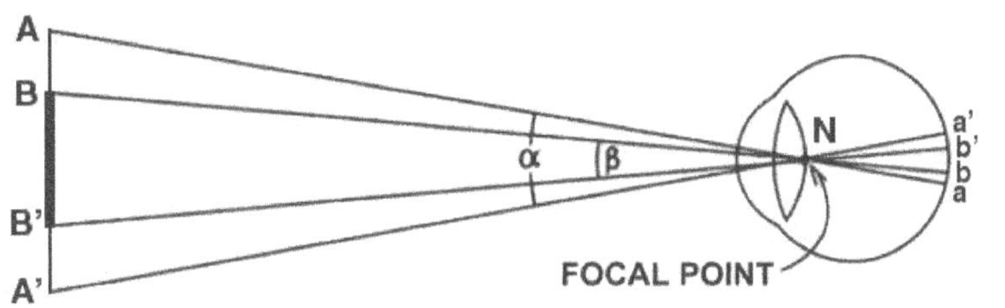

Tab. 37	Topographic resolution (optical physics).

[137] See Gian Paolo Paliaga, "The Vices of Refraction", Edizioni Minerva Medica, Turin 2008, page 45.

[138] Idem.

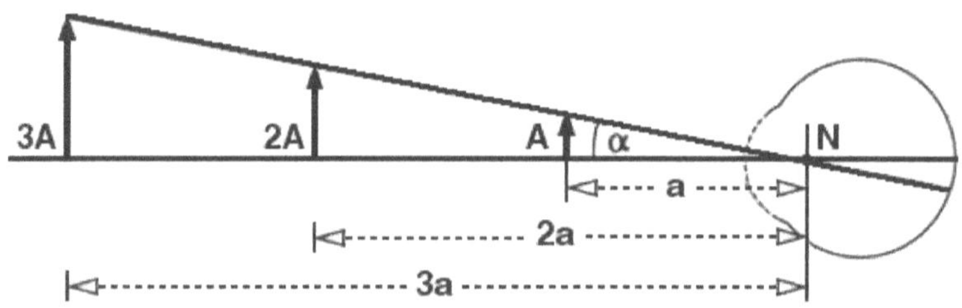

Tab. 38	Topographic resolution (optical physics).

It follows that in the context we have considered it is impossible (and scientifically incorrect) to speak of Geometric Resolution without considering the size of the imaged Structure. It is therefore entirely consequential to deduce that "objective" images are not necessarily satellite images that can be defined as High-Resolution.

Therefore, it must be said that the quality of a satellite image must be deduced from the ratio between the size of the imaged Structure and the Geometric Resolution obtained: this ratio is called "Topographic Resolution." To calculate the Topographic Resolution (TR), we must divide the side (S) of the imaged Structure (measured in Km) by the Geometric Resolution (GR) of the image (measured in meters), according to the following formula: TR = S:GR. The results derived from the ratio just indicated indicate values between 0.15 and 10.00.

To clarify the concept of Topographic Resolution, we provide the example of two Structures with different dimensions (Table 39): the photo above refers to a Structure in the Lybia Montes region, measuring 0.873 Km across, surveyed with a resolution of 5.78 meters; the photo below refers to a Structure in the Pavonis Mons region, measuring 7.0 Km across, surveyed with a resolution of 43.09 meters. As can be easily deduced from the calculation of the Topographic Resolution, these images, although taken with different levels of Geometric Resolution due to their different dimensions, present the same level of Topographic Resolution. Therefore, although there are two different GR values, the real value is the same, because it is related to the respective dimensional values, measured using the Topographic Resolution.

Definite Resolution (DR).

Given that the surveyed terrestrial Structures are part of our direct cultural heritage (because they are experienced directly through on-site visual observations and/or photographic imagery), it is possible to deduce a minimum hypothetical level of Geometric Resolution for any given surveyed Structure. Taking into account the levels of Geometric Resolution obtained (and the associated image quality) from satellite surveys conducted on Earth since 1972 by Landsat satellites, we used the Topographic Resolution to obtain the fixed number 0.26. This results in two sets of data: a hypothetical level of Geometric Resolution (which we will call Definite Resolution) and a *Quality Index.*

The first step is dividing the size of the surveyed Structure by the fixed number 0.26 (i.e., the Topographic Resolution), it will be possible to derive an optimal level of Geometric Resolution, which is the Definite Resolution.

As a second operation, we will compare the hypothetical Geometric Resolution (Defined Resolution) and the actual Geometric Resolution: the result will be the so-called *Quality Index.*

Let's now consider an example using a satellite survey from the Mariner 9 Mission, photo DAS 7686388 from 1972. The surveyed Structure measures 80 x 104 Km, has a Geometric Resolution of 446.42 meters, and a Topographic Resolution of 0.23. Once we divide the Structure's size by the fixed number 0.26 we obtain the optimal Geometric Resolution (such as the Defined Resolution) of 400 meters. The next step is to calculate a ratio between the DR of 400 meters and the actual GR of 446.42 meters, which gives us a negative result of -10.39%.

This means that the actual GR is lower than the optimal one of 10.39% and such a percentage represents the abovementioned *Quality Index.*

Pavonis Mons M9 Das 4402165	Lybia Montes MGS M02-03051
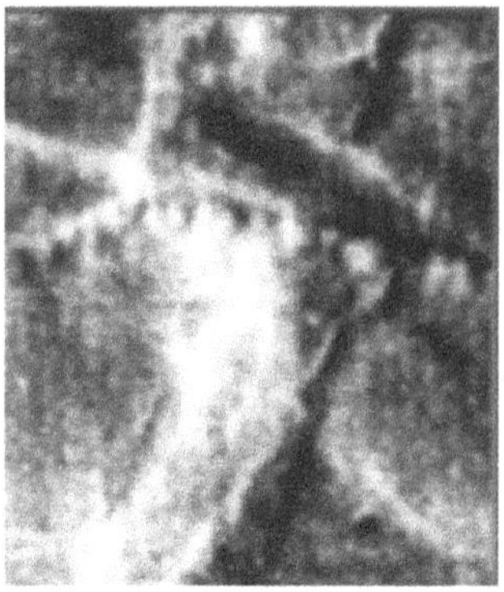	

RELATIONSHIP BETWEEN GEOMETRIC AND TOPOGRAPHIC RESOLUTION		
Dimensions of the structure (km)	Geometric resolution	Topographic resolution
Lybia Montes: 0,873	5,78	0,15
Pavonis Mons: 7,00	43,90	0,16

Tab. 39 Relationship between geometric and topographic resolutions.

6. Experimental Verification

The purpose of the experimental verification is to check the correspondence between the original images released by the satellite surveys and the planetary cartography composed by mosaicking the frames.

It includes the *reconfirmation process*, *modification methodologies* and *verification tests*, which carry out a detailed cartographic analysis of the various maps of Mars constructed on the basis of the satellite surveys of the missions Mariner 9 and Viking 1&2 (Table 40).

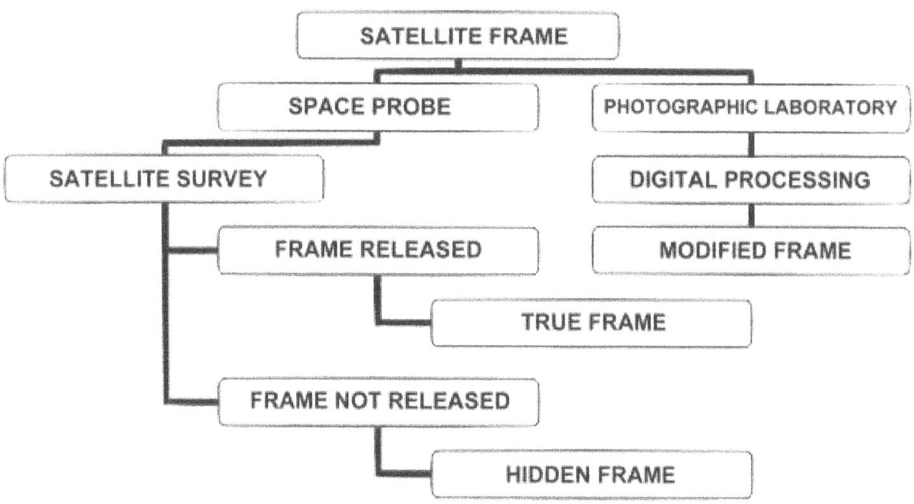

Tab. 40 Satellite image.

Reconfirmation process. The reconfirmation process takes into account the so-called 'temporal resolution', which represents the period of time between two successive shots of the same area. Our starting assumption is that a structure detected with a satellite camera is progressively redefined in detail as the shooting distance decreases. in this way, it is possible to establish an inversely proportional relationship between the decrease in the shooting distance (distance of the sensor from the detected structure) and the increase in the quality of the detected image. The purpose of this process is to check that a given image – analysed in its presumed detection path – is also a faithful representation of the observed scene **(Tables 41,42,43).**

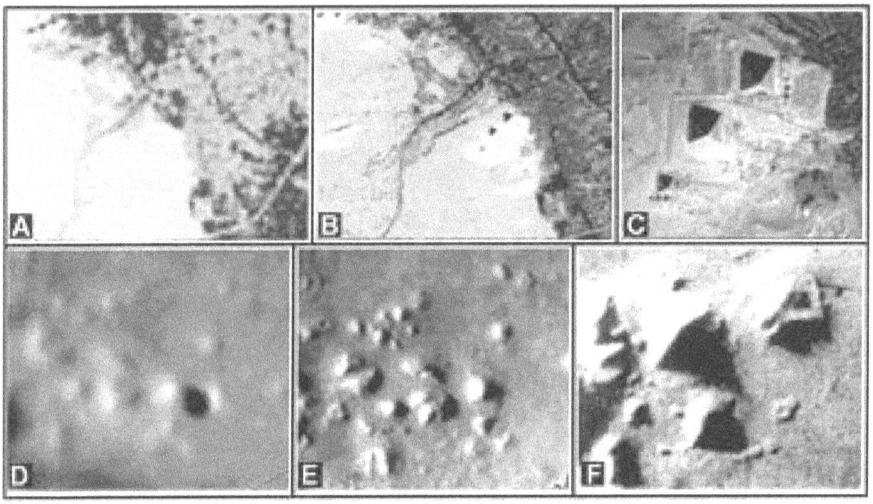

EARTH (Giza - Egypt)			MARS (Cydonia Colles - Arabia Terra)		
Mission Year	Intercepting Range Km	Geometric Resolution m	Mission Year Code	Intercepting Range km	Geometric Resolution m
Landsat 1 1972 A	920	79,00	Mariner 9-1972 Code 9162624 D	3.483	875,76
Landsat 4 1982 B	721	28,80	Viking 1-1977 Code 561A25 E	6.277	158,00
Ikonos 2 1999 C	681	1,00	Viking 1-1976 Code 35A72 F	1.873	47,13

Tab. 41 Reconfirmation process – Relationship between terrestrial and Martian structures. A,B,C) Graphical evolution of the pyramids of the Giza region in Egypt; D,E,F) Graphical evolution of the pyramids of Cydonia Colles, on Mars; attached diagram with satellite information.

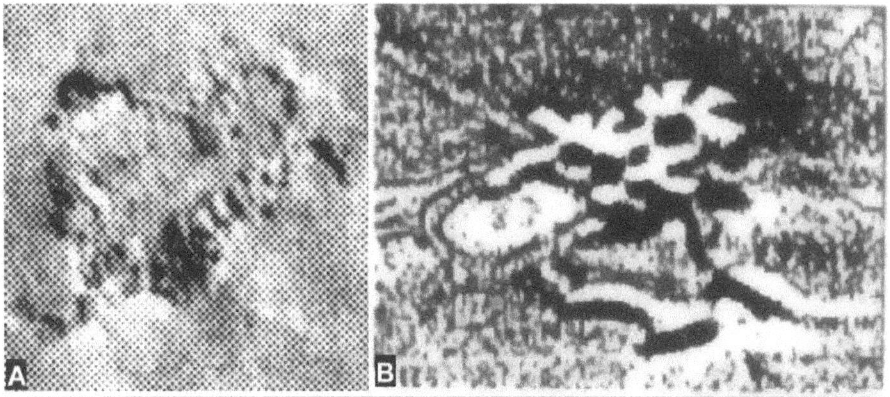

Tab. 42	Apollinaris Patera – Reconfirmation process – A) Photo 635A06 Viking 1 – ris. m 254 (1978); B) Photo Das 9341014 Mariner 9 Camera B – ris. m 48,2.

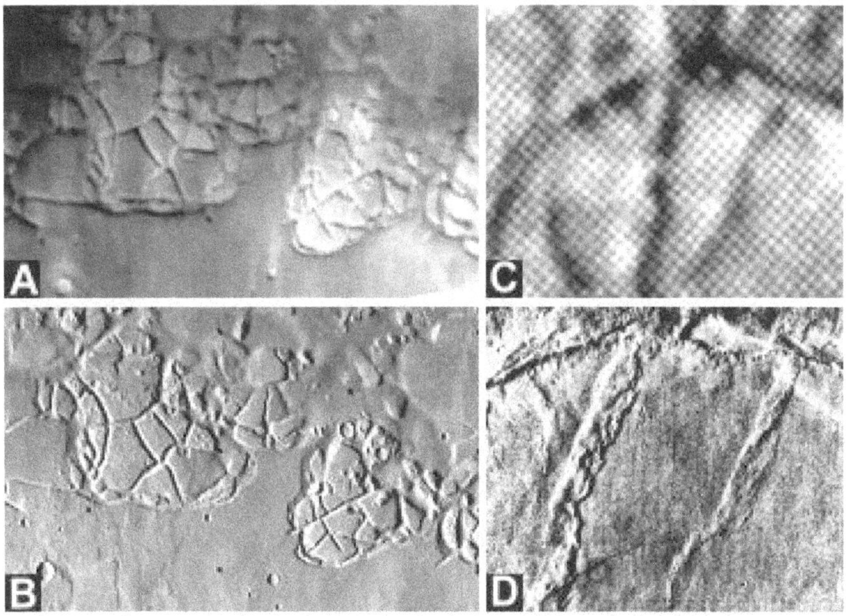

Tab. 43	Reconfirmation process (geometric resolution) – Galaxias Chaos – A) MGS M600806 2004 m 272.27; B) Viking 1 541A14 1977 – m 163; Pavonis Mons C) Mariner 9 Camera A – 1972 Das 7111128 m 535.20; D) Mariner 9 Camera B – Das 4402165 m 43.90.

In this context, experimentation must verify, in absolute terms, that the succession of resolution levels always corresponds to a progression of quality levels.

The overview of the images in sequence, allows a careful observer to grasp the gradual increases in quality of the images portrayed, placing them in a coherent context where it is possible to check a direct and correlative relationship between the various elements, the concatenation of which explains and justifies in a logical (hence scientific) manner, the elements present in each frame.

In the *actual process of reconfiguration*, the image presents the following characteristics: 1. It shows the same structures; 2. It produces a proportional increase in image quality; 3. Persistent structures are proportionally increased in quality; 4. It increases the possibility of distinguishing a greater number of details (Tables 44,45,46,47).

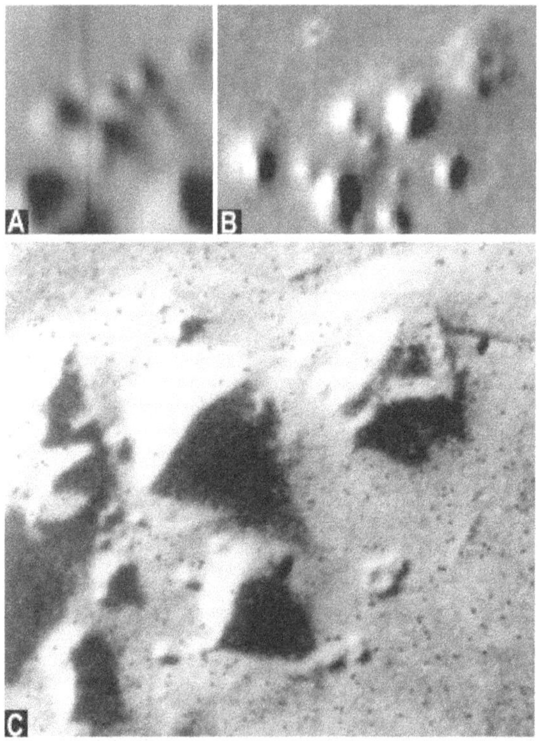

Tab. 44	Reconfirmation process of the 'Group of Pyramids' (Cydonia Colles); A) Mariner Mission 9 Das 9162624 (res. 875.76 metres); B) Viking 1 – code 561A25 (res. 189.565 metres); C) Viking 1 code 35A72 (res. 47.13 metres). The 'Pyramid Group' comprises a dozen structures, the most notable of which are the Akis Pyramid (located to the west) and the Ruined Pyramid (located to the east). The Akis Pyramid is a structure approximately 440 metres high, with a hexagonal base, a width of 2.3 km and a long axis of 2.8 km, a perimeter of 7.6 km and a total base area of 4.34 km2. The Ruined Pyramid presents a perfect geometric shape and is probably a pyramid structure with a rectangular base, partially demolished. The smaller side is about 2,000 metres, the larger 2,300 metres, and the height (prior to the probable demolition) is assumed to have been about 500 metres (today, no more than 400).

RECONFIRMATION PROCESS (Cydonia Colles)			
Technical data	Portion	Straightened portion	Explanatory notes
Viking 2 - 1977 Cod. 220S38 Res.m 728,0	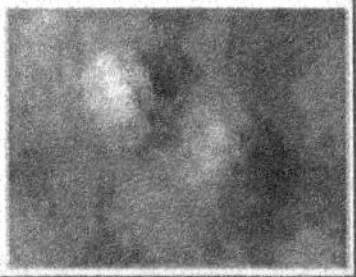	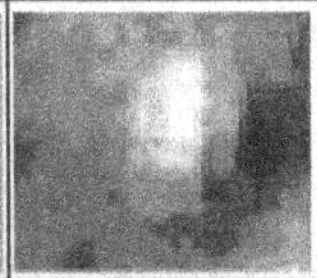	Invisible structure Position deduced from subsequent images
Viking 1 - 1978 Cod. 753A34 Res.m 232.0	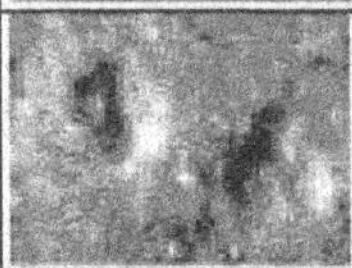	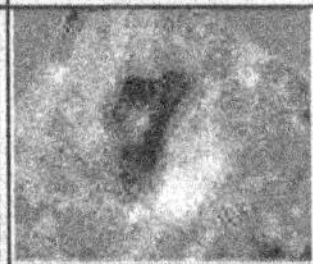	Structure just glimpsed (outline)
Viking 2 - 1978 Cod. 673B56 Res.m 169.0	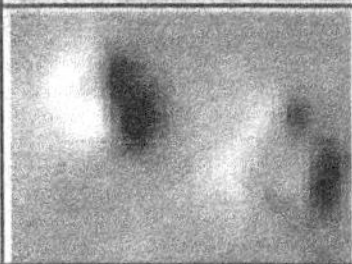	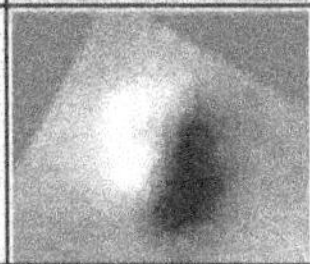	Structure with first contours
Viking 1 - 1977 Cod. 561A25 Res.m 158.0	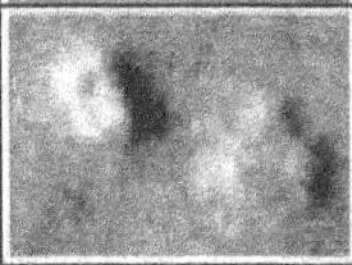	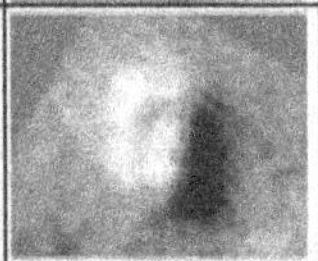	Structure surveyed with some details
Viking 1 - 1976 Cod. 35A72 Res.m 47.13	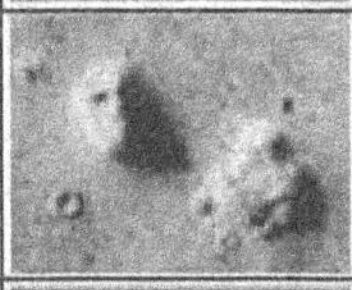	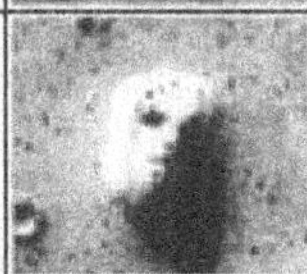	Structure detected: distorted by the sun being too low
Viking 1 - 1976 Cod. 70A13 Res.m 43.42	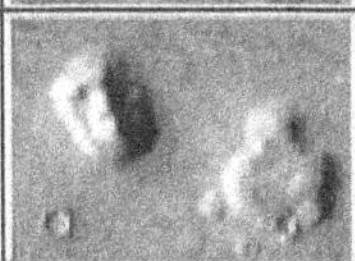	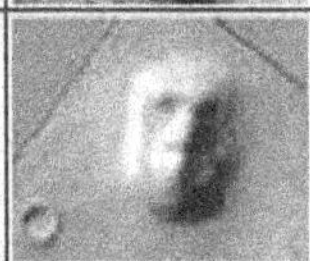	Structure faithfully captured with part of the left face
Mars Odyssey 2004 Cod. V12445004 Res.m 37.83	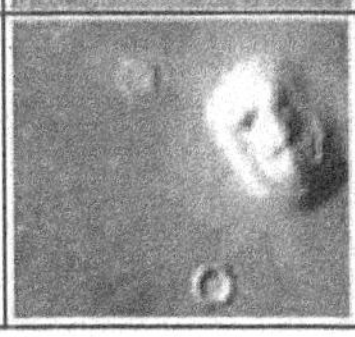		Reconfirmation and redefinition of the previous structure

Tab. 45	Reconfirmation process of an anthropomorphic structure (Cydonia Colles – Face on Mars).

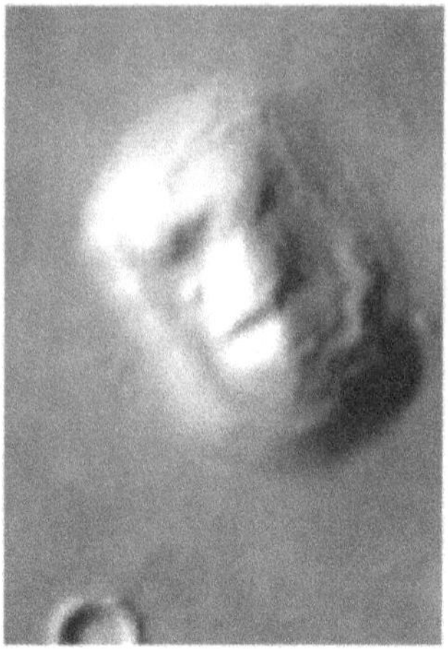 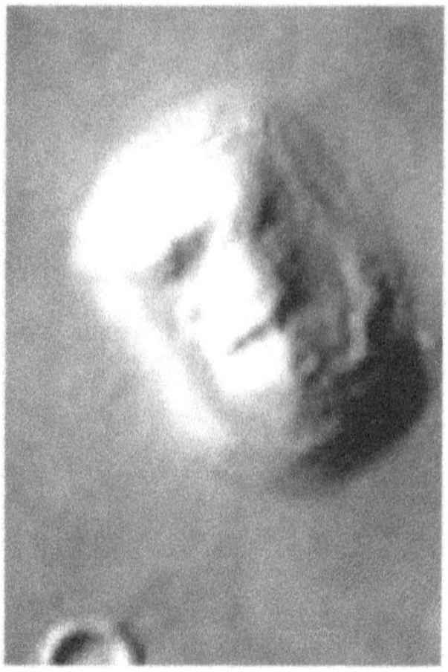

TWO IMAGES COMPARATED (Cydonia Colles)

Mission Code	Date	Frame time	Slant range	View angle	Geometric resolution
Viking 1 70A13	30 August 1976	16.84	1.725	62,65°	43,42
Mars Odyssey V12445004	4 October 2004	17.41	433,581	67,63356°	37,837

Tab. 46	Reconfirmation process – Cydonia Colles – Two images compared: Viking 1 – 70A13; Mars Odyssey V12445004.

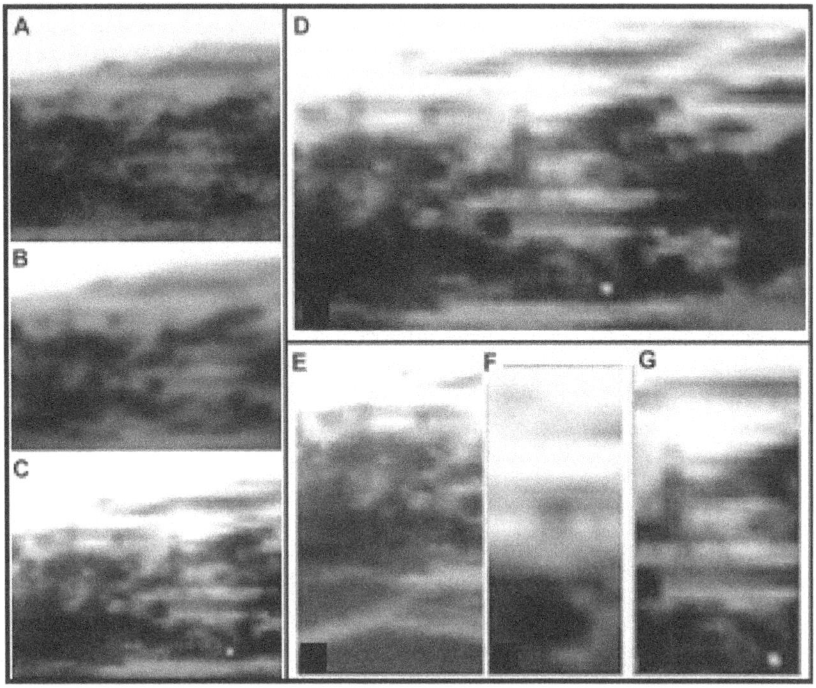

Tab. 47	Reconfirmation process – Chryse Planitia region (47.968°W – 2.480°N). Surface survey by the Viking Lander 1 probe (Code 12A235, 25 July 1976, survey distance 1,500 metres). Morphological data: hill 65 metres high, 500 metres wide. Structure with the appearance of a multi-storey building. A/B/C: Reconfiguration process of the structure; Operations on photo C; D-enhancement; E-western part; F-particles; G-eastern part.

Modification methodologies. Given the validity of the reconfirmation process, should the experimentation detect the presence of a frame that is not in line with a quality progression, we would be in presence of a non-homogeneous datum, which would attest to the certainty that the proposed images would not be authentic, but rather digital modifications of real images or virtual creations on a digital level.

Modification methodologies are distinguished by intervention sectors and types used. Sectors of intervention include *synthetic operations*, which treat a set of images (photomosaic), and *analytical operations*, which treat an isolated image (in some cases, included within a photomosaic).

The types of operations include erasure processes and modification processes.

Frame erasure consists of a graphic processing aimed at the total elimination of both the structures previously surveyed by other missions and the entire topographical context within which the structures in question are inserted. The result is the general modification of the morphology of the terrain for the extension of hundreds of kilometres, through the superimposition of an image or any other element that can hide the image itself, to the point of making it invisible (to the layman's eye), totally replacing the image with another specially constructed one. Another erasure technique consists of superimposing on a given image a series of vertical bands that are very close to each other, making it impossible to identify a given structure.

Modification of the frame can be implemented at the level of geometric or radiometric resolution processing.

Modification at the geometric level involves changing the size and increasing or decreasing the amplitude of the sides, one in relation to the other. These structural modifications are essentially carried out through the use of a high-band filter, which transmits all frequencies above a certain limiting frequency, which attenuates all the others. The purpose is to remove any visual clues as to the true size and shape of an object.

Modification at the radiometric level modifies the chromaticity with the use of contrast filters: one for background noise with one of the many commercially available computer programmes, another to obscure unwanted details that could not otherwise be removed, and finally, a relief filter, to add non-existent details, for example, terrain that is rougher and more 'natural' than a simple enlargement would make it appear (**Table 48**).

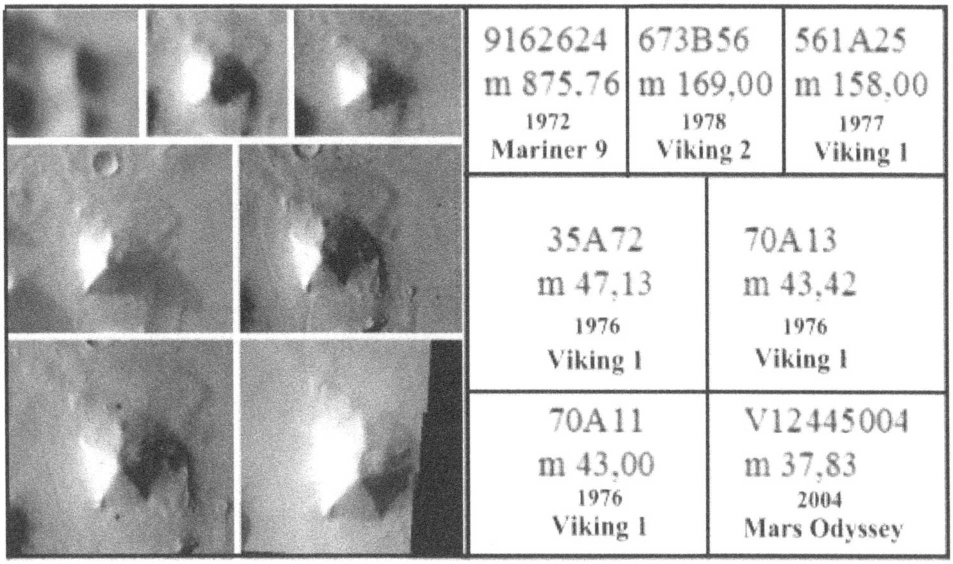

Tab. 48 Reconfirmation process (D&M Pyramid – Cydonia Colles). Explanation attached in the body of the photo.

The complex of such modifications includes the transformation of geometrically configured lines (traced and delimited), with clear and precise contours (so that they are recognisable as topographical structures), into slender and imprecise structures, devoid of limits and boundaries, in practice into fluid and undulating structures that can in some way be traced back to elements of a natural type. in some cases, such operations tend to pose images that are the result of shadows as real, transforming them into hypothetical and non-existent real and permanent structures.

Other operations concerning radiometric processing are the attenuation of the intensity of individual pixels, with an increase or decrease in brightness, and the negative effect, which results in a chromatically inverted image. in some cases, mirror rotation, also known as position inversion, is also used.

In the event that the reconfirmation process finds anomalies, the modified (or deleted) images can be restored and redefined thank to certain graphic recovery techniques, returning them as far as possible to the original situation (**Tables 114,119,120,121,123,124**).

Verification tests. in addition to the analysis of the progressive sequence of geometric resolution levels – known as the reconfirmation process – images can also be checked by means of photogrammetric analysis using verification tests, through which it is possible to both check the presence of authentic images and whether we are in the presence of natural or artificial structures.

The most common tests used in this field are the TM3D Test, Fractal Test, Shadow Test, Context Test, Bilateral Symmetry Test, Internal Relationship Test.[139]

The *TM3D Test* checks whether a given structure is such from whichever side you look at it, or whether only from certain angles; it then tries to check whether the image refers to a three-dimensional structure. When a given structure is considered to show a given configuration because it was taken from a certain angle, it is said to be an "optical illusion" and the effect shown is due to a "play of light and shadow"[140] (**Tables 49,50,51,52**).

---

[139] The result of even four of the listed tests can be predicted by pure chance is calculated at one in a billion.

[140] It should be noted that such expressions are an integral part of the *canonicity* of the so-called "orthodox" science, which uses them outside of a proper scientific context.

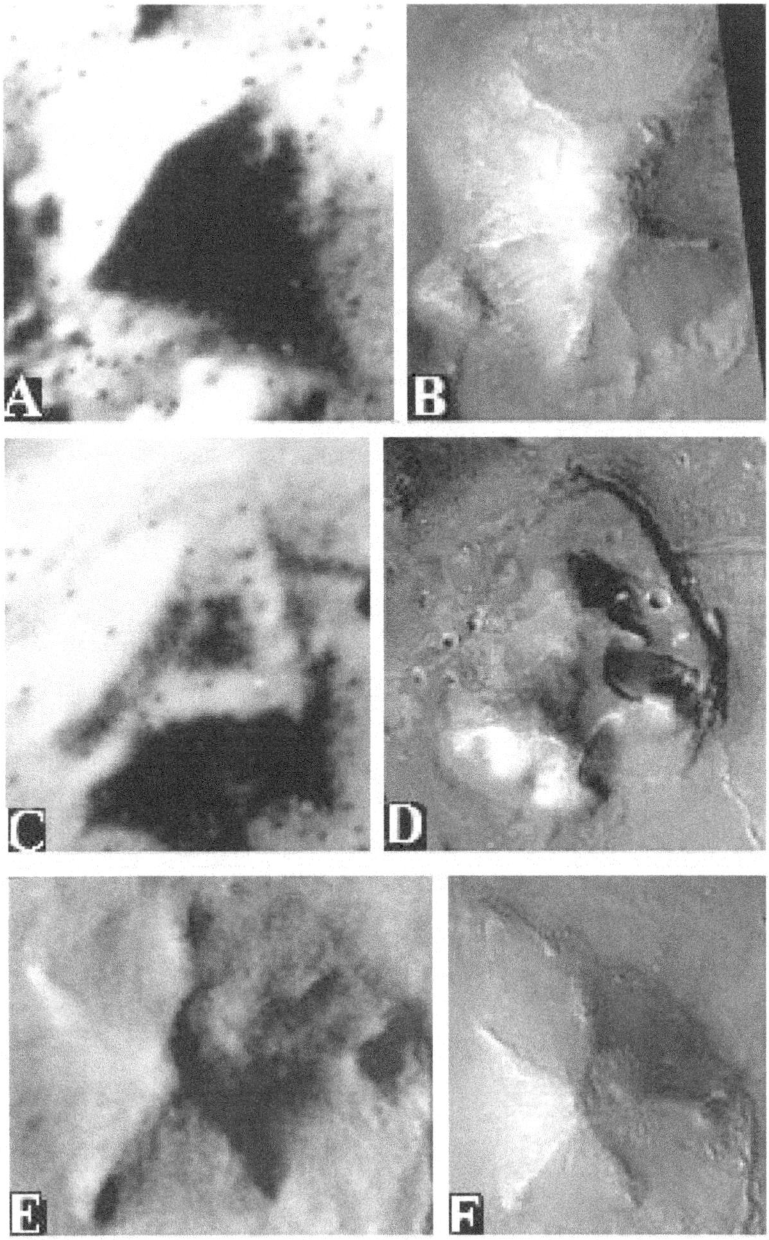

Tab. 49	Cydonia Colles - Reconfirmation and modification process: (B,D,F – MGS 1999); Akis Pyramid, Ruined Pyramid, D&M Pyramid; A,C,E: Viking 1 1976, 35A72; B-D-F). MGS 1999.

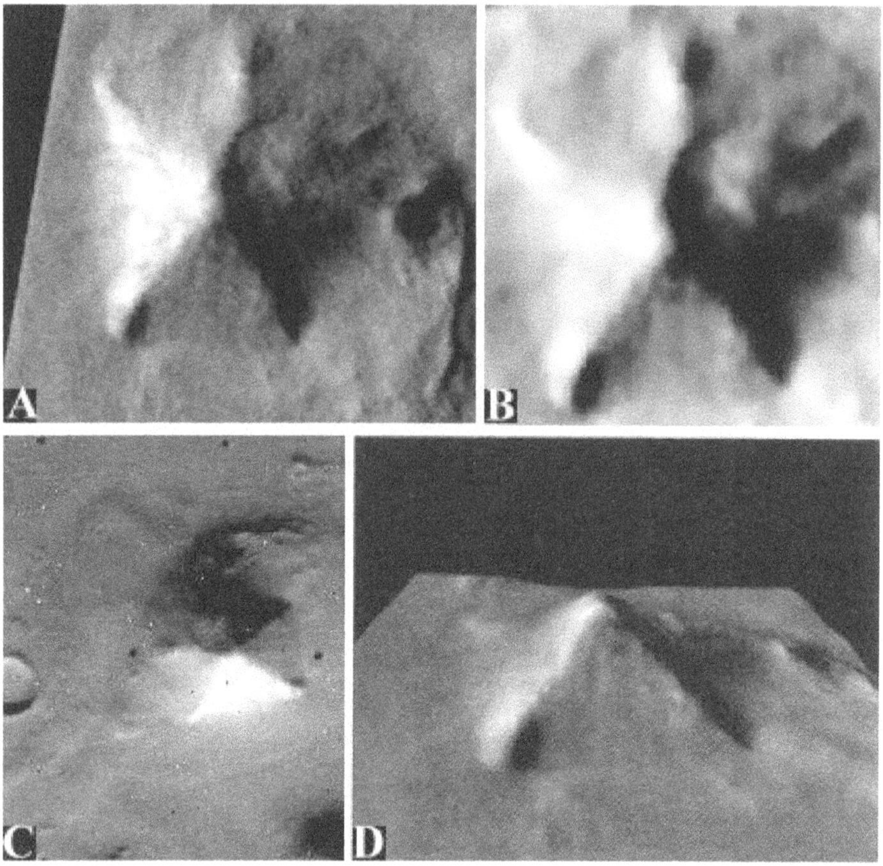

Tab. 50	3D reconstruction of Pyramid D&M (Cydonia Colles); A,B,C) Viking 1 satellite survey (1976); D) 3D reconstruction.

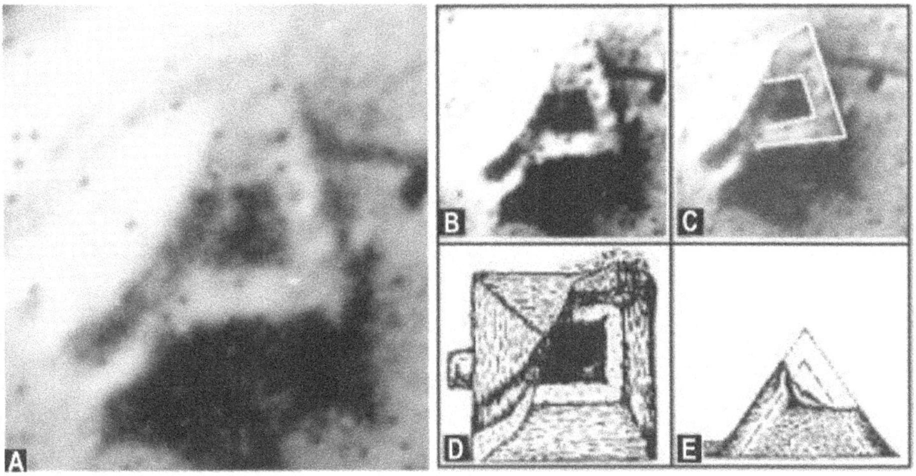

Tab. 51	3D reconstruction of the Ruined Pyramid (Cydonia Colles); from: Richard C. Hoagland, The Monuments of Mars, A City on the Edge of Forever, North Atlantic Book, Berkeley, California (USA), p. 86. The reconstruction is by Kynthia Lynn.

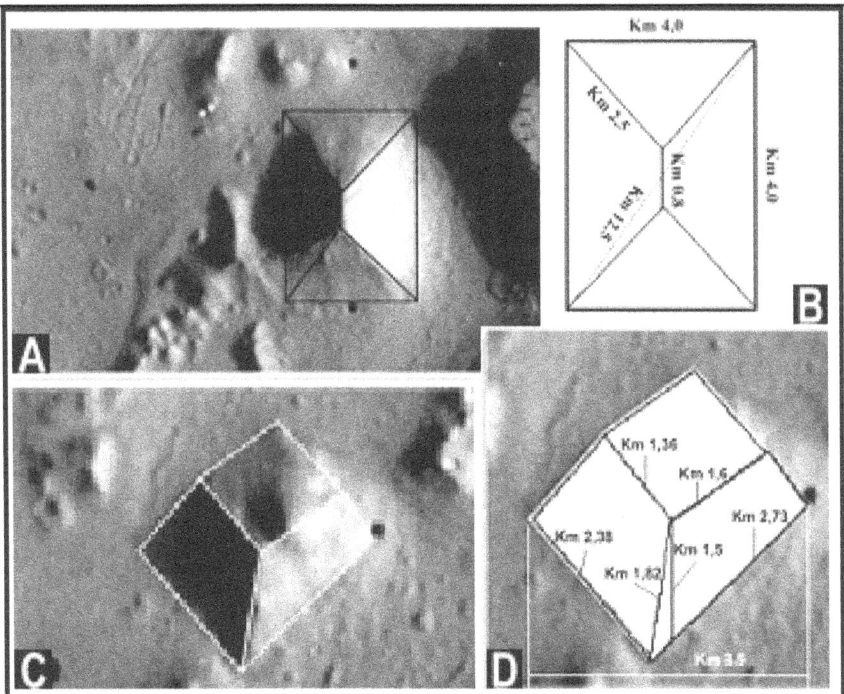

Tab. 52	3D processing – Cydonia Mensae (photo 219S16 – Viking 2) – A,B) Rectangular pyramid; C,D) Trapezoidal structure.

The *optical illusion* can be “absolute” and “relative”. It is said to be “absolute” when it indicates that the structure appearing in the frame does not exist in reality, it is said to be “relative” when the structure, which in the frame presents deformed parameters, actually exists, although the satellite image does not have the right features to show its exact physical conformation (**Tab 49A**).

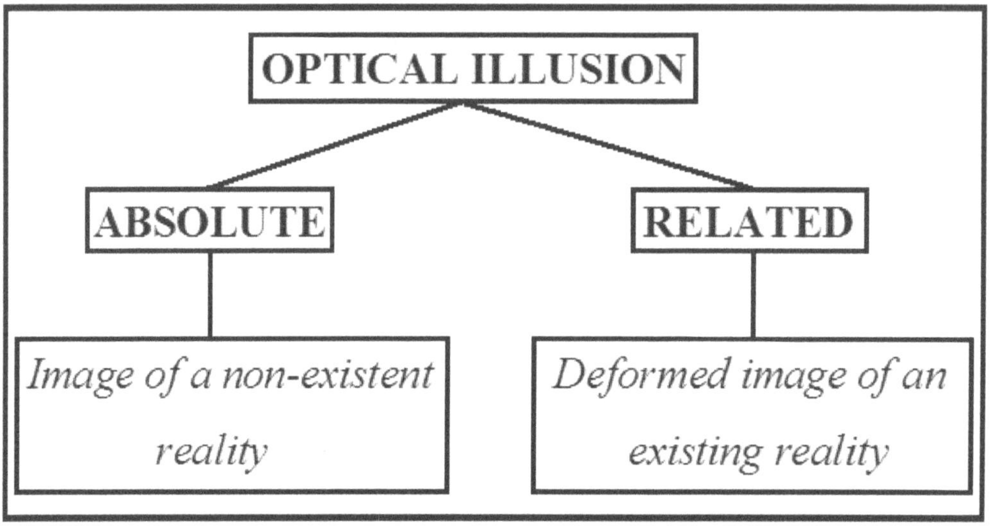

Tab. 49A	Diagram explaining the scientific meaning of the concept of optical illusion.

The *Fractal Test* checks whether certain structures, at first sight natural, are in fact artificial. Natural objects generally show a higher level of fractal content (i.e. fragmentary, broken lines), whereas artificial objects are more symmetrical and more regular (**Table 53**).

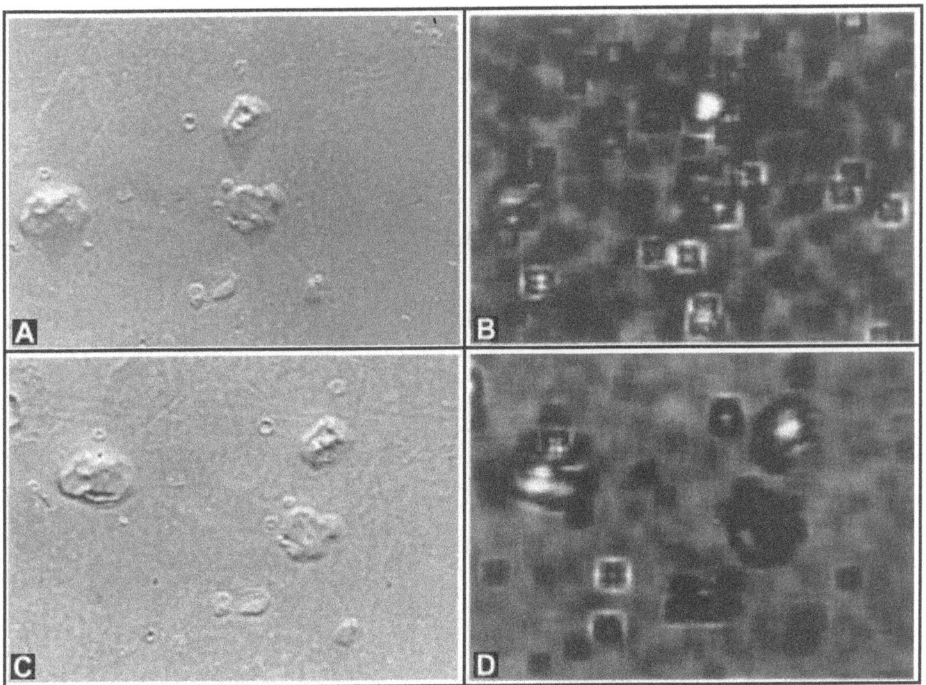

Tab. 53	Fractal analysis of the Cydonia Colles region (Mark Carlotto); from: Carlotto J. Mark, The Martian Enigmas: A Closer Look, North Atlantic Books, Berkeley (USA) 1991. A/B) image 35A72 Viking 1; C/D) image 70A13 Viking1.

The *shadow test* considers what is considered to be an element of secondary importance, namely the shadow, which is the opaque image created by a body that stands between the light source and the surface where this source is projected. The shadow is actually useful for identifying fundamental elements of a given structure, such as its size and shape, and consequently its height above the ground (**Tables 54,54A**).

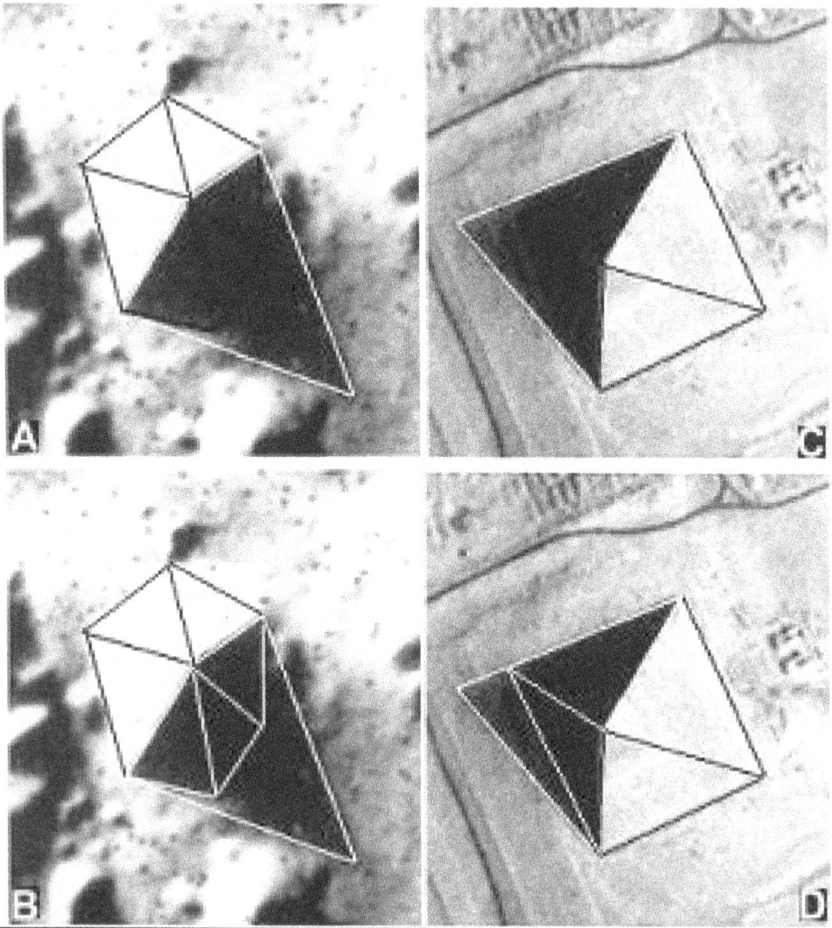

Tab. 54	Shadow analysis; A/B) Pyramid of Akis (Cydonia Colles); C/D) Pyramid of Cheops (Giza).

The *Context Test* seeks to find out whether there are other structures around a given one, which may suggest the existence of an artificial reality. in practice, a rock, which would appear to us as a natural structure, could be an accident, while the same element placed in the midst of other geologically inexplicable ones, could represent an element in favour of the artificiality of that structure considered (**Table 55**).

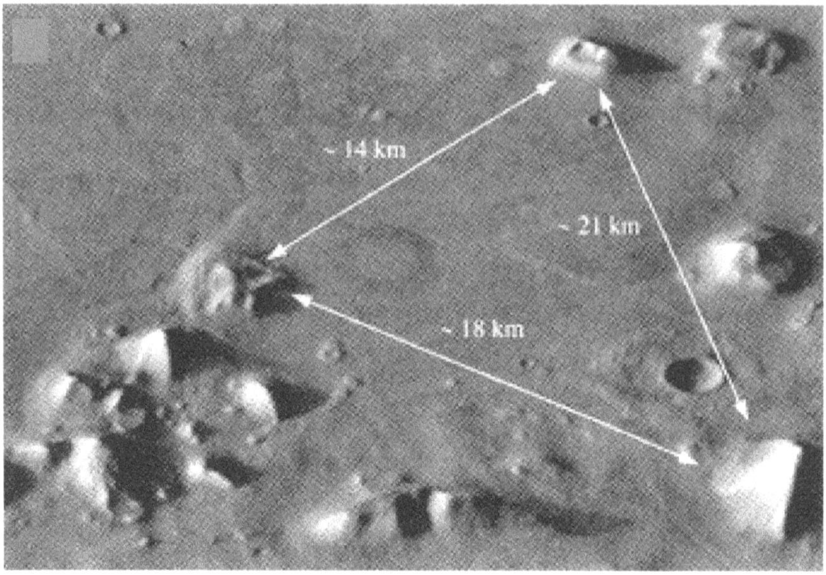

Tab. 55	Context Test – Cydonia Colles Region. The structure named 'Face on Mars' (NE position) placed in relation with other structures in the same area: in the SW position: the Ruined Pyramid and, next to it, the Akis Pyramid; in the SE position: the D&M Pyramid or Five-Pointed Pyramid.

The *Bilateral Symmetry Test* checks if an observed structure presents relevant elements of a geometric character, which characterise an element of non-natural origin. in particular, so-called bilateral symmetries are observed, which, in nature, are only found in crystalline formations. When it is possible to detect such symmetries, in the context of structures filmed by satellites, after having carried out all the other control tests and nevertheless having completed the analysis of the image according to the terms of photo-interpretation, we can affirm we are in the presence of an artificial structure (**Table 56**).

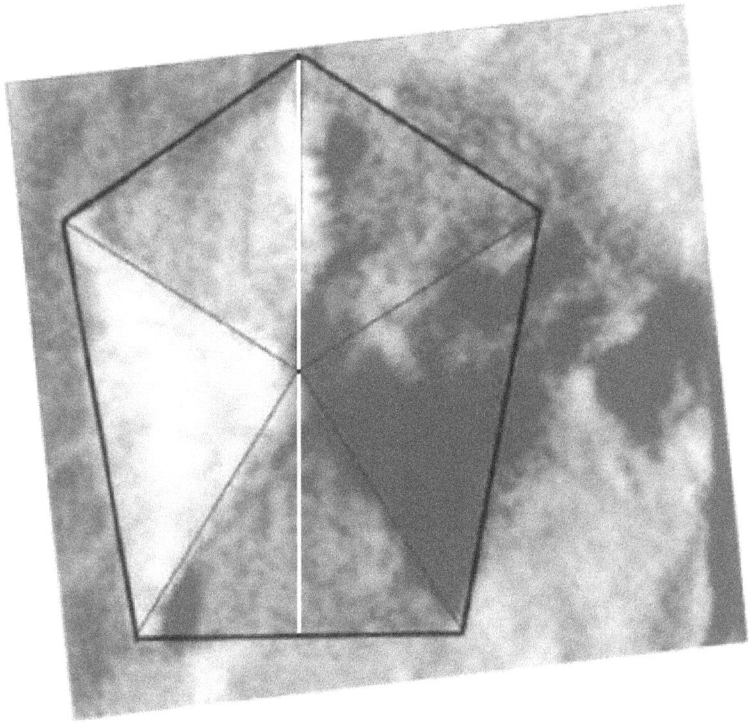

Tab. 56	Bilateral symmetry test – Cydonia Colles, D&M Pyramid, Viking 1 – 35A72.

The *Internal Relation Test*, takes into consideration an image which contains a given number of structures. If this image, subsequently taken with varied photogrammetric parameters (shooting distance, focal length, etc.), presents a change in the shape, of at least one of the structures presents an alteration, the same thing should be found in all the other structures in the context (**Tables 57,58**).

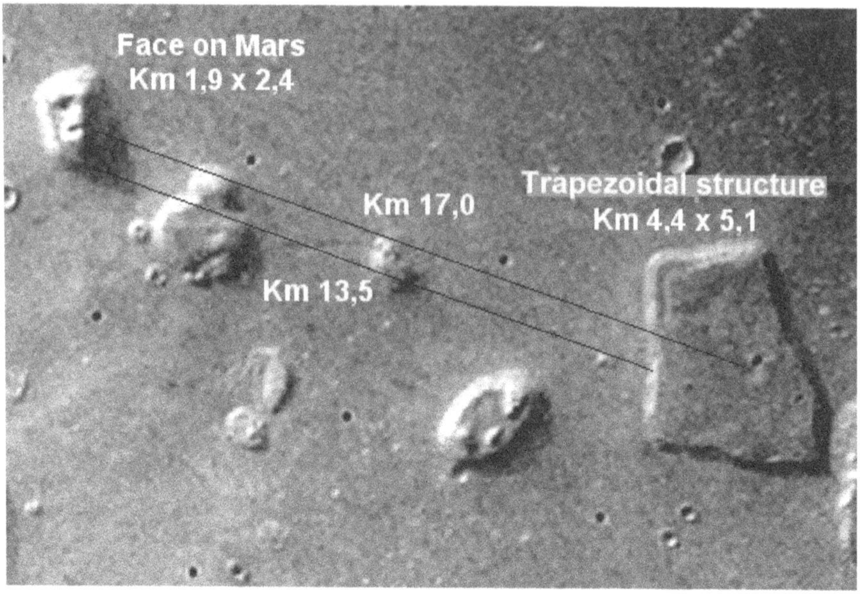

Tab. 57	Test of the internal relationship of Cydonia Colles: Face on Mars and a 'Trapezoidal Structure'.

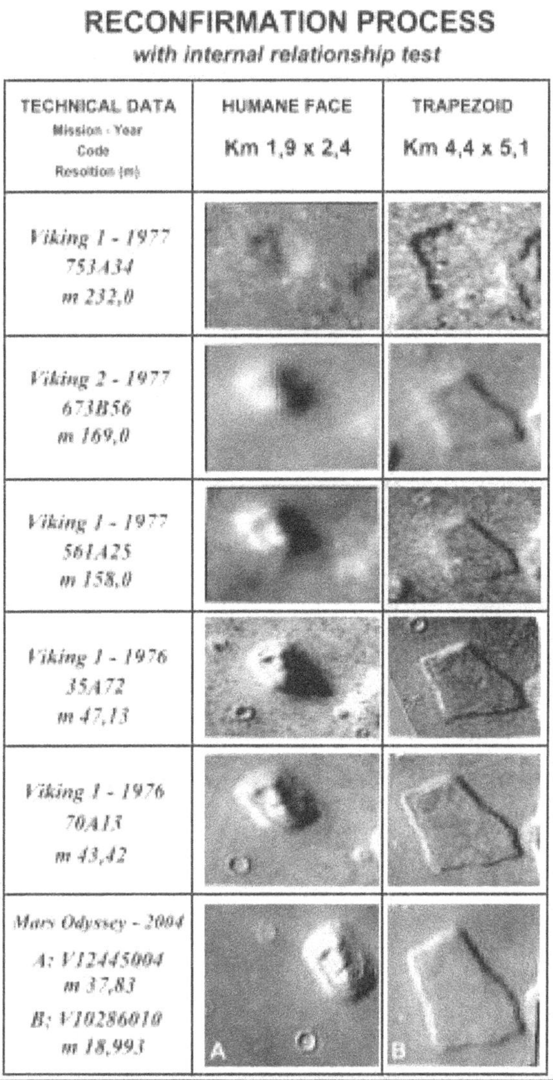

Tab. 58	Test of the internal relationship of Cydonia Colles: Face on Mars and a 'Trapezoidal Structure'. Reconfirmation process.

After subjecting the image to one or more verification tests, having ascertained whether a given genuine image had undergone a modification process, we would be able to state that we were in the presence of a modified image.

When the verification tests would then give a negative result, thus indicating that the analysed images were not genuine, (because they were different from the original photos), we would have to proceed to the graphic recovery of the concealed or modified image.

The work of the 'recovery technician' proceeds according to the following algorithm: analysis of the structures that have been damaged; checking of the original frames; recovering of the forged, modified or deleted structures; analysis of the portions of the map relating to the structures considered; relocation of the structures considered within the context of the maps, eliminating the rigged parts and replacing them with the authentic elements.

In order to find a modified or deleted image, within a frame whose coordinates coincide with the portion of the surface from which the real scene has been 'removed' to make space for a fake one, it is useful to make use of so-called 'photographic control (and reference) points' through which it is possible to narrow down the area concerned until the portion that is presumed to have suffered artistic damage is identified.

Once you have precisely identified the portion of the surface that coincides with the portion of the rigged image, you can: 1. Place the original image within the rigged frame; 2. Superimpose the two elements, precisely defining the contours; 3. Measuring the structure using a metrical scale.

In the practical application of graphic retrieval techniques, the representation of one surface on another is obtained when a bi-univocal correspondence is established between the points of the two surfaces, i.e. when each point of one corresponds to a unique and determined point of the other, and reciprocally, through the similarity relationship provided by the scale of the representation.

In the following operation, it would be necessary to examine in detail the nature of the two surfaces in order to ascertain whether they could be developed one on top of the other and, if so, the similarity of the corresponding figures would be ensured, like a flexible but inextensible veil, i.e. without elongation or contraction, overlapping or breakage. Once the necessary bi-univocal correspondence between the reference points in the cartographic representations had been established, it would be possible to assess the extent of the deformations made and, above all, to know what methods, if any, were used for this purpose.

Part III
Analysis of Aerographic Maps

1. Cartography of Mars

The surface of the planet Mars was mapped between 1971 and 1972 by the Mariner 9 mission[141] and between 1975 and 1983 by the Viking 1 and 2 missions.[142] (**Tables 59, 60,61**).

Currently, the classification of remotely captured images is based on a process that makes possible to create thematic maps of the surface. The surface of Mars has been divided into thirty cartographic quadrangles, numbered from 1 to 30 with the prefix MC (Mars Chart). This is the format approved by the International Astronomical Union and commonly used by all the mapping agencies. This division considers the construction of the terrestrial maps, in which a *graphic representation* of the actual surface is created, scaled down and with the use of conventional signs, including objects, land masses, regions and aerographic zones (Table **62**).

SATELLITE SURVEYS OF MARS (1965 – 1982)				
Years	Space probe	Surface Surveys%	Number of images	Geometric res. (max in m)
1965	Mariner 4	1	21½	1.250
1969	Mariner 6	10	428	1.000
1969	Mariner 7	10	749	1.000
1971-72	Mariner 9/A	85	3.820	430
1971-72	Mariner 9/B	54	3.509	37
1976-80	Viking 1&2 Orbiter*	97	50.028	6
1976-82	Viking 1&2 Lander**	------	3.983	0,1

* 2%: 6-8 m; 10%: up to 25 m; 39%: up to 100 m; 49%: over 100 m.
** minimum resolution: 13,3 metres.

Tab. 59 Mars satellite survey (1965 – 1982).

[141] The imaging equipment of the Mariner 9 consisted of two cameras. The first, Camera A, was fitted with a WAC (Wide Angle Camera) with a lens with a focal length of 52.267 mm. The second, Camera B, was a NAC (Narrow Angle Camera) with a wide-angle lens with a focal length of 500.636 mm.

[142] It was composed of two orbiting modules with two cameras with a focal length of 475 mm. Between 1976 and 1980, they mapped 97% of the Martian surface.

IMAGES FROM MISSION MARINER 9 AND VIKING 1&2 (1971 - 1982)					
SPACE MISSIONS	YEARS	ORBITAL SURVEYS	YEARS	SURFACE SURVEYS	TOTAL
MARINER 9	1971-72	7.329	-------	-------	7.329
Camera A	1971-72	3.820	-------	-------	3.820
Camera B	1971-72	3.509	-------	-------	3.509
VIKING 1 & 2	1976-80	50.028	1976-82	3.983	54.011
Viking 1	1976-80	24.701	1976-82	2.090	26.791
Viking 2	1976-78	15.752	1976-80	1.893	17.645
Viking 2S	1976-78	9.575	-------	-------	9.575
Totali	1971-80	57.357	1976-82	3.983	61.340

Tab. 60 Images of the Mariner 9 and Viking 1&2 missions (1971 – 1982).

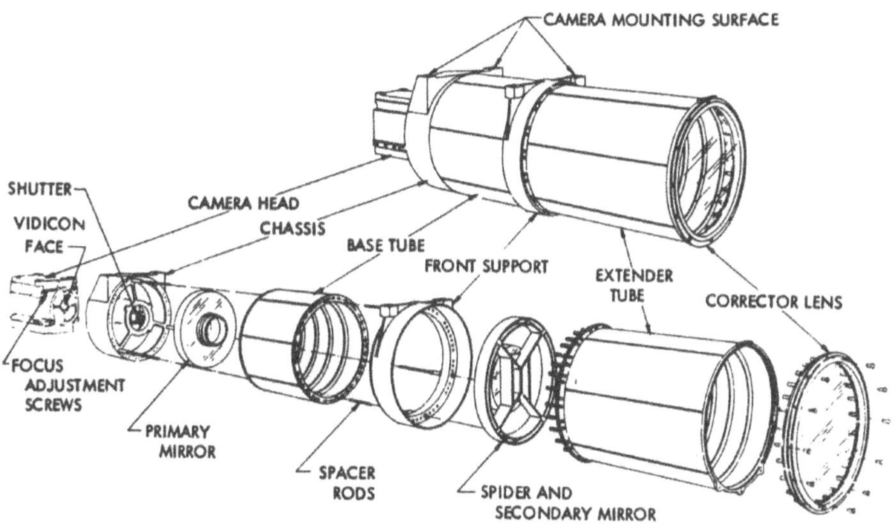

Tab. 61 Diagram of the structure of the narrow-angle (telephoto) 'Camera B' of the Mariner 9 probe.

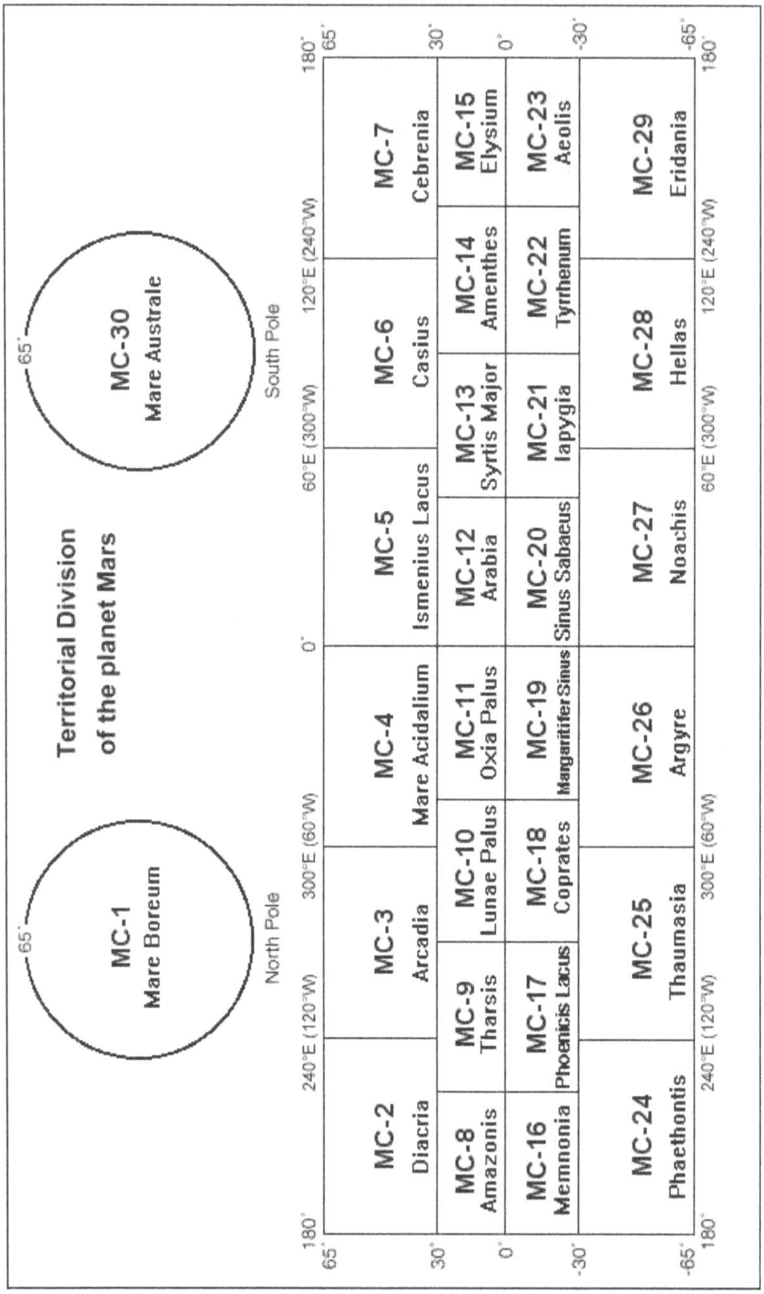

Tab. 62	Territorial division of the planet Mars.

To accurately reproduce the surface features of the planet, the spherical surface must be converted into a flat one, by means of *projections* based on geometric procedures and complex mathematical calculations. Of the many types of projections in use today, three have been applied to the Martian Cartography.

From latitude 30° North to latitude 30° South, the Mercator projection is used, resulting in two bands of 8 (eight) quadrangles each. For the latitudes 30° North to 65° North and from 30° South to 65° South, 12 (twelve) quadrangles use the Lambert Conformal Conic Projection. The two polar regions, between 65° and 90°, use a polar stereographic projection.

Sixteen quadrangles are mapped using the *Mercator projection*, eight north of the Equator and eight south of the equator.

Since the quadrangles are numbered from north to south, the sixteen Mercator projection quadrangles lie between the 8th and 23rd quadrangles. Each of these central quadrangles, which straddle the Equator, is divided into four sectors, according to their orientation, i.e., N-W, N-E, S-W, S-E.

While the quadrangles are identified by a number and have been given a unique name (e.g., MC-11 Oxia Palus), the sectors have not been given any topographical designation. The quadrangles that use the Mercator projection are *MC-8 Amazonis. MC-9 Tharsis. MC-10 Lunae Palus. MC-11 Oxia Palus. MC-12 Arabia. MC-13 Syrtis Major. MC-14 Amenthes. MC-15 Elysium. MC-16 Memnonia. MC-17 Phoenicis Lacus. MC-18 Coprates. MC-19 Margaritifer Sinus. MC-20 Sinus Sabaeus. MC-21 Iapygia. MC-22 Mare Tyrrenhum. MC-23 Aeolis.*

The twelve quadrangles that use the *Lambert Projection* are numbered 2 to 7 (North of the Equator) and 24 to 29 (south of the equator). They are *MC-2 Diacria. MC-3 Arcadia. MC-4 Mare Acidalium. MC-5 Ismenius Lacus. MC-6 Casius. MC-7 Cebrenia. MC-24 Phaetontis. MC-25 Thaumasia. MC-26 Argyre. MC-27 Noachis. MC-28 Hellas. MC-29 Eridania.* These quadrangles are further divided, some into four, others into five sectors. Two more identifiers are added to those given above (N-W, N-E, S-W, S-E) relating to the central parts of the map, either south of the equator (SC: South-Central) or North (NC: North-Central).

The two *polar regions* are included in sectors identified by the numbers 1 and 30 respectively, comprising the north pole and the south pole. These sectors are circles whose circumferences coincide, in the case of MC-1 (*Mare Boreum*) with the northern sides of the Lambert quadrangles (2 to 7) and in the case of MC-30 (*Mare Australe*) with the southern sides of the Lambert Quadrangles (24 to 29).

The Reference's meridian of Mars (corresponding to the *Greenwich Meridian* on Earth) was defined in 1840 on the first topographic map produced by Beer and Mädler. Until then, due to the poor quality of the telescopes used, it had not been possible to distinguish details less than several thousands of kilometres in length.

Planetary Cartography essentially uses a series of separate images and combines them into a mosaic, known as a *photomosaic* or *photographic mosaic*.

The accuracy of the photomosaics depends on the photographic scale and the criteria used in their construction. The flattened-out images are assembled systematically to create the final, detailed photographic work.

In the mosaic process, two fundamental elements must be taken into consideration: compliance with the tolerances defined for the orientation of each single frame and the dimensions of the image or their geometric congruence (dimensions and orientation). Assembling the digital data of the individual images, the internal radiometric congruence or contrast and brightness, will then be guaranteed.

In creating photomosaics, the images are placed in order latitudinally and longitudinally, then joined together to give the observer a continuous view of the area being mapped. The central part of the images shows less radial and height distortion. The general aim of the mosaic methodology is to take these factors into account.

This phase involves the definition of the lines along which the images will be connected, chosen from those natural lines that delimit the passage between surfaces with high radiometric differences that make it possible to mask the joins between the images with different levels of contrast.

In some cases, this result can be achieved by applying a *smoothing* filter along the overlapping edges of the images, which results in a more visually pleasing image, but certainly NOT one that is more correct or accurate! This makes the transition between images in the mosaic more homogeneous and gradual.

By putting the photos together, processed in this way, a precise and detailed map of the entire surface of a planet or part of it, can be produced.

A set of maps of Mars were produced according to planetographic mapping, in a manner that now largely defies astronomical logic, where the structures shown have similarities to the terrestrial ones. A geological map of Mars was created (**Tabs 63A,63B**). A number of historical maps relating to Mars' past were then created (**Tabs. 64,65**).

Three photographic and cartographic representations of the surface of Mars were created between 1972 and 1981, based on data from the Mariner 9 and Viking 1 and 2.

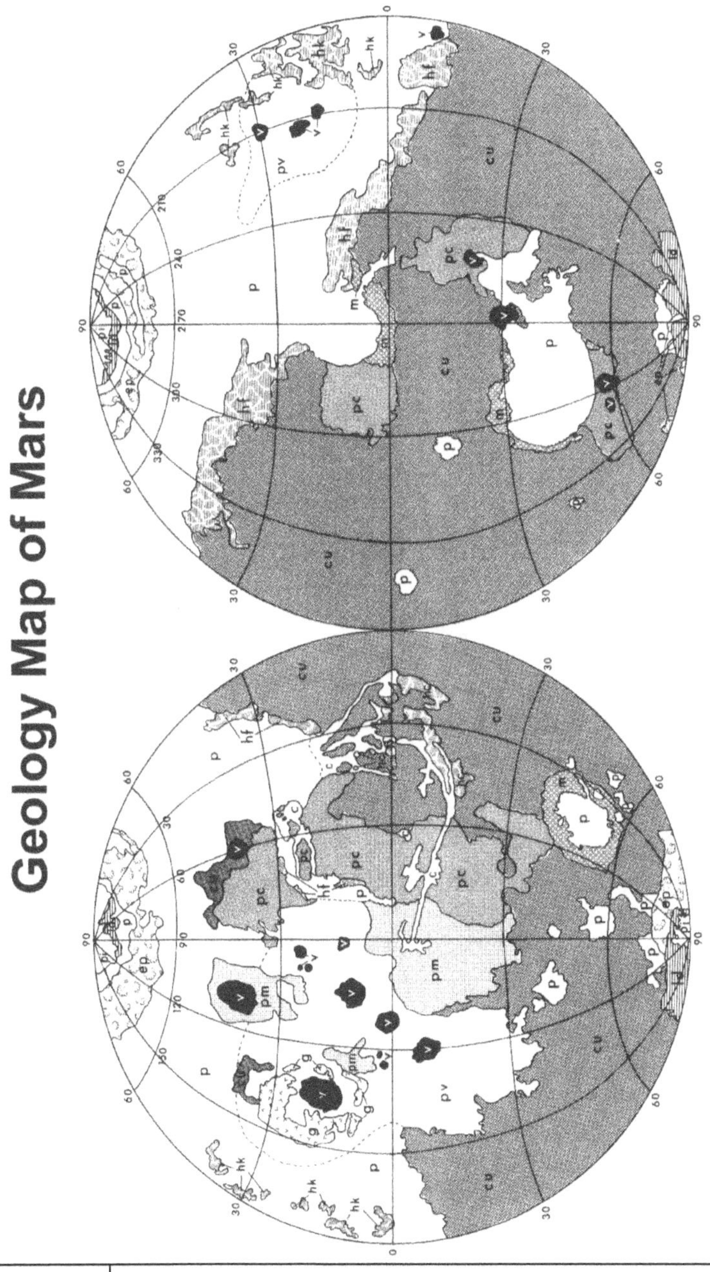

Tab. 63A	Geology Map of Mars; Thomas A. Mutch, The Geology of Mars, Pricenton University Press, New Jersey, 197; p. 58.

POLAR UNITS

pi	**Permanent ice**
ld	**Layered deposits**—thin (less than 100 m) continuous laminae at B-frame resolution. Appear as shallow benches alternating with steeper slopes. Probably exposed by wind erosion of layered dust deposits.
ep	**Etched plains**—Irregular, pitted surface. Thick, unlayered deposits eroded by wind.
	VOLCANIC UNITS
v	**Volcanic constructs**—shields, domes or cones.
pv	**Volcanic plains**—few craters, lobate scarps interpreted as flow fronts.
pm	**Moderately cratered plains**—lacks volcanic features.
pc	**Cratered plains**—most densely cratered plains unit, contains ridges resembling those of the lunar maria, also contains older eroded volcanic features.
	MODIFIED UNITS
hc	**Hummocky terrain, chaotic**—chaotic terrain, disrupted and tilted blocks of local surface material adjacent to channels or in closed basins.
hf	**Hummocky terrain, fretted**—marginal unit to northern scarp, located between cratered terrain and undivided plains. Large blocks near contact with cratered terrain, smaller blocks out into plains.
hk	**Hummocky terrain, knobby**—isolated regions of 10 km knobs, not clearly related to areas of chaotic or fretted hummocky terrain.
c	**Channel deposit**—smooth floors of channels, probably composed of alluvial, wind blown, or mass-wasted wall materials.
p	**Plains, undivided**—sparsely to moderately cratered at A frame resolution, B frames show irregular ridges, scarps and channels, general scoured appearance.
g	**Grooved terrain**—arcuate regions of linear mountains adjacent to Olympus Mons, 1-5 km wide and typically 100 km long, outer zone has finer texture.
	ANCIENT UNITS
cu	**Cratered terrain, undivided**—densely to moderately cratered uplands, most ancient of all surfaces.
m	**Mountainous terrain**—rugged basin margin material, probably eroded basin ejecta.

Tab. 63B	Geology Map of Mars; Thomas A. Mutch, The Geology of Mars, Pricenton University Press, New Jersey, 197; p. 341.

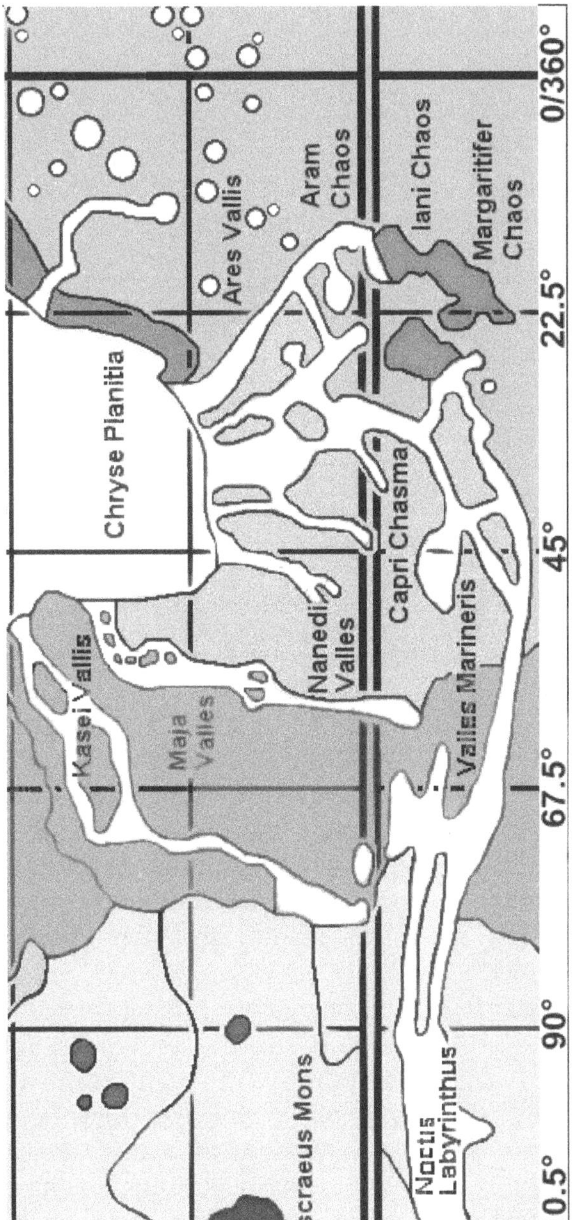

Tab. 64	Cartography of the river region of Mars (Gianni Viola).

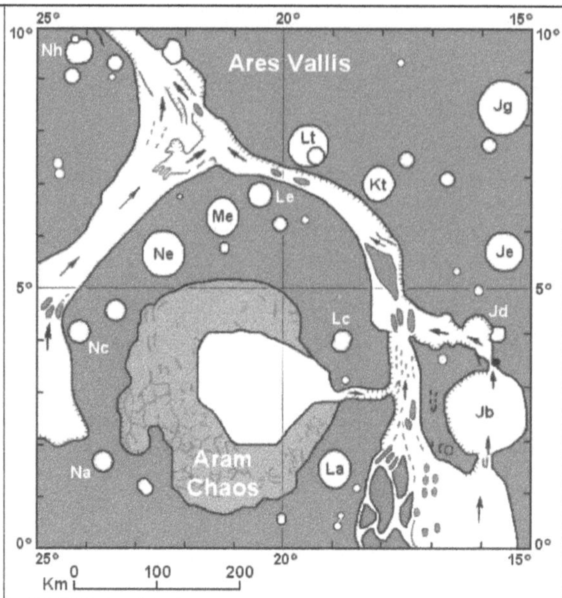

Tab. 65	Cartography of the Martian region of Aram Chaos.

1. *"Mars Equatorial Mosaic"* (1972 - code 211-4294).

We find written on a NASA website[143]: In today's digital world, we obtain images quickly and easily and take it for granted. Just a few decades ago, downloading images from spacecraft was not so simple. Before, there were no computers or software capable of combining digital images. They were printed on photographic paper, cut by hand and stuck in place on a large blackboard, according to a detailed diagram of the probe's photographic coverage of the planet.

In this particular field of activity, since NASA's inception 60 years ago, the list of people who have worked successfully is long, from engineers to pilots to researchers and many others. Although astronauts seem to enjoy the most success and visibility among enthusiasts, there is the work of many "unsung heroes", not only the technicians in charge of space missions, but also the various individuals who, after acquiring satellite images, studied, analysed and assembled the various images coming from space into photomosaics. These works were created both by NASA's Jet Propulsion Laboratory, which produced 96 mosaic plates of selected areas of the Martian surface during its long scientific activity, and by the United States Geological Survey, which created many others.

[143] Image of Mars of the day – Mariner Mainer 9 – NASA.

However, despite this impressive output, the large photomosaic called "Mars Equatorial Mosaic" (code 211 – 4294), created from a series of images from the Mariner 9 mission, remained unsurpassed **(Tabs 66,67).**

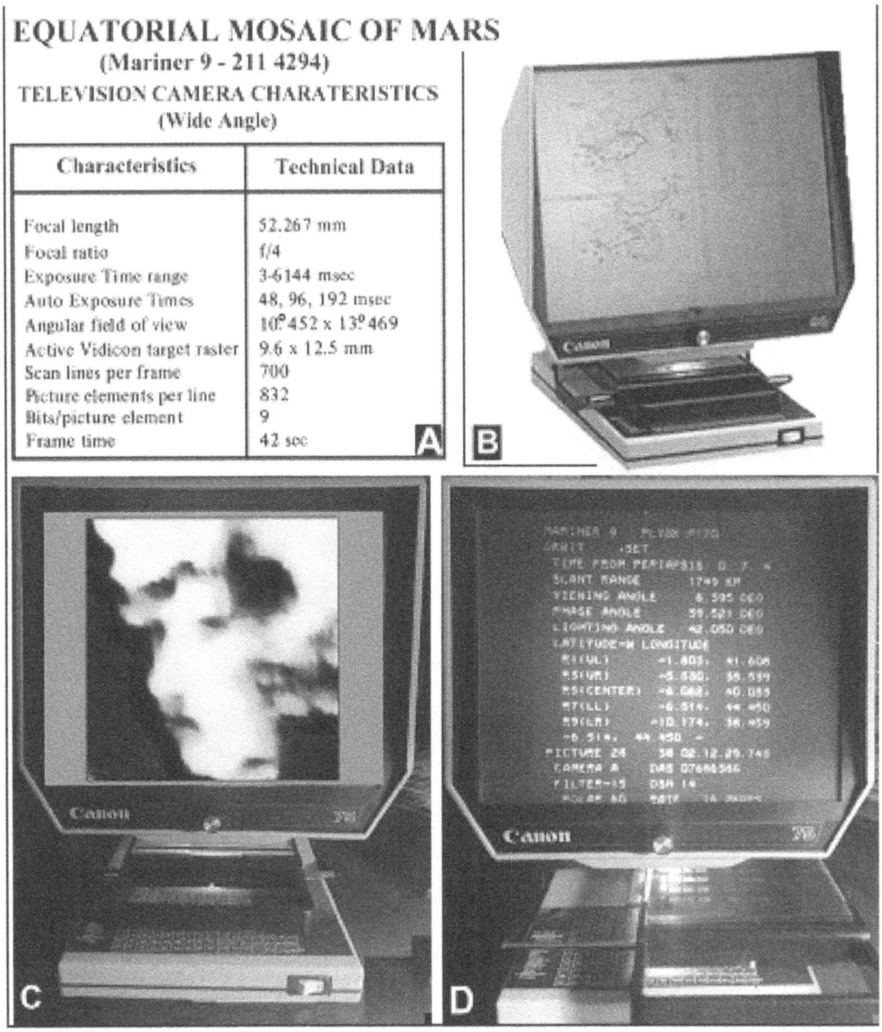

EQUATORIAL MOSAIC OF MARS
(Mariner 9 - 211 4294)
TELEVISION CAMERA CHARATERISTICS
(Wide Angle)

Characteristics	Technical Data
Focal length	52.267 mm
Focal ratio	f/4
Exposure Time range	3-6144 msec
Auto Exposure Times	48, 96, 192 msec
Angular field of view	10°452 x 13°469
Active Vidicon target raster	9.6 x 12.5 mm
Scan lines per frame	700
Picture elements per line	832
Bits/picture element	9
Frame time	42 sec

Tab. 66	A - Equatorial mosaic of Mars (Mariner 9 – 211-4294) – Camera characteristics (Wide-angle A). B - Microfiche viewer with 42x lens. Cards measuring 105x148 mm. Initial tool for studying images from Mariner 9. C - On the monitor: Capri Chasma Region., DAS 7686388 (see Tables 119,120 and 121). Detail: Antropohtaphic structure – Km 80x104. D - Mariner 9 – Scheda informativa della microfiche (MM'71/MTC/MTVS Video Roll 4201-1; Das 7686388.

MOSAIC OF EQUATORIAL REGION OF MARS
MARINER 9   211-4294   P13113

MC	LON	LAT	ID DAS Time	FRAME NUMBER	TAPE_ID FILE #	number MARS MAP 1:5.000.000	HEIGHT KM	WIDTH KM	TEXT
10-1307	47,0°W	3,0°N	7614918	4199-060	020 060	78	495	403	68A20 M05163
10-1307	49,0°W	1,0°S	7614848	4199-054	020 058	44	487	409	68A18 M05161
18-1184	49,0°W	25,0°S	7686108	4201-021	022 121	61	392	323	70A0732 05215
18-1208	49,0°W	6,0°S	7614498	4199-045	020 055	20	420	323	68A15 M05158
18-1208	52,0°W	11,0°S	7614428	4199-039	020 053	30	407	314	68A13 M05156
18-1208	46,0°W	20,0°S	7686178	4201-027	022 123	52	393	314	70A0932 05217
19-1209	37,0°W	3,0°N	7686808	4201-060	022 134	79	494	403	70A2032 05228
19-1381	27,0°W	20,0°S	7829958	4206-030	020 186	39	394	315	74A1032 05347
19-1381	30,0°W	6,0°S	7758278	4203-045	020 123	69	421	324	72A1532 05287
19-1381	32,0°W	11,0°S	7758208	4203-039	020 121	60	408	315	72A1332 05285
19-1381	37,0°W	20,0°S	7758068	4203-027	020 117	42	394	314	72°0932 05281
19-1381	40,0°W	6,0°S	7686388	4201-045	022 129	71	420	323	70A1532 05223
19-1381	42,0°W	11,0°S	7686318	4201-039	022 127	62	407	314	70A1332 05221
19-1381	44,0°W	16,0°S	7686248	4201-033	022 125	51	398	312	70A1132 05219

Tab. 67 Classification codes for some Mariner 9 images used to compose the Equatorial Mosaic 211-4294 (P13113), from 1972 (author: JPL NASA).

This mission provided amazing images comprising 30 million bits of information, collected between November 1971 and October 1972. The photogrammetric analyses covered the 3,820 orbital shots taken by Camera A (wide-angle camera, known as 'low resolution') and the 3,509 taken by Camera B (telephoto camera, known as 'high resolution').

The Mars Equatorial Mosaic was created using 774 satellite images, representing 20 per cent of the total images obtained by Camera A between late January and mid-March 1972. This mosaic represents a 360-degree panoramic view of the entire equatorial region of Mars, between 30° Lat N and 30° Lat S. **(Tab 68).**

The central part of the photo mosaic comprises about 200 images and is located between 20° North Latitude and 40° South Latitude and between 10° and 120° West Longitude **(Tab 69).**

OFFICE OF PUBLIC INFORMATION
JET PROPULSION LABORATORY
CALIFORNIA INSTITUTE OF TECHNOLOGY
NATIONAL AERONAUTICS AND SPACE ADMINISTRATION
TELEPHONE 213-354-5011

Mariner 9
Mars Equatorial Mosaic
JPL 211-4294

PHOTO CAPTION

June 7, 1972

North-Top

NORTH - TOP

A 360 degree panoramic view of the entire equatorial region of Mars is seen in this mosaic of pictures taken by Mariner 9 from late January until mid-March, 1972. Several hundred individual TV photo frames were scaled to size for this mosaic, which extends from 30° North Latitude to 30° South Latitude. The equator bisects the mosaic horizontally. The mosaic covers 50% of Mars, some 27 million square miles. Visible at left center is the complex of newly-discovered giant volcanic mountains and their summit craters. The largest of these, -- Nix Olympica -- which measures 310 miles across at the base, is at upper left center. Nine huge volcanoes have been seen in Mariner 9 pictures. The center section of the mosaic contains an enormous canyon -- 2500 miles long, 75 miles wide and, at its maximum depth, nearly 20,000 feet deep. On Earth, this Mars feature would extend from Los Angeles to New York. Its maximum depth is four times that of the Grand Canyon in western America. Mariner 9 has mapped 85% of the surface of Mars, including the area seen in this mosaic and the South Pole. During the next nine weeks of operation it is planned to map the remaining 15% at the North Pole.

Tab. 68	Caption of the Equatorial Photomosaic 211-4294 (original document).

NATIONAL AERONAUTICS AND SPACE ADMINISTRATION

Technical Memorandum 33-585

Volume I, Addendum 1

Mariner Mars 1971 Television Picture Catalog

Experiment Design and Picture Data

Dennis Crosby

JET PROPULSION LABORATORY

CALIFORNIA INSTITUTE OF TECHNOLOGY

PASADENA, CALIFORNIA

August 15, 1974

Tab. 69	NASA, Technical Memorandum 33-585 volume I, Addendum 1, Mariner Mars 1971 Television Picture Catalog, Experiment Design and Picture Catalog, Crosby – JPL California Institute of Technology – Pasadena, California (USA) – 15 Agosto 1974.

It should be noted that the Mars Equatorial Mosaic is the largest mosaic graphic representation in the history of space science **(Tab 70A).**

But who actually created this masterpiece, this work of art?

The author of the original photomosaic (211-4294) was Patricia Conklin. This researcher worked at the US Geological Survey in Flagstaff (USA), but for this project she was transferred to NASA's Bioscience and Planetology Section at the Jet Propulsion Laboratory (part of the California Institute of Technology)[144].

As stated on a NASA website[145] dated 30 March 2018, Patricia Conklin "was one of the people at JPL who assembled the Mariner 9 photos into large mosaics", whereas it would have been correct to specify that she – as can be seen from the attached photo **(Tab 70B)** was the author of the photomosaic, as she clearly appears in the photo to be arranging the precious frames of the mosaic.[146]

---

[144] In 1971, researcher Patricia Conkiln contributed to the article 'The Surface of Mars – Uncratered terrain', published in the Journal of Geophysical Research on 10 January 1971 (Vol. 76 – No. 2).

[145] Technology Then and Now.

[146] Here, Conklin worked as a planetary cartographer, doing all the work manually. Secondly, with regard to the rectification, or simply the correction of the photos, these operations were carried out by Robert Skinner, a photographic technician in the astrogeology department at the University of Flagstaff, under the direction of Dr. Laurence Soderblom, a researcher on the Mariner 9 project (**Table 71**).

Tab. 70A	Equatorial Mosaic of Mars (Mariner 9 – 211-4294) – Central portion.

Tab. 70B	Patricia Conklin – planetary cartographer JPL-NASA, author of the "Equatorial Photomosaic Mariner 9 - 211-4294" (the photo is dated May 16, 1972).

For this reason, it can be said without fear of contradiction that researcher Patricia Conkiln was the most important figure involved in this activity, even though, strangely, there is no official mention of this fact.

From the moment that the manual work of assembling photos into photomosaics was completely abandoned, it was also possible to verify the end of graphic creativity in a field where the exclusive use of computer monitors effectively blocked the application of photogrammetric techniques in cartography.

It should be noted that, in order to be viewed in its entirety, a photomosaic must necessarily be produced in paper format, since digitization, although necessary for graphic analysis, prevents it from being viewed correctly as a whole.

Tab. 71	Arizona Daily Sun August 30, 1972: The press announces the construction of the Equatorial Photomosaic of Mars, elaborated with images of Mariner 9.

2. **The 1979 *Atlas of Mars* Map Series** (1 volume), (**Tabs. 72,186**) with a scale of 1:5,000,000 and composed of images from the Mariner 9 mission, divided into a series of 30 charts, with partial radiometric correction and in some cases structural modifications, so that only some of the images are as originally captured.

This Atlas contains charts in various formats: *topographic*, *shaded relief* and *controlled photomosaic* versions.

The first satellite images used for the mapping of a planet (in this case, Mars) were from the Mariner 9 mission. As in the Viking mission, three years later, this was achieved through the application of verifiable scientific principles, for example by providing the global context into which each image would be inserted.[147]

[147] What has been done with the images from subsequent missions is a total decontextualization and a criminal fragmentation of information. Even an expert in the geomorphology of Mars would find it very difficult to identify the portion of the surface to which a given image corresponds. The same operation would be completely impossible for those ignorant of the relative scientific concepts.

UNITED STATES
DEPARTMENT OF THE INTERIOR
GEOLOGICAL SURVEY

INTERAGENCY REPORT: ASTROGEOLOGY 56

MARS CARTOGRAPHY:
The 1:5,000,000 Map Series

by

R.M. Batson

October, 1973

Prepared under JPL Contract WO-8122

This report is preliminary and has not been edited or reviewed for conformity with U.S. Geological Survey standards and nomenclature.

Prepared by the Geological Survey for the National Aeronautics and Space Administration

Tab. 72 USGS – Mars Cartography – The 1:5.000.000 – Map series – R.M. Batson – JPL-NASA October 1973.

The charts produced based on data from *Mariner 9* was called the *Mars Chart* (MC) and served as the basis for creating the Atlas of Mars in 1979, in addition to two charts with a scale of 1:25,000,000 (Tab.72A) including a *shaded relief* map and a topographic map, along with a whole series of twenty 1:5,000,000 *shaded relief* maps. Other mosaics have been created from filtered, corrected and enhanced photos, using more precise scales and positioning.[148] A photographic mosaic was also created on a globe of Mars.

[148] See "Mars Equatorial Mosaic" – pp. 194-202.

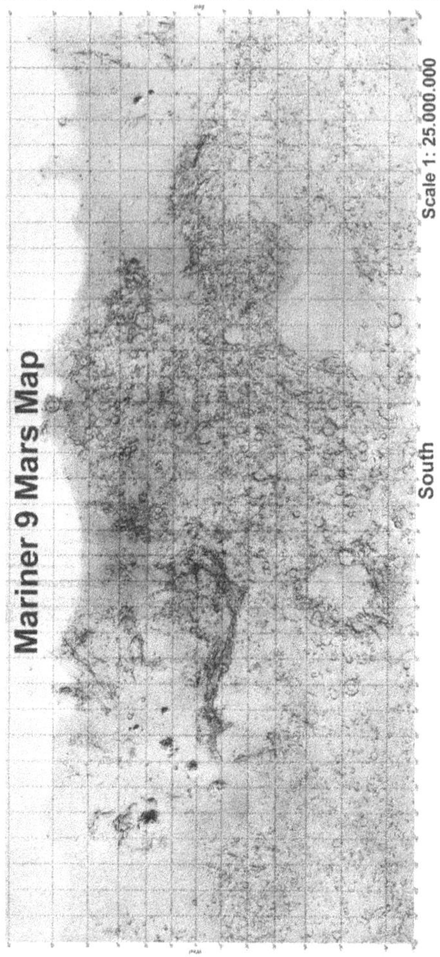

Tab. 72A	Preliminary map of Mars (Mariner 9); Prepared by the Department of Interior and United States Geological Survey (USGS), in cooperation with the Jet Propulsion Laboratory, California Institute of Technology, National Aeronautics and Space Administration (NASA); Mercator projection; Scale 1: 25,000,000.

3. **The 1979–81 *Atlas of Mars* (topographic series), (Tabs.73,74,75,76)** on a scale of 1:2,000,000, consists of 140 charts created using images from the Viking 1 and 2 missions. The images have been radiometrically corrected and, in some cases, modified or deleted. Therefore, they do not match the original images taken on the previous Mariner 9 mission.

Images from the Viking missions are available in two versions: *pristine* images with no geometric and radiometric corrections and *mosaic* images that are geodetically controlled.

Using pristine imagery, the USGS created a digital map of the Martian surface, using some map projection models. Thus, it used a sinusoidal equivalent, where each line could be matched to a parallel of latitude, compressing its length and, where necessary, the cosine of the latitude itself. This way, each pixel of the image is identifiable by absolute latitude and longitude geographic coordinates, thus obtaining a global map reference system. As for the polar zones, the polar stereographic map projection was used.

Two sets of controlled photomosaic maps were produced using the Viking images, one at a scale of 1:2,000,000, consisting of 140 maps, the other at a scale of 1:500,000, consisting of 126 maps. in particular, the first was drawn up according to the subdivision of the controlled photomosaics of the MC series, the second according to the MTM projection system, where each of the 18 zones of the MTM covers 20 degrees of longitude, while the maps north and south of 72° of latitude are in polar stereographic projection. High-resolution photomosaics covering some areas of the planet are also part of this set and it was precisely through the MTM set that the MC maps were revised.

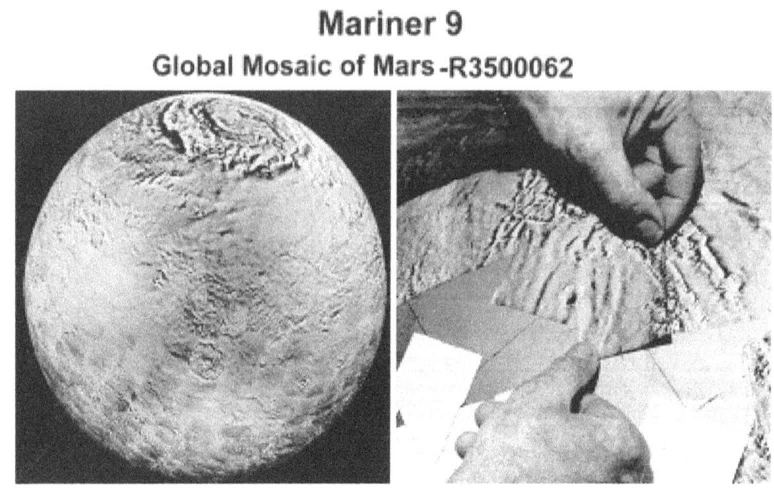

Tab. 73 Global mosaic of Mars (Mariner 9, R3500062).

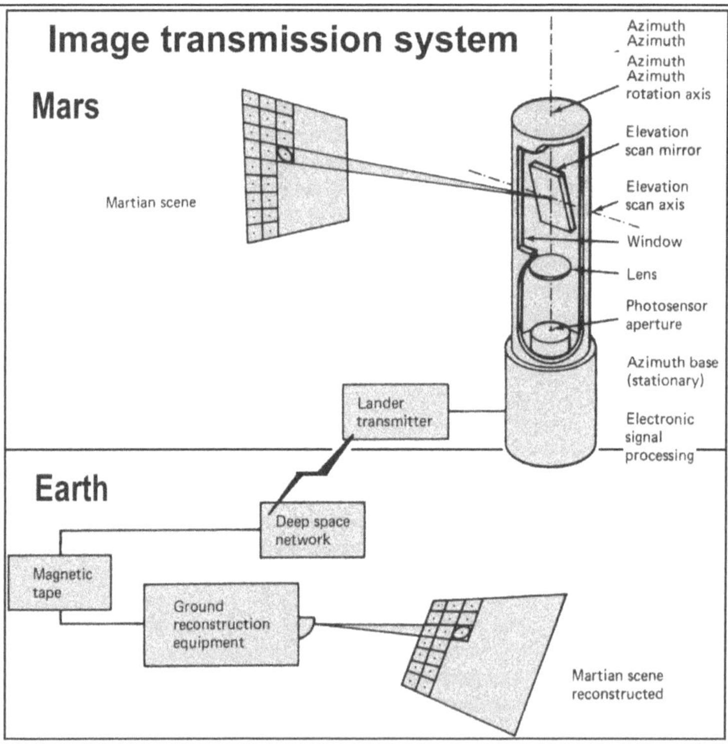

Tab. 74 Image transmission system between Viking probes and Earth.

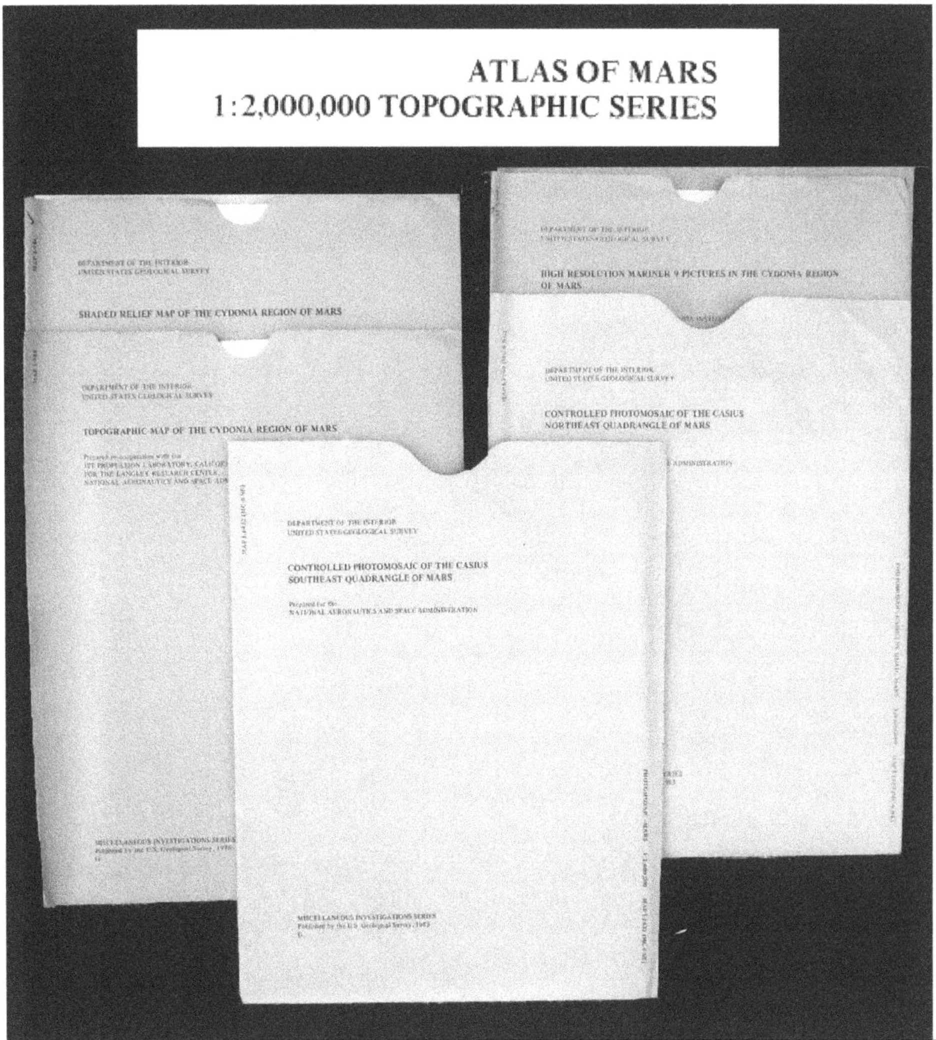

Tab. 75	Controlled Photomosaic of Mars – Published by the United States Geological Survey (USGS – USA); 140 Maps.

Tab. 76	Viking Orbiter Image of Mars (CD).

Analysing the Equatorial Photomosaic (Point 1—Regional Division of the Planet) it appears that, in some well-documented cases or where the need arose (and we will explain this point in detail later), some images have not been included in the corresponding cartography (Points 2–3). Sometimes, they have been modified or even deleted completely and been replaced with others created from scratch.

Now that the original images have been so extensively modified, the Equatorial Photomosaic has now acquired the value of a very important "archive", as a permanent record of a series of images that are preserved in their full authentic photographic integrity.

In detail, we have the following situation: 1) in the 1979 Atlas of Mars, half of the structures detected by Mariner 9 were deleted 2) in the 1979–81 Atlas of Mars (composed of images from the Viking missions), the work of erasing the part containing the remaining original images (from Mariner 9), previously left untouched in the 1979 Atlas of Mars, was completed.

This forms the basis of the cartographic material that was subjected to qualitative (photo-interpretative) and quantitative (photogrammetric) analysis, following the methodology that is analytically illustrated in the eight sheets attached to each of the eight[149] topographic regions that were examined.

We divided the entire territory covered by the Equatorial Photomosaic into two zones A and B (**Tabs. 77,78,79**).

[149] It must be said that this tampering with planetary cartography is a very serious blow to scientific integrity and respect for the truth, especially if we consider that this documentation serves as a basis of study for students, teachers and enthusiasts of the subject and for writing degree theses.

Tab. 77	Equatorial photomosaic: portions analysed (Zone A, Zone B).

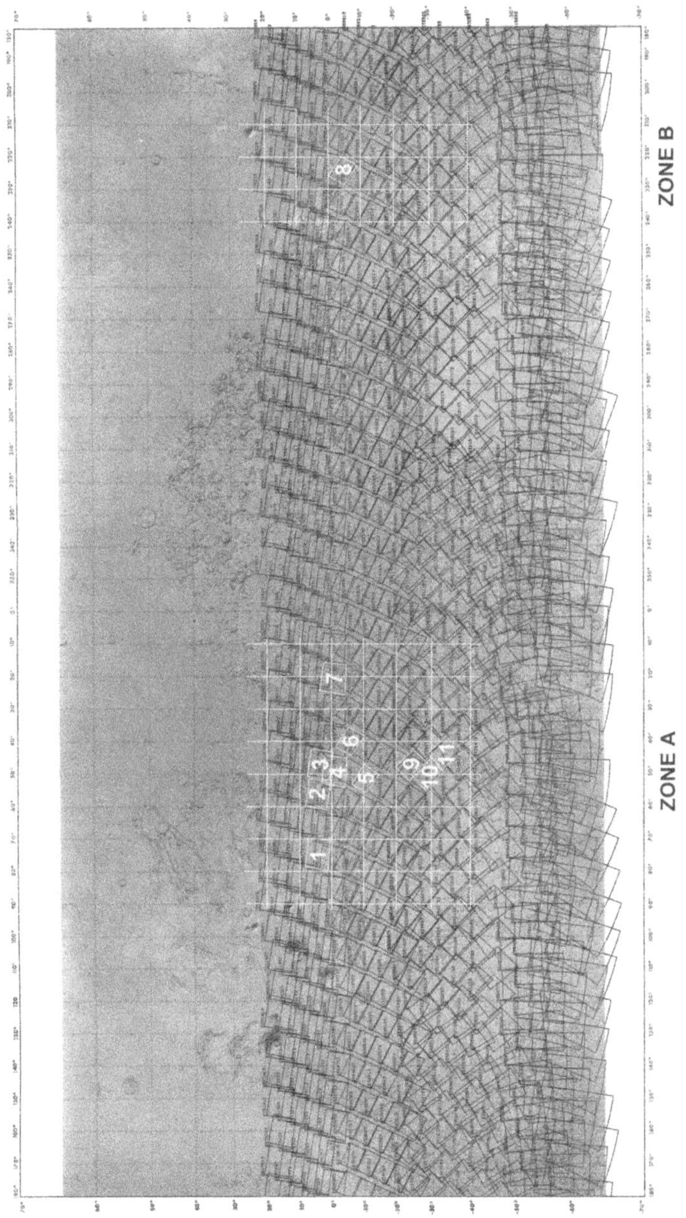

Tab. 78 — Equatorial photomosaic: portions analysed (Zone A, Zone B).

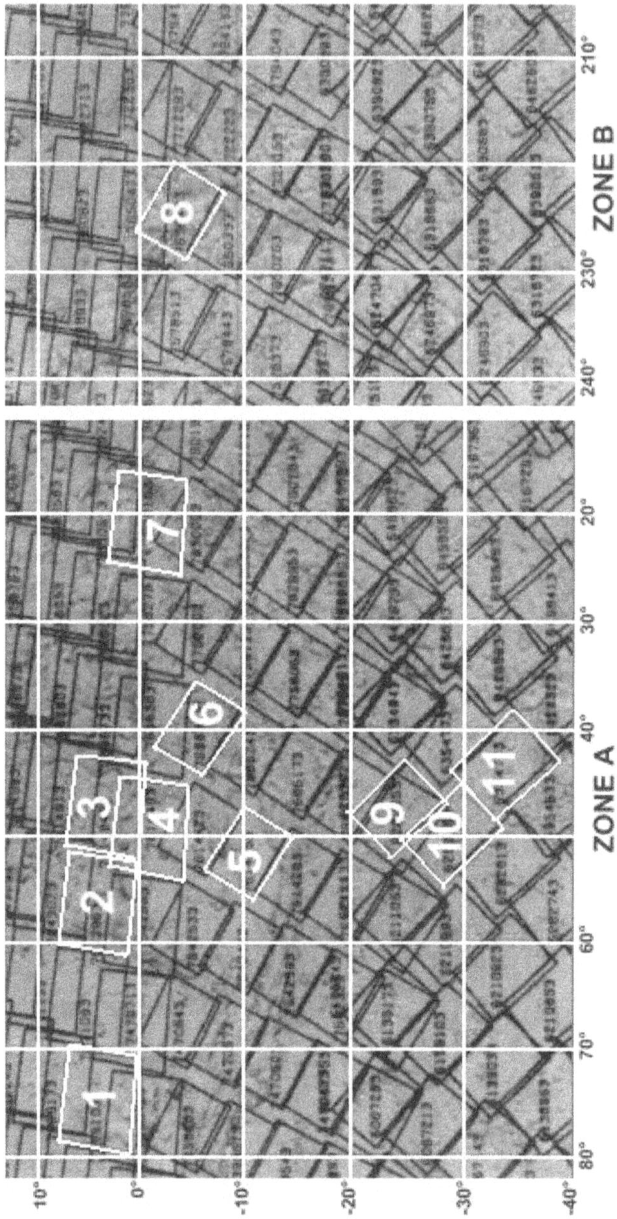

Tab. 79	Equatorial photomosaic: localization of analyzed frames, with reduced grid (Zone A, Zone B).

Zone A (**Tab. 80A**) includes regions of the MC-10 Quadrangles (Echus Fossae; Nanedi Valles), MC-18 (Ophir Planum; Aurorae Planum), MC-19 (Capri Chasma; Iani Chaos) and MC-26 (Nirgal Vallis), while *Zone B* (**Tab. 80B)** includes regions of the MC-23 Quadrangle (Aeolis Mensae).

ZONE A

Tab. 80A	Equatorial photomosaic: Zone A.

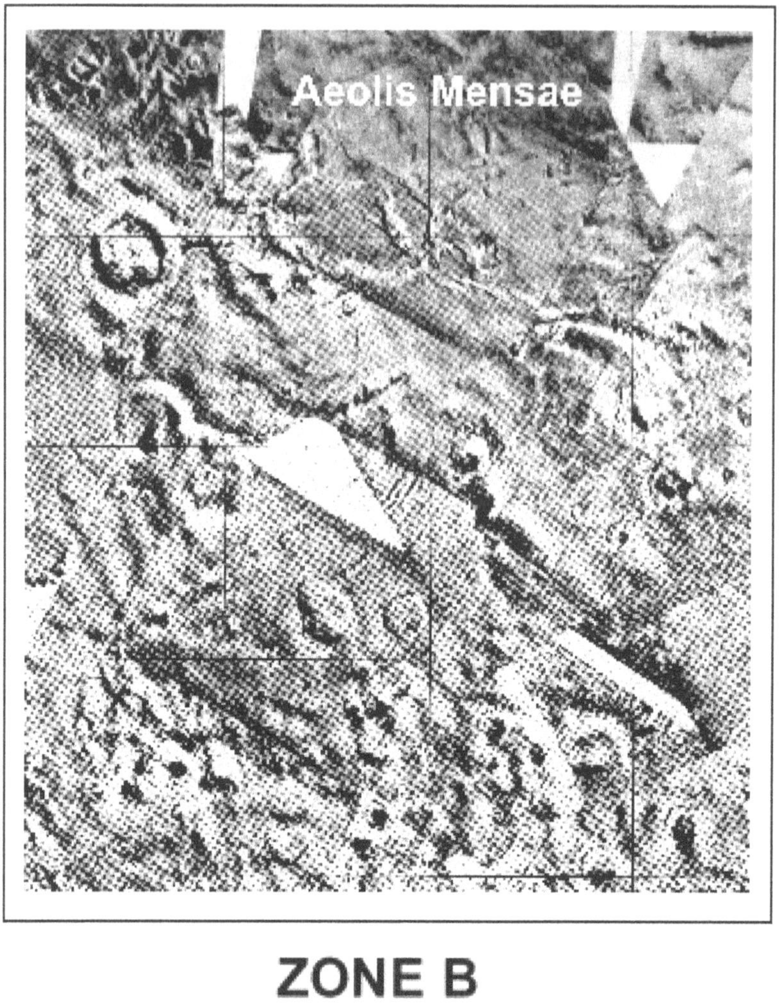

Tab. 80B	Equatorial photomosaic: Zone B.

Altogether, we analysed eight structures in five quadrants (MC) (Table **81**), 10 sectors (Table **82**) and eight regions, distributed on Mars according to the attached table (Table **83**) and with the general data tabulated for a quick search (Table **84**). Finally, for a better understanding of the data, we have summarised the main terminologies in use in the planetographic field (Table **85**).

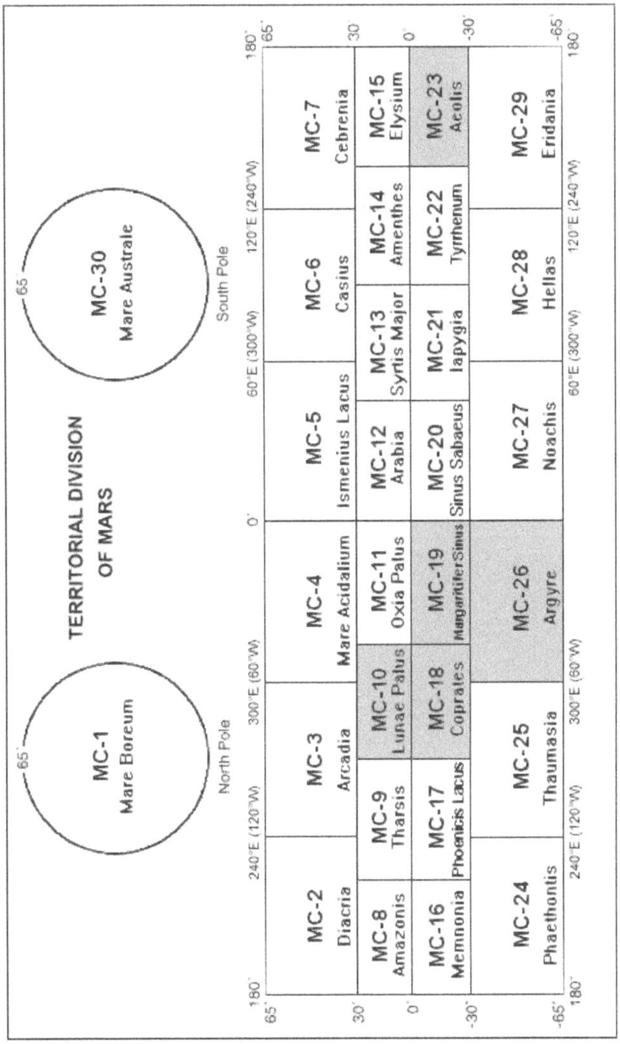

Tab. 81	Quadrant (MC) analyzed.

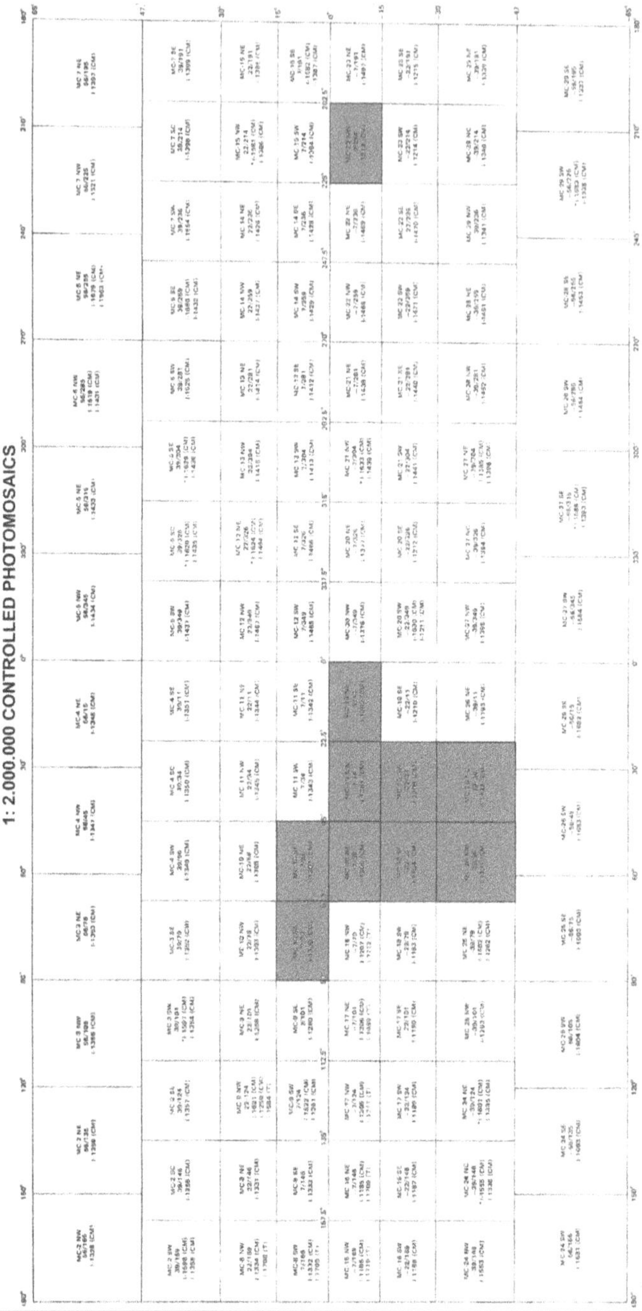

Tab. 82 Quadrant sectors (MC), analysed.

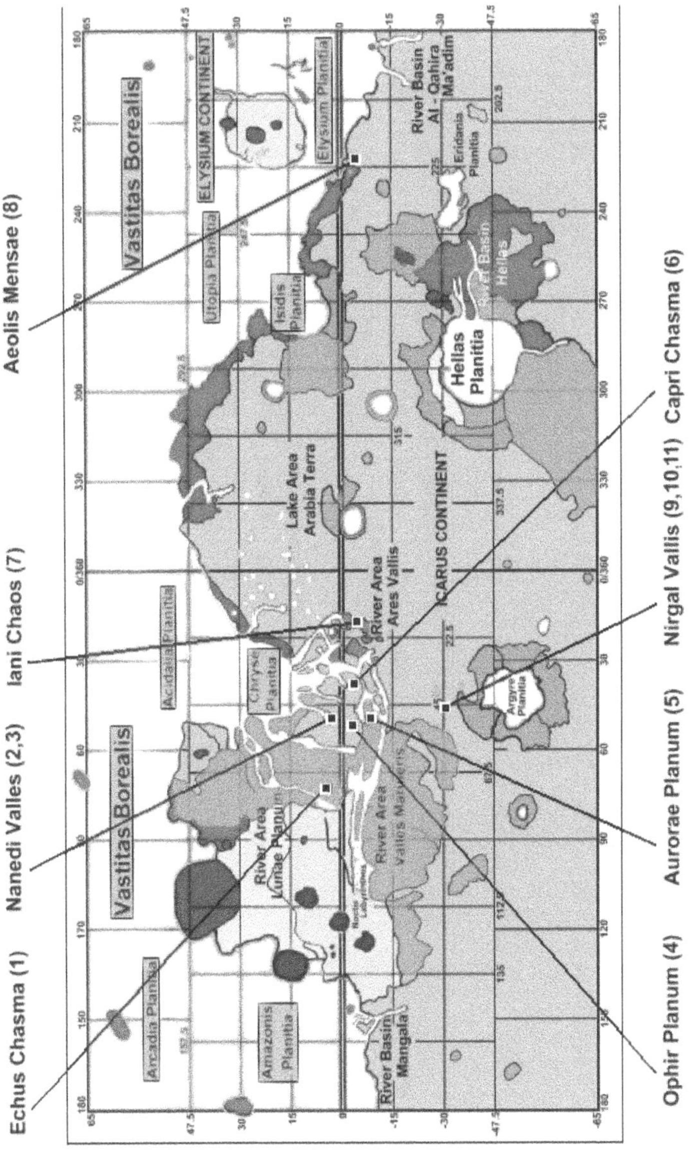

Tab. 83	Localization of the analyzed regions.

EQUATORIAL MOSAIC (Tab 83)

Mariner 9 - 211.4294

N.	MARINER 9 DAS CODE	MARTIAN REGION	CARTOGRAPHIC LOCATION	ANALYSED STRUCTURES Km	GEOMETRIC RESOLUTION m
1	7399173	Echus Fossae	MC 10 1306 SW	48.4x159.2	549.48
2	7543023			40x40	544,38
3	7614913	Nanedi Valles	MC 10 1307 SE	60x60	541,82
4	7614843	Ophir Planum	MC 18 1208 NE	68x80	535,20
5	7614423	Aurorae Planum	MC 18 1208 NE	79x79x88	436,22
6	7686383	Capri Chasma	MC 19 1381 NW	80x104	446,42
7	7830513	Iani Chaos	MC 19 1382 NE	85x155 13.7x16.0	537,20
8	7650473	Aeolis Mensae	MC 23 1213 NW	70x90	450,50
9	6283023		MC 18 1184 SE		424,48
10	6282953		MC 26 1191 NW		424,25
11	6354773	Nirgal Vallis	MC 26 1192 NC	48x68	430,35

Tab. 84 Legend of Tav. 83.

ANALYSIS SHEETS – LEGEND	
MAP: paper photomosaic depicting the surface of Mars. **CODES:** naming of frames. **DETECTIONS DETAILS:** photogrammetric data from a satellite survey. **FRAME:** satellite detection. **PHOTOMOSAIC:** composition of frames. **PHOTOGRAPHIC CONTROL POINTS:** reference points to delect structures not present in the cartographic representation. **QUADRANGLES – MC:** units of the territorial division of Mars (1-30). **REGION:** name given to the portion of surface taken from an analysed frame. **DETECTION SCHEME:** graphical representation of a satellite survey. **STRUCTURE:** shape detected within a frame.	
Tab. 85	Detailed explanation of some scientific terms used in the text (Mariner 9).

The algorithm for the analysis of Mars mapping consists of the following points:

1. Identifying a given structure, observed within the Equatorial Photomosaic.
2. Searching for the frame code of the Mariner 9 mission.
3. Identifying the coordinates of the frame and searching for it on the map.
4. In the event that the search is unsuccessful—i.e., if the structure is found to have changed or disappeared—the structure should be relocated to the precise position where it should have been, had there been no digital modification.

Satellite images are analysed and classified according to interdisciplinary geomatic synthesis (**Tab.85A**).

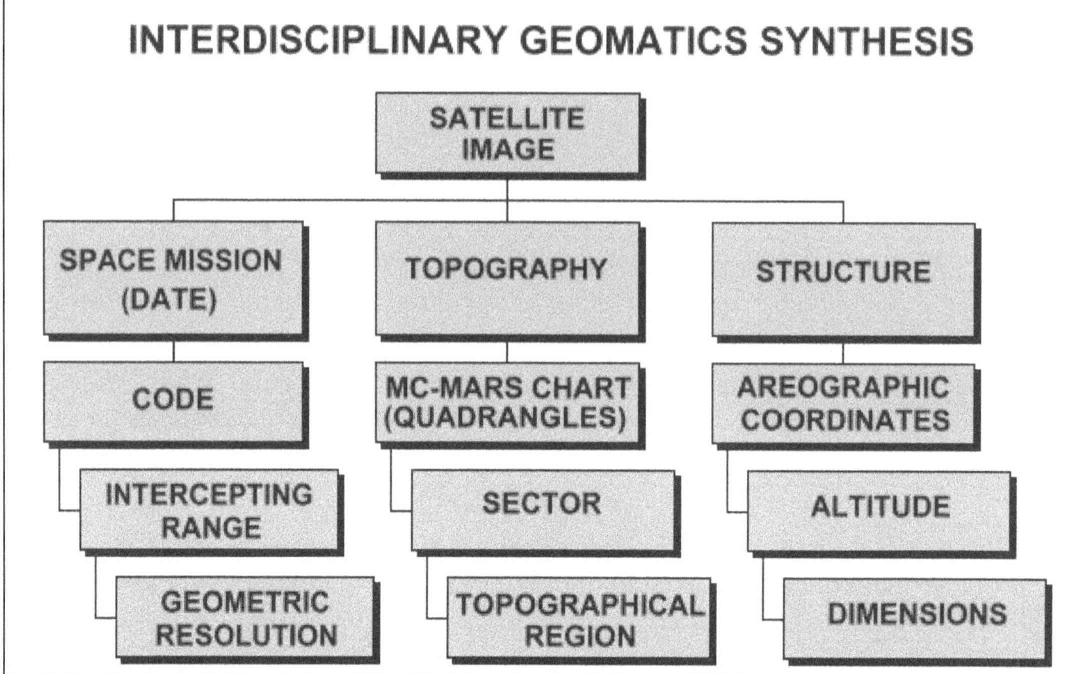

Tab. 85A	Interdisciplinary Geomatics Synthesis.

GENERAL COMPENDIUM OF THE PAST OF MARS[150]			
AREOGRAPHIC PART	TOTAL FIGURE	NORTHERN HEMISPHERE	SOUTHERNE HEMISPHERE
MARS	144.371.391 100,0	72.185.696 100,0	72.185.696 100,0
Land Surface	92.397.690 64,0	21.872.265 30,3	70.525.425 97,7
CONTINENTS	91.675.833 63,5	21.222.594 29,4	70.453.239 97,6
ICARUS[151]	85.179.120 59,0	14.725.881 20,4	70.453.239 97,6
ELYSIUM	4.908.627 3,4	4.908.627 6,8	
BOREUM	1.588.086 1,1	1.588.086 2,2	
ISLANDS	721.856 0,5	649.671 0,9	72.185 0,1
Water Surface	51.973.700 36,0	50.313.431 69,7	1.660.271 2,3
OCEANS	18.335.165 12,6	18.335.165 23,4	
BOREALIS	18.335.165 12,6	18.335.165 23,4	
MINOR SEAS	33.638.535 23,3	31.978.266 44,3	1.660.271 2,3
ELYSIUM	9.528.511 6,6	8.178.643 11,3	1.349.868 1,9
ARCADIA	5.919.227 4,1	5.919.227 8,2	
UTOPIA	5.486.112 3,8	5.486.112 7,6	
ACIDALIA	4.908.627 3,4	4.908.627 6,8	
AMAZONIS	3.753.656 2,6	3.443.259 4,8	310.397 0,4
ISIDIS	2.309.942 1,6	2.309.942 3,2	
CHRYSE	1.732.456 1,2	1.732.456 2,3	

[150] Data are given in square kilometres.

[151] In the southern part of the Icarus continent, there were three lake formations of considerable size (the Eridania of 242,000 sq. km, the Argyre of 500,000 sq. km and the Hellas of 2,800,000 sq. km) that could be considered true inland seas. The water surface figure (36 %) added to the figures for the two aforementioned lakes (2.45 %), gives a total of over 38.45 % water surface area. in the southern part of the planet, the outer seas, Elysium, and Amazonia, and the three inland seas Hellas, Argyre and Eridania, added up to about 5.2 million square kilometres, corresponding to about 7.2 % of the total surface area of the southern hemisphere.

2. Region of Echus Fossae

Map MC 10 1306 SW

4.23° N – 74.56° W (alt. km 6.0)

In quadrant MC- 10 Lunae Palus (**Table 86A**) we take into consideration picture No.19 of the attached list (**Table 86B**), with the following identification number (Das 7399173).

In **table 87A** we observe the pattern of detection and in table **87B** the original frame.

In **table 88A** the detected structure is inserted in the context of Equatorial Photomosaic 211-4294; the asterisk (*) indicates the detected structure.

It could be described as a human face placed in profile, having the following dimensions: 48.4x159.2 km (**Table 87B**).

Located in the region of Echus Fossae (**Table 88B**), where the cartographic position is indicated, the original structure is completely disappeared on map MC.10.1306.SW (**Table 89A**).

A cartographic analysis shows that, in the corresponding part occupied by the original structure, there is the frame 682A25 attributed to the Viking 1 mission (**Table 89B**).

To superimpose the original image on the "false" one, taking into account the original photo (Table 87A), we fix some photographic control points (**Table 90A**), then we find them on map MC.10.1306.SW (**Table 90B**).

In **table 91** we have inserted the detailed data of the frame Das 7399173.

Stable elements of experimentation/photographic points of reference: (2.0-5.0°N – 73.5-78.40°W); 1) N: P/Side dishes of Echus Chasma; 2) E: Craters Re e Rd; 3) SE: Crater Rb; 4) SW: Crater Tc.

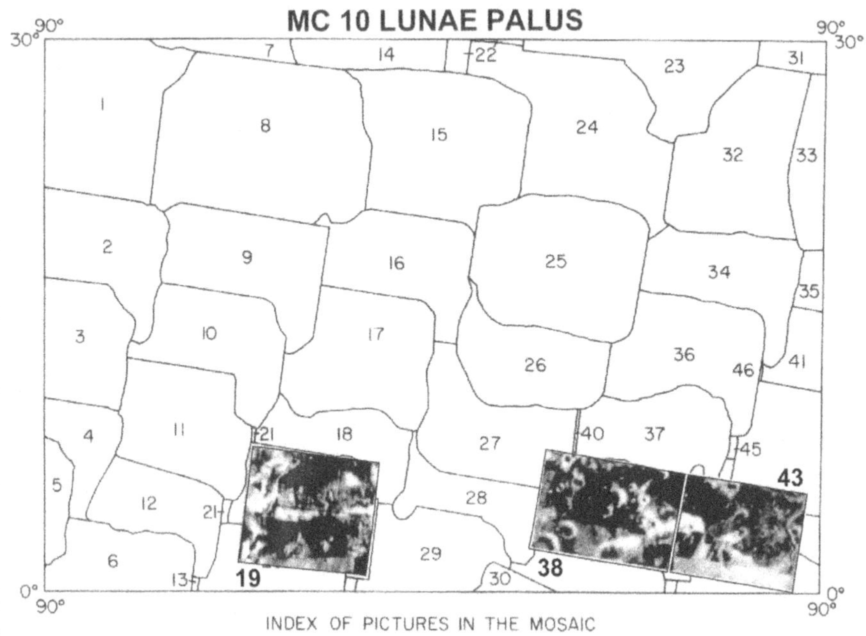

Tab. 86A	MC 10 Lunae Palus: Index of images Mariner 9.

IDENTIFICATION NUMBERS OF PICTURES IN THE MOSAIC

Index No.	DAS No.	Index No.	DAS No.
1	7255813	24	7471693
2	7255463	25	7471343
3	7255393	26	7471273
4	7255673	27	7471203
5	7255253	28	7471133
6	7327143	29	7471063
7	7255953	30	6139733
8	7327773	31	8946669
9	7327423	32	7543583
10	7327353	33	7615473
11	7327283	34	7543233
12	7327213	35	7615123
13	5380793	36	7543163
14	8730859	37	7543093
15	7399733	■ 38	7543023
16	7399383	39	7542953
17	7399313	40	7471553
18	7399243	41	7615053
■ 19	7399173	42	7614983
20	7399103	■ 43	7614913
21	7327633	44	7614843
22	8802819	45	7543443
23	8874779	46	7613513

Tab. 86B	MC 10 Lunae Palus: image identification number.

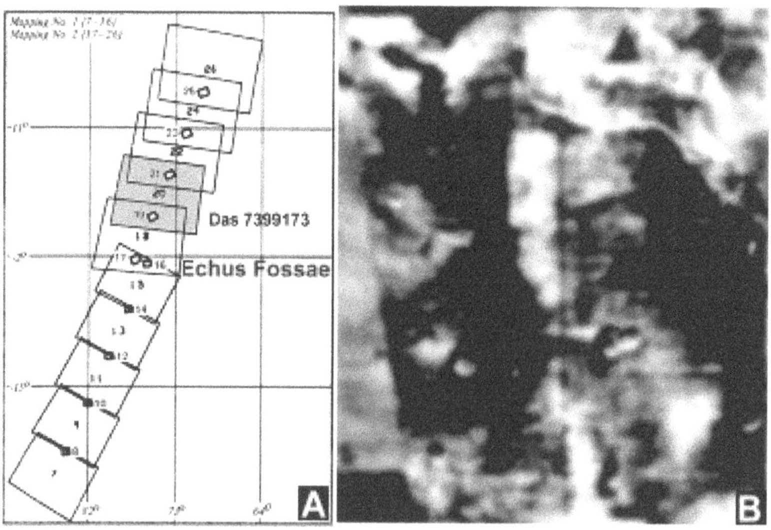

Tab. 87	Echus Fossae: Photo Das 7399173: A) detection scheme; B) original frame.

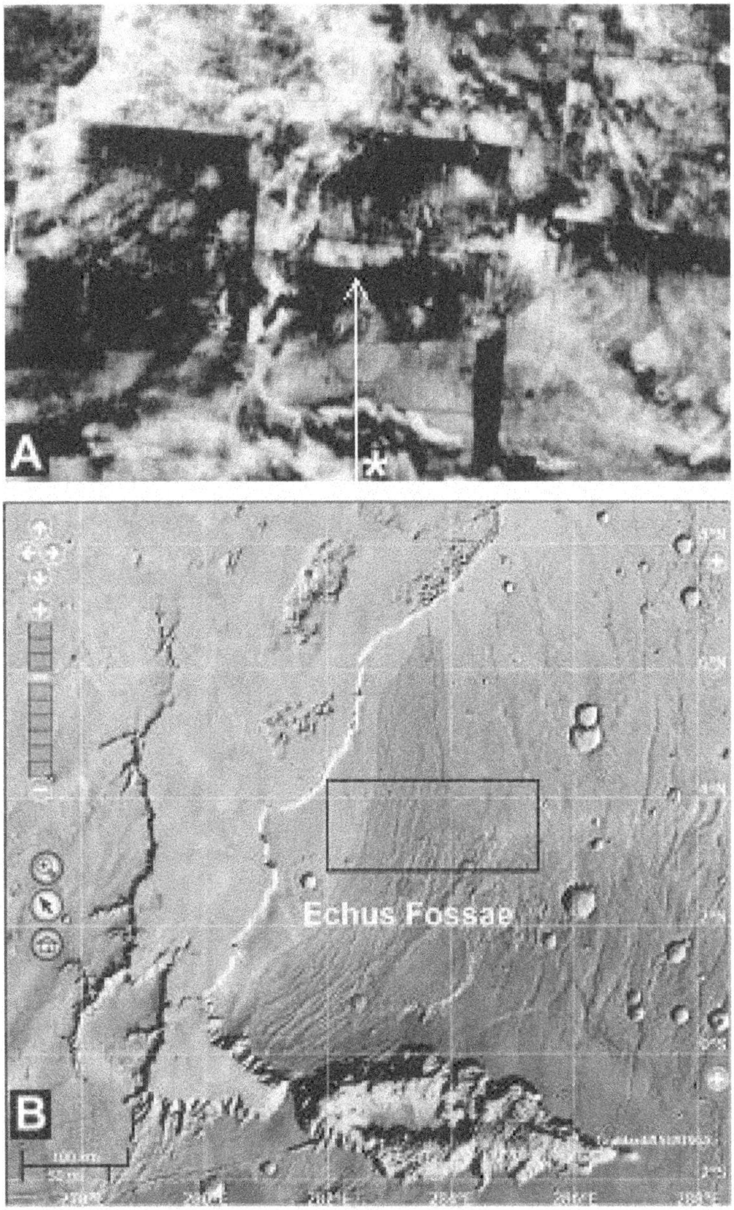

Tab. 88	Echus Fossae: A) Photomosaic 211-4294 (portion) with indication (*) of the detected structure; B) Cartographic position of the detected structure.

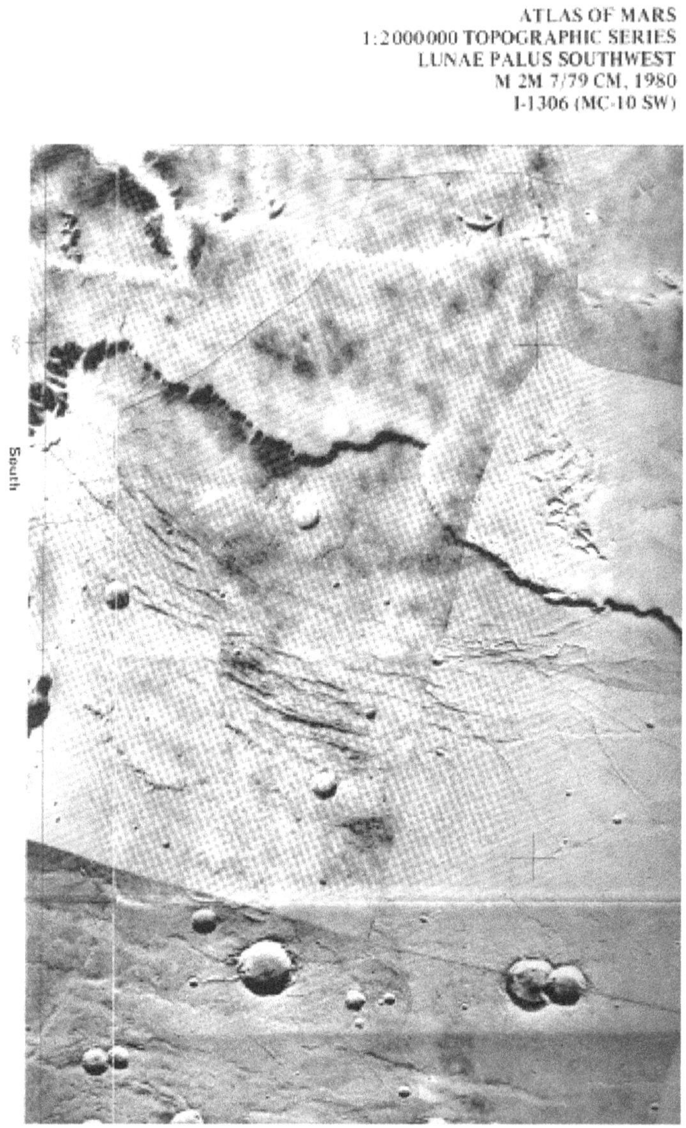

Tab. 89A	Echus Fossae – Map MC 10 1306 SW (portion), where the structure is analysed.

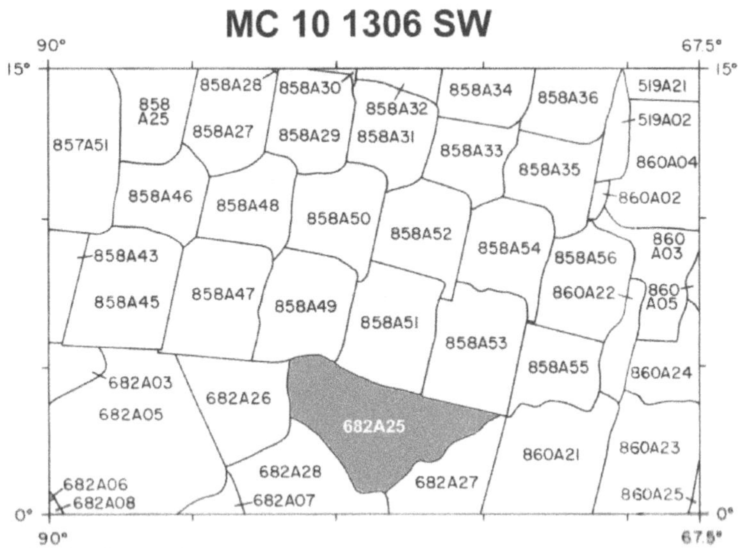

Tab. 89B	Echus Fossae – Map MC 10 1306 SW – Index of Viking images.

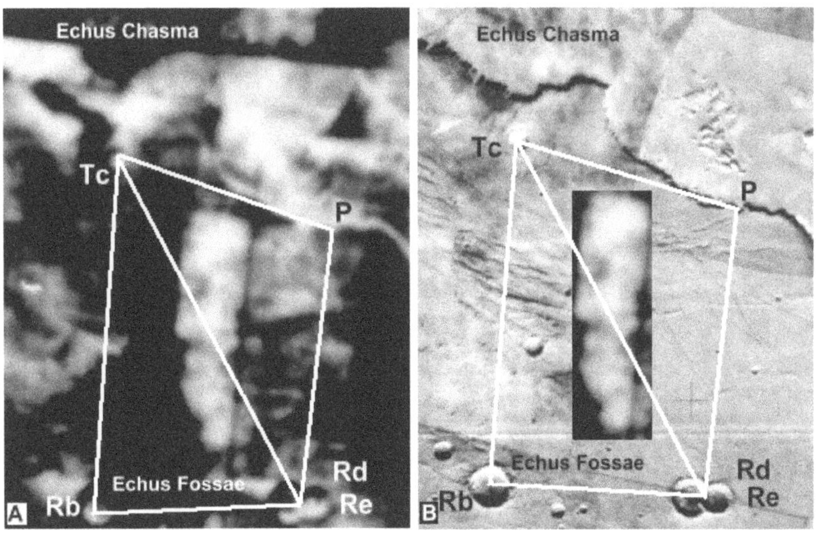

Tab. 90	Echus Fossae: A) Photomosaic 211-4294 (portion), with photographic control points; B) Carta MC 10 1306 SW (portion), discovery of photographic control points and relocation of the detected structure.

Mariner 9 Das 7399173

INST	TYPE	TIME (GMT) D H M S	PERI TIME H M S	SPACECRAFT LAT	SPACECRAFT LONG-W	HGT	PLATFORM CONE	PLATFORM CLOCK	INTERCEPTING LAT	INTERCEPTING LONG-W	INTERCEPTING RANGE	VIEW ANGLE	LIGHT ANGLE	PHASE ANGLE	DAS REFERENCE TIME
TETRAD NO 1															
1	B	33 23 10 16	-0 33 38	-63.32	167.81	3272	112.23	145.77	-59.05	171.05	3293	9.22	66.68	67.76	7,396,688
2	A	33 23 10 58	-0 32 56	-63.52	165.75	3217	112.08	145.85	-59.75	156.68	3232	7.76	64.91	67.98	7,396,723
3	B	33 23 11 40	-0 32 14	-63.68	163.63	3163	114.09	149.02	-58.52	169.47	3200	12.10	65.56	65.90	7,396,758
4	B	33 23 13 4	-0 30 50	-63.90	159.20	3057	117.44	152.32	-58.87	169.88	3112	14.94	65.74	62.55	7,396,828
SINGLE B NO 1															
5	B	33 23 24 15	-0 19 38	-58.81	121.53	2289	125.63	156.59	-56.27	116.95	2305	8.71	47.54	54.36	7,397,388
SINGLE B NO 2															
6	B	33 23 29 52	-0 14 2	-51.40	106.84	1991	141.58	192.18	-56.61	117.11	2077	20.98	47.87	38.41	7,397,668
MAPPING NO 1															
7	A	33 23 45 58	0 2 3	-19.15	82.81	1656	137.42	249.34	-23.56	85.42	1695	15.11	37.29	42.65	7,398,473
8	B	33 23 46 40	0 2 45	-17.59	82.13	1662	137.42	249.31	-21.44	84.19	1691	12.96	37.97	42.57	7,398,508
9	A	33 23 47 22	0 3 27	-16.04	81.46	1670	137.41	249.30	-18.90	83.10	1687	9.82	38.60	42.67	7,398,543
10	B	33 23 48 4	0 4 9	-14.48	80.81	1680	137.43	249.30	-16.80	81.95	1690	7.70	39.54	42.56	7,398,578
11	A	33 23 48 46	0 4 51	-12.93	80.17	1691	137.31	249.20	-14.24	81.00	1695	4.61	40.40	42.76	7,398,613
12	B	33 23 49 28	0 5 33	-11.39	79.55	1704	137.18	249.17	-12.13	79.99	1706	7.58	41.47	42.81	7,398,648
13	A	33 23 50 10	0 6 15	-9.85	78.94	1719	137.06	249.10	-9.58	79.10	1719	.91	42.57	43.01	7,398,683
14	B	33 23 50 52	0 6 57	-8.31	78.34	1736	136.96	249.10	-7.48	78.11	1737	2.54	43.84	43.03	7,398,718
15	A	33 23 51 34	0 7 39	-6.79	77.75	1754	136.89	249.18	-4.96	77.22	1759	5.56	45.16	43.18	7,398,753
16	B	33 23 52 16	0 8 21	-5.28	77.17	1774	136.94	249.34	-2.91	76.16	1784	7.46	46.67	43.05	7,398,788
MAPPING NO 2															
17	B	33 23 57 52	0 13 57	6.34	72.84	1986	114.94	282.03	-2.60	77.42	2122	26.21	46.83	65.05	7,399,068
18	A	33 23 58 34	0 14 39	7.72	72.33	2019	115.10	282.13	-.17	76.55	2127	23.30	48.29	64.97	7,399,103
19	B	33 23 59 16	0 15 21	9.09	71.82	2053	115.28	282.18	1.77	75.40	2142	21.07	50.00	64.71	7,399,138
■ 20	A	33 23 59 58	0 16 3	10.43	71.31	2089	115.48	282.13	4.23	74.56	2154	18.02	51.56	64.59	7,399,173 ■
21	B	34 0 0 40	0 16 45	11.76	70.80	2125	115.56	282.07	6.16	73.50	2176	15.87	53.25	64.42	7,399,208
22	A	34 0 1 22	0 17 27	13.06	70.30	2163	115.55	281.94	8.56	72.82	2198	13.06	54.74	64.52	7,399,243
23	B	34 0 2 4	0 18 9	14.35	69.80	2202	115.47	281.85	10.43	71.86	2227	11.11	56.37	64.52	7,399,278
24	A	34 0 2 46	0 18 51	15.62	69.30	2242	115.40	281.75	12.77	71.21	2257	8.48	57.87	64.67	7,399,313
25	B	34 0 3 28	0 19 33	16.86	68.81	2283	115.32	281.69	14.59	70.22	2292	6.56	59.54	64.67	7,399,348
26	A	34 0 4 10	0 20 15	18.09	68.31	2325	115.25	281.60	16.92	69.54	2328	4.05	61.11	64.81	7,399,383
TETRAD NO 2															
27	A	34 0 8 22	0 24 27	25.02	65.34	2595	99.85	303.99	7.28	68.48	2952	38.74	60.15	80.22	7,399,593
28	A	34 0 9 46	0 25 51	27.17	64.35	2691	100.09	299.07	15.55	68.28	2858	26.63	63.03	79.98	7,399,663
29	B	34 0 10 28	0 26 33	28.21	63.85	2741	100.23	298.99	17.20	67.09	2886	24.82	64.70	79.75	7,399,698
30	A	34 0 11 10	0 27 15	29.23	63.34	2790	100.28	294.05	23.26	68.15	2852	16.13	66.10	79.79	7,399,733
DYAD															
31	A	34 0 13 58	0 30 3	33.13	61.30	2996	107.10	270.50	54.43	67.55	3471	42.94	80.80	72.97	7,399,873
32	B	34 0 14 40	0 30 45	34.06	60.78	3048	107.13	270.44	56.66	65.90	3566	44.53	82.78	72.86	7,399,908
~~33~~	A	34 0 15 22	0 31 27	34.96	60.26	3101	107.15	270.50	60.02	64.96	3716	48.10	84.86	72.91	~~7,399,943~~

JPL TECHNICAL MEMORANDUM 33-585, VOL. II 6-355

Tab. 91 Echus Fossae: Photo Das 7399173 (78) – Detailed satellite survey information.

3. Region of Nanedi Valles

Map MC 10 1307 SE

37.35° N - 54.53° W (alt. km 3.0)

In quadrant MC- 10 Lunae Palus (**Table 86A**) we take into consideration the images n. 38 and 43 of the attached list (**Table 86B**), with the following identification numbers (Das 7543023, 7614913).

In **tables 92A and 93A** we look at the detection patterns and in **tables 92B** and **93B** we have the original frames.

In **table 94A** the detected structure is inserted in the context of Equatorial Photomosaic 211-4294; the asterisk (*) indicates the detected structure. It could be described as two human faces placed frontally and vertically aligned, having the following dimensions: 40x40 km and 60x60 (**Table 99A**).

Located in the region of Nanedi Valles (**Table 94B**), where the cartographic position is indicated, the original structure has completely disappeared on map MC.10.1307.SE (**Table 96A**).

From a cartographic analysis, it appears that, in the corresponding part occupied by the original structure, there is the frame 860A34 attributed to the mission Viking 1 (**Table 96B**).

To superimpose the original image on the "false" one, taking into account the original photo (**Table 95A**), we fix some photographic control points (**Tables 97A, B, C**), then we find them on map MC.10.1307.SE (**Tables 97D, E, F**). In the original photo (**Tables 98A,C,E**) we placed other photographic control points, then found in paper MC.10.1307.SE (**Tables 98B,D,F**). Another operation, the same as the previous one, has been carried out - always with the original photo (**Tables 99A,C**), through the highlighting of the coordinates 3.27 - 52.96 of W, then found on the map MC.10.1307.SE (**Tables 99B,D**).

The structures of Nanedi Valles, given the remarkable dimensions (**Table 95A**): km 40x40 and 60x60, were modified/deleted also on the pages of the Atlas of Mars of 1979 (**Table 95B**).

In **Tables 100 and 101** we have inserted the detailed data of the frames Das 7543023 and 7614913.

Stable elements of experimentation/photographic points of reference: (2.0-3.0°N - 50.0-55.0°W); 1) SW: Crater Mutch; 2) NE: Crater Camichel (Db); 3) NE: Crater Dcb; 4) E: 1 –Fa; 2 – Ek; 3 – Ekc; 4 – Dj; 5 – Ck; 6 –

Bk; 7 – Ej; 8 – Dh; 9 – Fgh; 10 – Fg; 11; – Dgb; 12 – Cga; 13; - Cg; 14; - Bf; 15; - Ef; 16; - Fe; 17; - Eo; 18 - Dg.

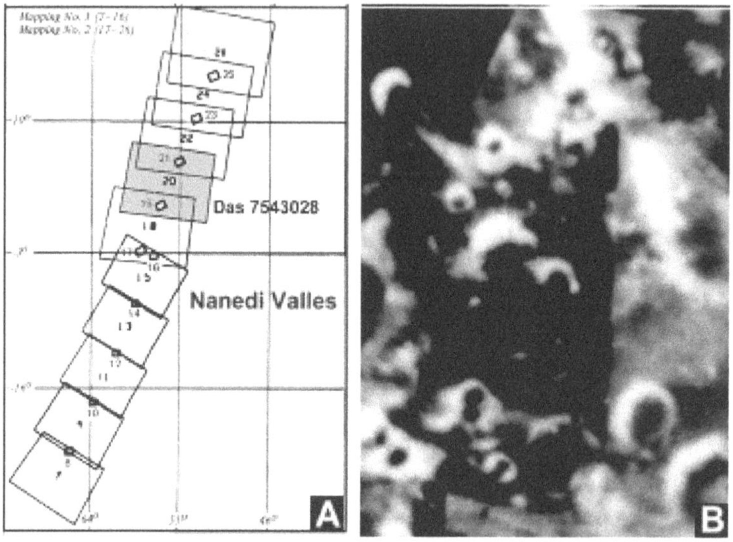

Tab. 92	Nanedi Valles – Photo Das 7543028: A) Detection scheme; B) Original frame.

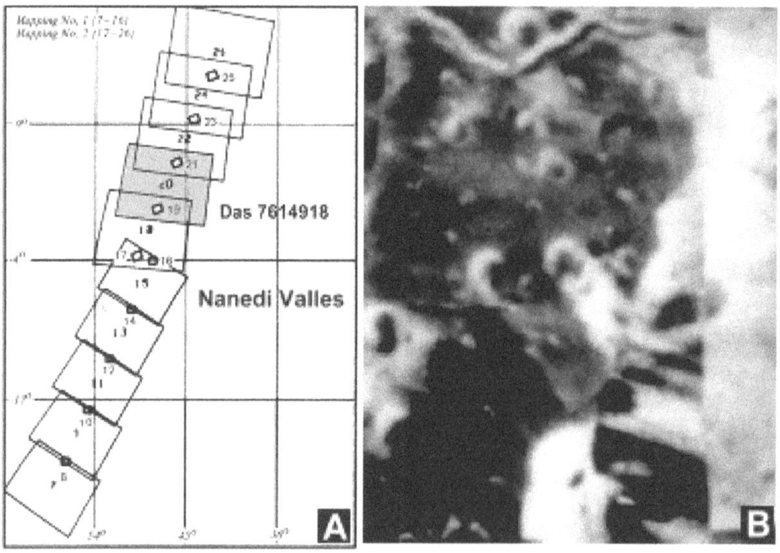

Tab. 93	Nanedi Valles – Photo Das 7614918: A) Detection scheme; B) Original frame.

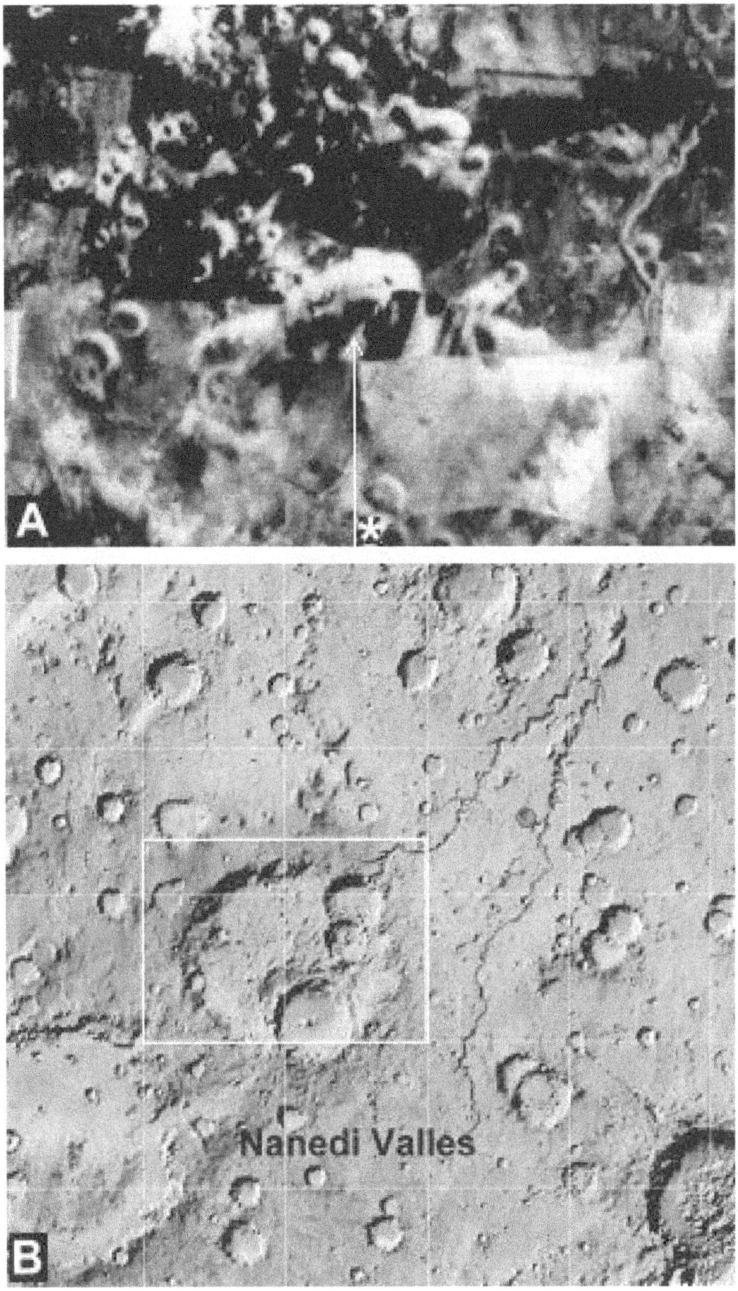

Tab. 94	Nanedi Valles: A) Photomosaic 211-4294 (portion) with indication (*) of the detected structure; B) Cartographic position of the detected structure.

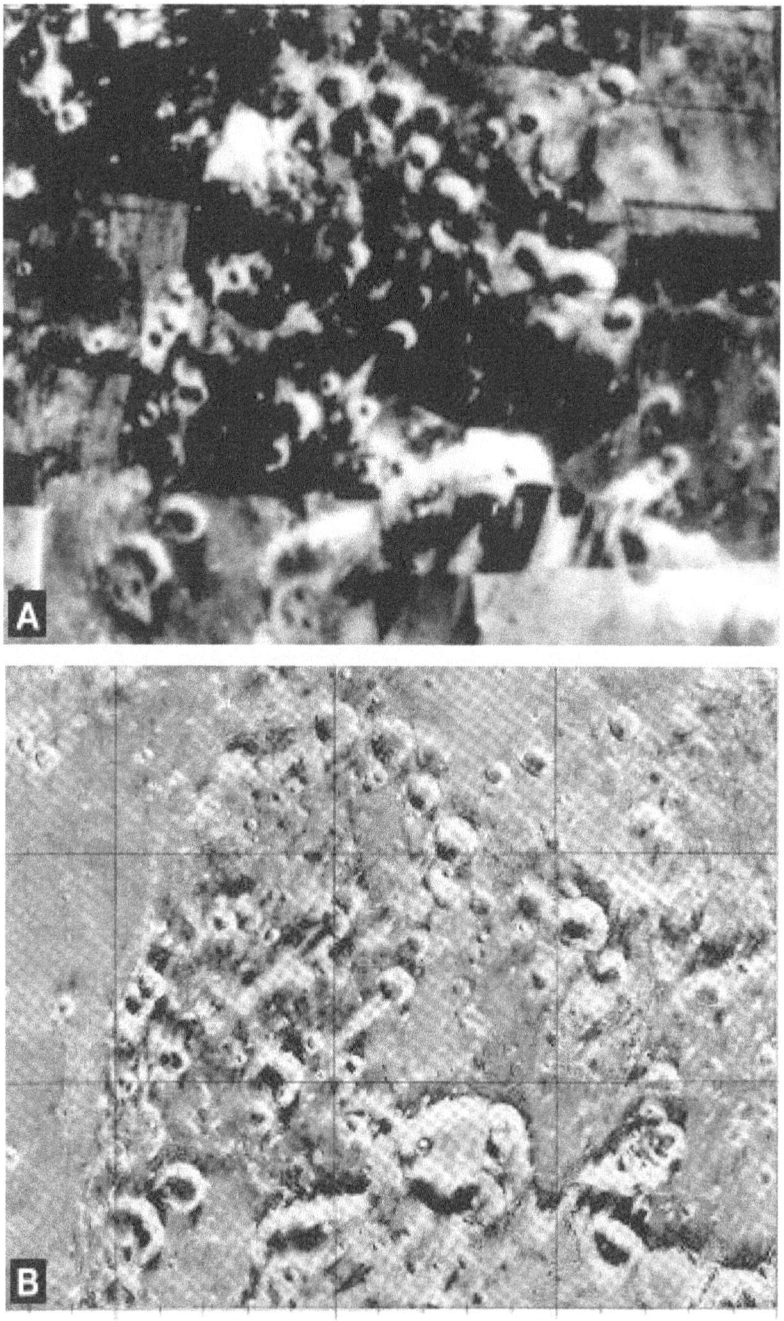

Tab. 95	Nanedi Valles: A) Photomosaic 211-4294 (portion); B) Modified photographic cartography (Atlas of Mars 1979).

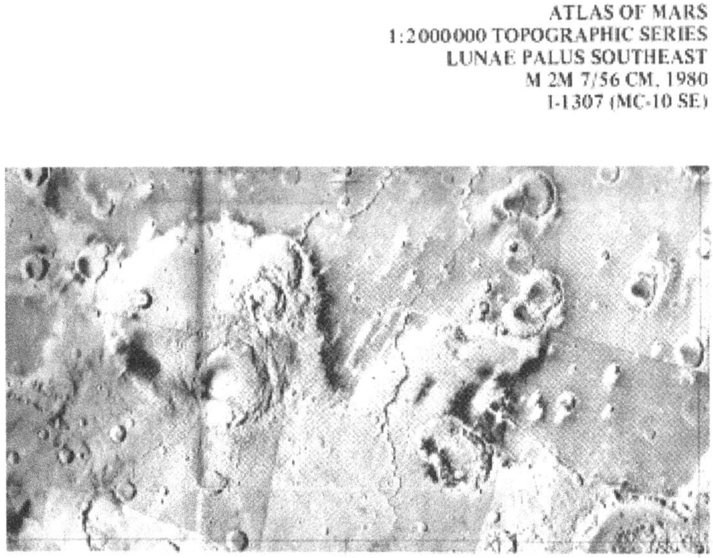

Tab. 96A	Nanedi Valles – Map MC 10 1307 SE (portion), where the structure analyzed falls.

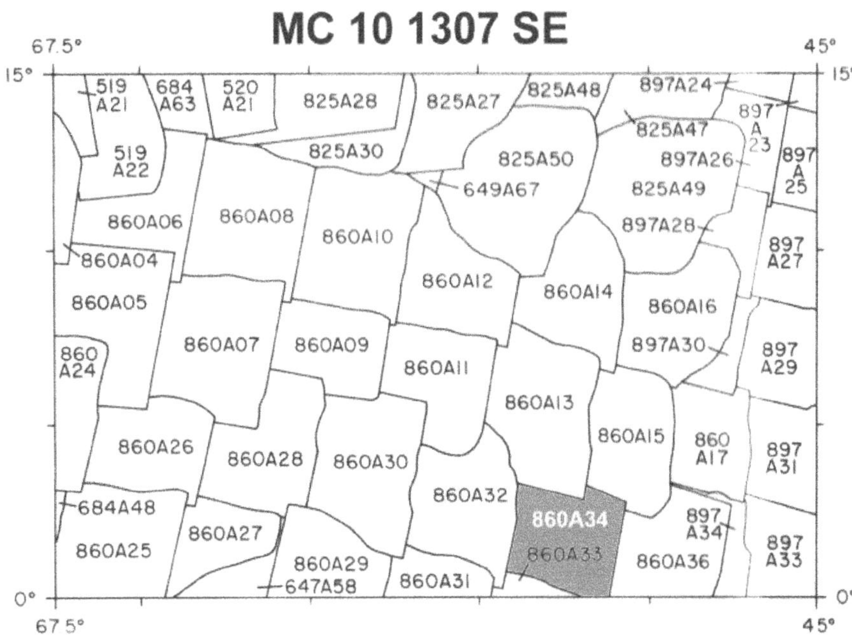

Tab. 96B	Nanedi Valles – Map MC 10 1307 SE – Index of Viking images.

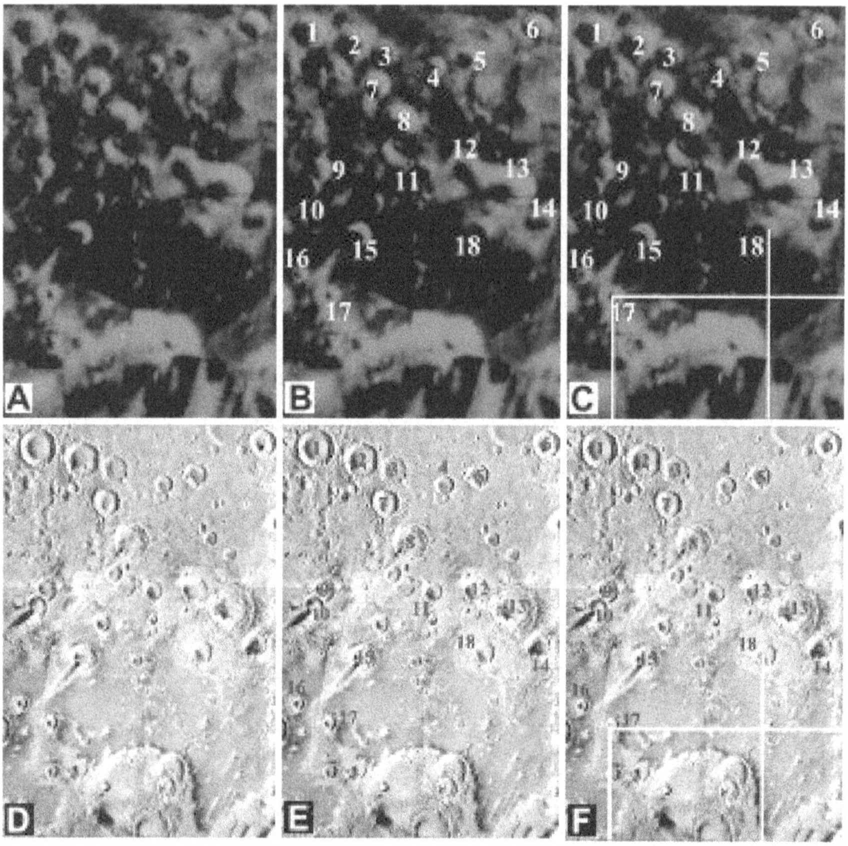

Tab. 97	Nanedi Valles: A,B,C) Photomosaic 211-4294 (portion), with photographic control points; D,E,F) Map MC 10 1307 SE (portion), with the discovery of photographic control points.

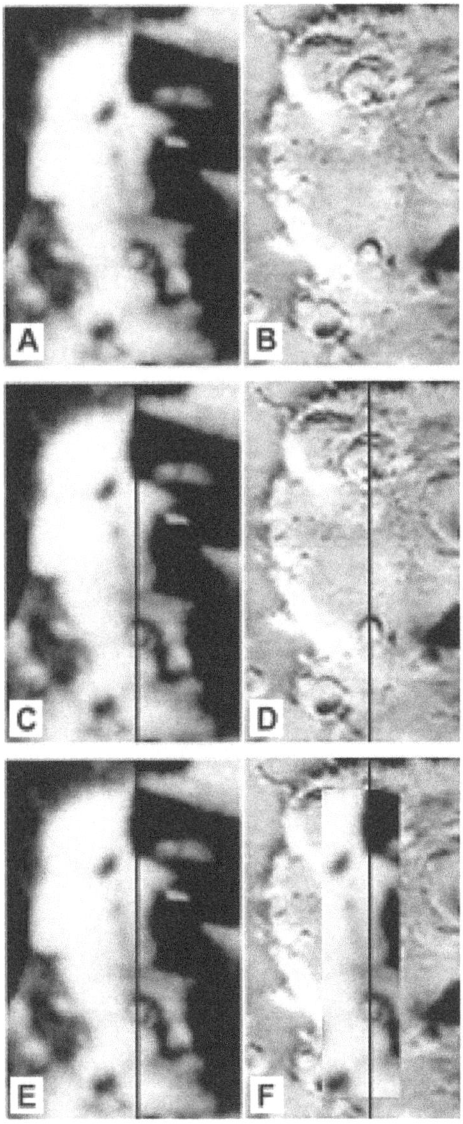

Tab. 98	Nanedi Valles: A,C,F) Detected structure – Photomosaic 211-4294 (portion); B,D,F) Corresponding position on Map MC.10.1307.SE (portion) – Identification and correspondence of a photographic control point: an alleged "eye" of the structure at the bottom.

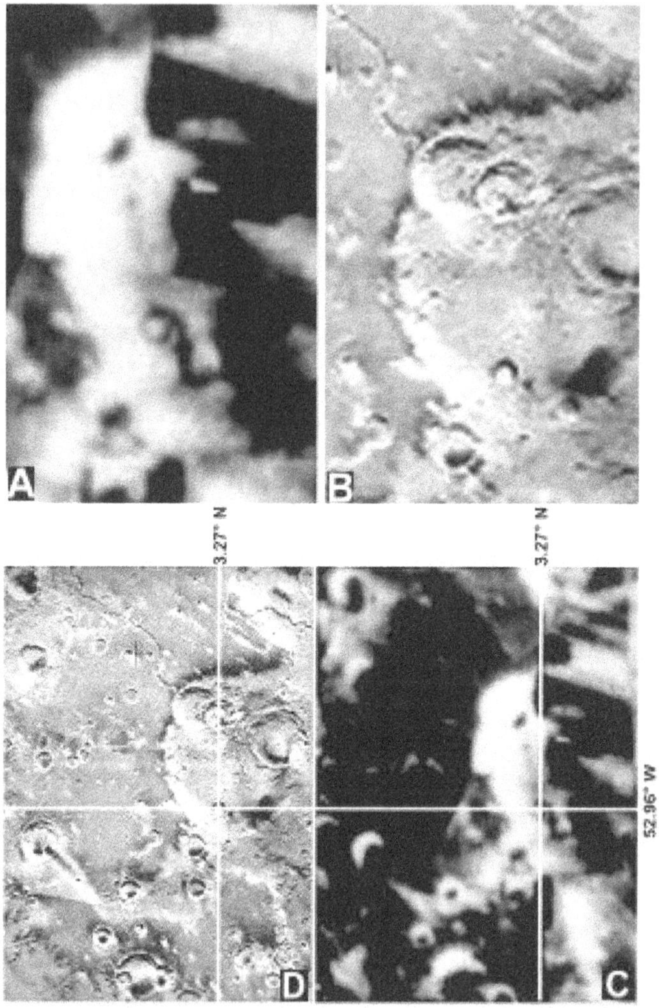

Tab. 99	Nanedi Valles: A) Detected structure – Photomosaic 211-4294 (portion); B) Corresponding part on Map MC 10 1307 SE (portion); C) Detected structure – Photomosaic 211-4294 (portion) – Indication of a photographic control point (3.27 N; 52.96 W); D) Map MC10-1307.SE (portion): photographic control point found.

Mariner 9 Das 7543023

INST		TIME (GMT)				PERI TIME			SPACECRAFT			PLATFORM		INTERCEPTING			VIEW	LIGHT	PHASE	DAS REFERENCE
	TYPE	D	H	M	S	H	M	S	LAT	LONG-W	HGT	CONE	CLOCK	LAT	LONG-W	RANGE	ANGLE	ANGLE	ANGLE	TIME
TETRAD NO 1																				
1	B	35	23	7	15	-0	33	53	-63.28	149.99	3291	99.78	158.52	-38.70	147.58	3866	46.01	57.75	80.21	7,540,538
2	A	35	23	7	57	-0	33	11	-63.48	147.95	3237	99 79	158 57	-40.01	144.87	3770	44.57	56.05	80.27	7,540,573
3	B	35	23	8	39	-0	32	29	-63.66	145.84	3183	100.55	163.91	-33.71	148.29	4011	54.94	56.76	79.44	7,540,608
4	B	35	23	10	3	-0	31	5	-63.90	141.45	3076	103.78	168.02	-33.37	150.86	3968	57.23	58.49	76.21	7,540,678
SINGLE B NO 1																				
5	B	35	23	21	15	-0	19	53	-59.09	103.69	2305	132.28	162.05	-61.09	107.73	2315	6.07	53.49	47.71	7,541,238
SINGLE B NO 2																				
6	B	35	23	26	51	-0	14	17	-51.82	88.81	2004	135.34	159.87	-50.54	81.21	2037	13.15	44.30	44.65	7,541,518
MAPPING NO 1																				
7	A	35	23	42	57	0	1	48	-19.74	64.49	1656	138.96	247 27	-23.99	67.10	1593	14.71	36.23	41.11	7,542,323
8	B	35	23	43	39	0	2	30	-18.18	63.80	1661	138.61	247.13	-21.84	66.06	1689	12.72	36.62	41.38	7,542,358
9	A	35	23	44	21	0	3	12	-16.63	63.13	1669	138.61	247.30	-19.36	64.95	1685	9.74	37.18	41.46	7,542,393
10	B	35	23	45	3	0	3	54	-15.07	62.47	1678	138.83	247.49	-17.32	63.71	1688	7.65	38.13	41.19	7,542,428
11	A	35	23	45	45	0	4	36	-13.52	61.83	1688	139 01	247.60	-14.83	62.58	1692	4.49	39.07	41.06	7,542,463
12	B	35	23	46	27	0	5	18	-11.97	61.20	1701	139.14	247.43	-12.67	61.43	1702	2.20	40.21	40.85	7,542,498
13	A	35	23	47	9	0	6	0	-10 43	60.59	1715	139.09	247.16	-10.05	60.49	1715	1.15	41.28	40.98	7,542,533
14	B	35	23	47	51	0	6	42	-8.89	59.99	1731	139.04	246.91	-7.86	59.48	1733	3.30	42.54	40.95	7,542,568
15	A	35	23	48	33	0	7	24	-7.37	59.40	1749	138.94	246.90	-5.30	58.59	1756	6.48	43.81	41.13	7,542,603
16	B	35	23	49	15	0	8	6	-5.85	58.82	1768	138.84	246.96	-3.21	57.63	1780	8.41	45.21	41.15	7,542,638
MAPPING NO 2																				
17	B	35	23	54	51	0	13	42	5.81	54.48	1976	116.98	281.02	-2 72	58.98	2102	25.32	45.29	63.01	7,542,918
18	A	35	23	55	33	0	14	24	7 19	53.96	2009	117.24	281.01	-.21	58 07	2105	22.16	46.80	62.82	7,542,953
19	B	35	23	55	15	0	15	6	8.56	53.45	2042	117.40	280.99	1 77	56.96	2121	19.89	48.47	62.59	7,542,988
■ 20	A	35	23	56	57	0	15	48	9.91	52.94	2077	117.61	280.98	4.20	56.11	2134	16.89	50.03	62.46	7,543,023 ■
21	B	35	23	57	39	0	16	30	11.25	52.44	2113	117.62	280.89	6 14	55.10	2157	14.77	51.67	62.37	7,543,058
22	A	35	23	58	21	0	17	12	12.56	51.94	2151	117.57	280 83	8.50	54.43	2180	12.10	53.14	62.50	7,543,093
23	B	35	23	59	3	0	17	54	13.85	51.44	2189	117.51	280.81	13.35	53.44	2210	10.17	54.79	62.48	7,543,128
24	A	35	23	59	45	0	18	36	15.13	50.94	2229	117 48	280.81	12.66	52.74	2240	7.59	56.33	62.59	7,543,163
25	B	36	0	0	27	0	19	18	16.38	50 44	2269	117.43	280.82	14.45	51.71	2276	5.68	58.02	62.56	7,543,198
26	A	36	0	1	9	0	20	0	17.62	49.95	2311	117.41	280 81	16.76	50.99	2313	3.73	59.63	62.66	7,543,233
TETRAD NO 2																				
27	A	36	0	5	21	0	24	12	24.59	46.99	2579	100.76	304.07	6.24	50.46	2962	40.08	57.96	79.30	7,543,443
28	A	36	0	5	45	0	25	36	26.75	46.00	2675	100.95	297.82	15.63	50.81	2837	26.35	60.63	79.12	7,543,513
29	B	36	0	7	27	0	26	18	27.80	45.50	2724	101.04	297.57	17.42	49 74	2862	24.25	62.35	78.95	7,543,548
30	A	36	0	8	9	0	27	0	28.83	45.00	2773	101.10	291.58	24.20	51.25	2833	15.95	63.68	78.97	7,543,583
DYAD																				
31	A	36	0	10	57	0	29	48	32.76	42.96	2977	105.87	269.96	52.89	53.94	3450	42.83	76.44	74.25	7,543,723
32	B	36	0	11	39	0	30	30	33.69	42.45	3030	105.75	269.87	55.09	52.92	3539	44.22	78.26	74.24	7,543,758
~~33~~	A	36	0	12	21	0	31	12	34.61	41.93	3083	105.69	269.81	58.48	53.02	3689	47.81	80.18	74.38	~~7,543,793~~

JPL TECHNICAL MEMORANDUM 33-585, VOL. II

6-363

Tab. 100	Nanedi Valles: Photo Das 7543023 (28) – Detailed satellite survey information.

Mariner 9 Das 7614913

INST	TYPE	TIME (GMT) D H M S	PERI TIME H M S	SPACECRAFT LAT	SPACECRAFT LONG-W	HGT	PLATFORM CONE	PLATFORM CLOCK	INTERCEPTING LAT	INTERCEPTING LONG-W	INTERCEPTING RANGE	VIEW ANGLE	LIGHT ANGLE	PHASE ANGLE	DAS REFERENCE TIME
TETRAD NO 1															
1	B	36 23 5 3	-0 34 10	-63.19	141.41	3314	115.93	115.98	-86.78	69.62	3945	48.03	78.75	64.06	7,612,428
2	A	36 23 5 45	-0 33 28	-63.41	139.40	3259	115.93	115.98	-84.81	51.85	3933	49.69	77.44	64.14	7,612,463
3	B	36 23 6 27	-0 32 46	-63.60	137.32	3205	119.15	114.42	-85.13	10.59	4009	54.15	80.43	60.84	7,612,498
4	B	36 23 7 51	-0 31 22	-63.88	132.97	3098	121.33	115.45	-84.44	22.54	3862	53.22	79.20	58.65	7,612,568
SINGLE B NO 1															
5	B	36 23 19 3	-0 20 10	-59.40	95.15	2323	132.27	161.83	-61.67	100.58	2338	8.57	54.65	47.72	7,613,128
SINGLE B NO 2															
6	B	36 23 24 39	-0 14 34	-52.27	80.06	2018	134.02	171.48	-51.55	81.24	2020	2.75	43.73	45.97	7,613,408
MAPPING NO 1															
7	A	36 23 40 45	0 1 31	-20.40	55.38	1655	139.96	245.87	-24.65	57.97	1692	14.70	35.71	40.11	7,614,213
8	B	36 23 41 27	0 2 13	-18.84	54.68	1660	139.85	245.94	-22.57	56.79	1688	12.72	36.16	40.14	7,614,248
9	A	36 23 42 9	0 2 55	-17.28	54.00	1666	139.75	245.99	-20.05	55.72	1682	9.68	36.58	40.32	7,614,283
10	B	36 23 42 51	0 3 37	-15.73	53.34	1675	139.67	246.02	-17.96	54.62	1685	7.66	37.32	40.32	7,614,318
11	A	36 23 43 33	0 4 19	-14.17	52.69	1685	139.63	245.93	-15.40	53.60	1688	4.53	38.06	40.44	7,614,353
12	B	36 23 44 15	0 5 1	-12.62	52.06	1697	139.58	245.85	-13.27	52.55	1698	2.40	39.06	40.41	7,614,388
13	A	36 23 44 57	0 5 43	-11.08	51.44	1710	139.69	245.86	-10.74	51.51	1710	1.03	40.16	40.37	7,614,423
14	B	36 23 45 39	0 6 25	-9.54	50.83	1725	139.84	245.88	-8.64	50.40	1727	2.95	41.47	40.15	7,614,458
15	A	36 23 46 21	0 7 7	-8.01	50.24	1742	140.03	245.93	-6.10	49.36	1749	6.15	42.83	40.07	7,614,493
16	B	36 23 47 3	0 7 49	-6.49	49.65	1761	140.12	245.94	-3.99	48.28	1773	8.29	44.31	39.87	7,614,528
MAPPING NO 2															
17	B	36 23 52 39	0 13 25	5.21	45.29	1964	117.90	280.84	-3.52	49.86	2096	25.91	44.16	62.09	7,614,808
18	A	36 23 53 21	0 14 7	6.60	44.78	1996	118.09	280.77	-.98	49.00	2098	22.79	45.58	61.98	7,614,843
19	B	36 23 54 3	0 14 49	7.98	44.26	2029	118.29	280.80	.99	47.84	2112	20.49	47.26	61.70	7,614,878
■ 20	A	36 23 54 45	0 15 31	9.34	43.76	2064	118.48	280.81	3.42	46.99	2124	17.90	48.81	61.58	7,614,913 ■
21	B	36 23 55 27	0 16 13	10.68	43.25	2099	118.54	280.72	5.38	45.96	2146	15.34	50.46	61.45	7,614,948
22	A	36 23 56 9	0 16 55	12.00	42.75	2136	118.56	280.72	7.73	45.23	2168	12.61	51.96	61.51	7,614,983
23	B	36 23 56 51	0 17 37	13.30	42.25	2174	118.53	280.74	9.56	44.24	2197	10.73	53.60	61.49	7,615,018
24	A	36 23 57 33	0 18 19	14.58	41.75	2213	118.44	280.80	11.84	43.54	2227	8.20	55.12	61.63	7,615,053
25	B	36 23 58 15	0 19 1	15.85	41.26	2253	118.40	280.83	13.63	42.51	2261	6.31	56.81	61.59	7,615,088
26	A	36 23 58 57	0 19 43	17.09	40.76	2294	118.36	280.92	15.88	41.77	2297	3.83	58.40	61.71	7,615,123
TETRAD NO 2															
27	A	37 0 3 9	0 23 55	24.11	37.80	2560	101.09	304.02	5.23	41.68	2968	41.34	56.36	78.98	7,615,333
28	A	37 0 4 33	0 25 18	26.29	36.81	2655	101.29	297.71	14.75	42.06	2833	27.56	59.00	78.78	7,615,403
29	B	37 0 5 15	0 26 0	27.35	36.32	2704	101.46	297.48	16.59	40.92	2855	25.34	60.79	78.53	7,615,438
30	A	37 0 5 57	0 26 42	28.39	35.82	2753	101.56	291.56	23.31	42.38	2822	17.13	62.15	78.51	7,615,473
DYAD															
31	A	37 0 8 45	0 29 31	32.35	33.79	2957	112.69	283.50	40.61	30.51	3039	18.34	78.65	67.38	7,615,613
32	B	37 0 9 27	0 30 13	33.29	33.28	3009	112.83	283.54	42.20	28.61	3109	20.15	80.67	67.16	7,615,648
33	A	37 0 10 9	0 30 55	34.21	32.76	3061	112.89	283.52	44.58	27.14	3195	23.14	82.59	67.17	7,615,683

JPL TECHNICAL MEMORANDUM 33-585, VOL. II

6-367

Tab. 101	Nanedi Valles: Photo Das 7614913 (18) – Detailed satellite survey information.

4. Region of Ophir Planum

Map MC 18 1208 NE

3.0° S – 52.0° W – (alt. km 3.5)

In quadrant MC- 18 Coprates (**Table 102A**) we take into consideration picture No.9 of the attached list (**Table 102B**) with the following identification number (Das7614843).

In **table 103A** we observe the pattern of detection and in **table 103B** the original frame.

In **table 104A** the structure detected is inserted in the context of Equatorial Photomosaic 211-4294; the asterisk (*) indicates the structure detected.

It could be described as a human face in profile, occupying the entire area of a crater, with the following dimensions: km 68 x 80 (**Table 106B**).

Located in the region of Ophir Planum (**Table 104B**), where the cartographic position is indicated, the original structure is completely disappeared on map MC.18.1208.NE (**Table 105A**).

A cartographic analysis shows that, in the corresponding part occupied by the original structure, there is the frame 649A61 attributed to the Viking 1 mission (**Table 105B**).

To superimpose the original image on the "false" one, taking into account the original photo (**Plate 104A**), we fix some photographic control points (**Table 106B**), then we find them on map MC.18.1208.NE (**Table 106A**).

The image of Mariner 9 (**Table 107A of 1972**) was later reconfirmed in 2004 by Mars Odyssey (**Table 107B**), where it appears as a two-faced human face in profile.

In **table 108** we have inserted the detailed data of the frame Das 7614843.

Stable elements of experimentation/photographic points of reference: (0.0-5.0 – 50.0-55.0 – W); 1) Crater Dx; 2) Contour Craters (possibly Secondary Craters): NC Ez, NE Dz, SW Ex, Ew, SE Dx, Dw, Cw.

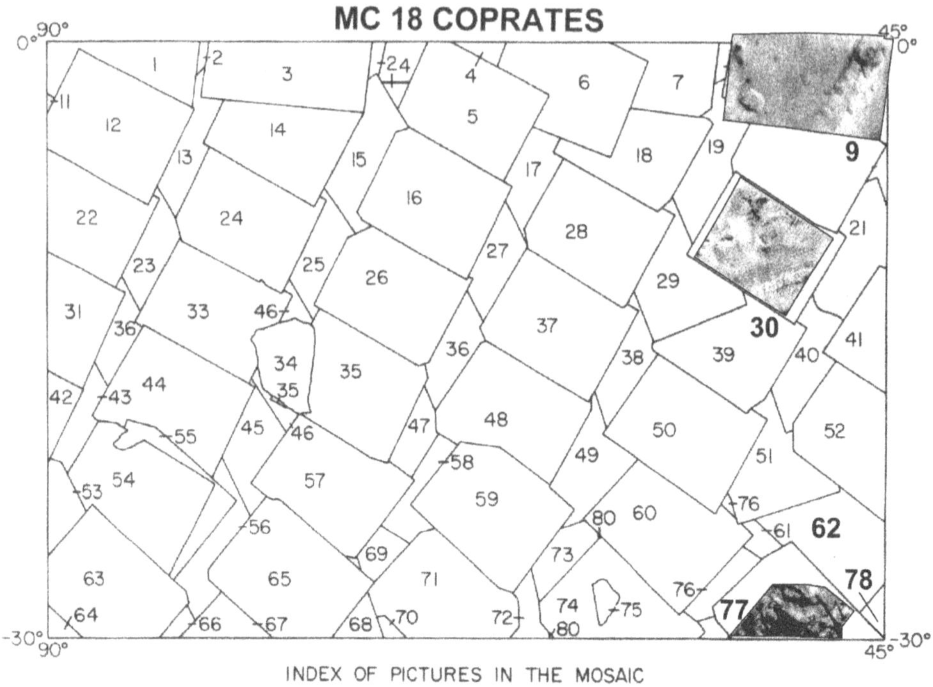

Tab. 102A	MC 18 Coprates – Index of images Mariner 9.

IDENTIFICATION NUMBERS OF PICTURES IN THE MOSAIC

Index No.	DAS No.	Index No.	DAS No.	Index No.	DAS No.
1	7327143	28	7542533	55	5995393
2	5380793	29	9017229	56	8873239
3	7399103	■ 30	7614423	57	6067353
4	7471063	31	7326653	58	7542323
5	7470713	32	8801419	59	6139243
6	6139733	33	7398613	60	6211133
7	7542953	34	12010324	61	7686103
8	6355403	35	7470503	■ 62	6283023
■ 9	7614843	36	8945269	63	5995323
10	9089189	37	7542463	64	5995253
11	8729599	38	9017159	65	6067283
12	7326793	39	7614353	66	8873169
13	8801559	40	9089049	67	6067213
14	7398753	41	7686243	68	8945059
15	8873519	42	7326583	69	8945129
16	7470643	43	8801349	70	6139103
17	8945409	44	7398543	71	6139173
18	7542603	45	8873309	72	9016949
19	9017299	46	8873379	73	9017019
20	7614493	47	8945199	74	6211063
21	9089119	48	7542393	75	9088839
22	7326723	49	9017089	76	9088909
23	8801489	50	7614283	■ 77	6282953
24	7398683	51	9088979	■ 78	9160729
25	8873449	52	7686173	79	6211623
26	7470573	53	8801279	80	5524783
27	8945339	54	7398473		

Tab. 102B	MC 18 Coprates – Image identification number.

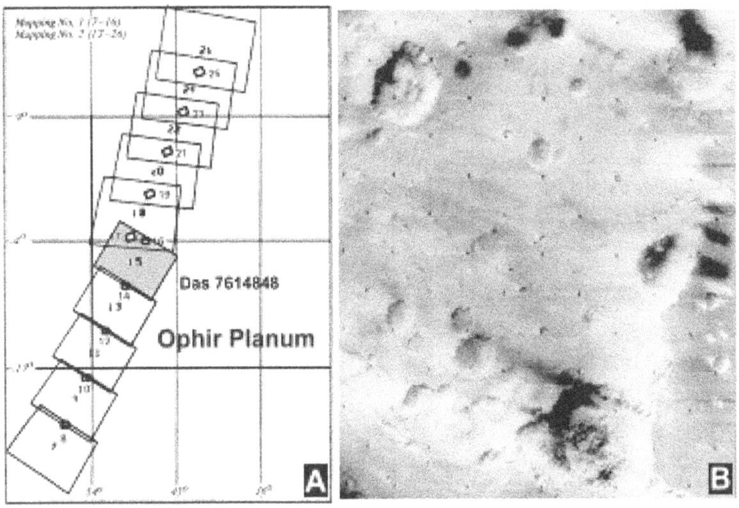

Tab. 103	Ophir Planum – Photo Das 7614848: A) Detection scheme; B) Original frame.

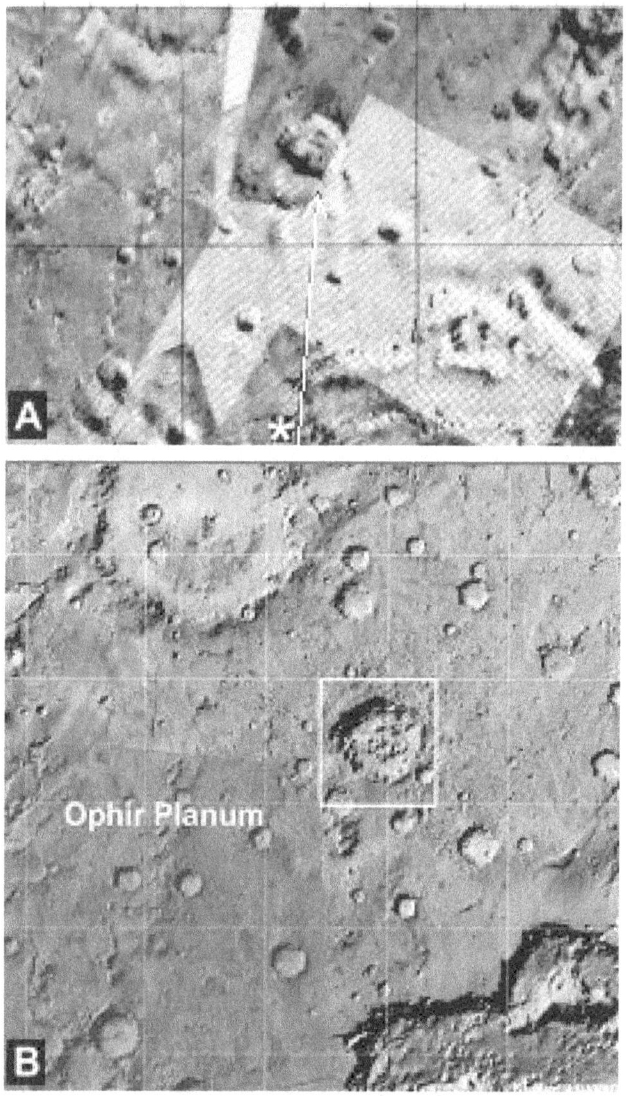

Tab. 104	Ophir Planum: A) Photomosaic 211-4294 (portion) with indication (*) of the detected structure; B) Cartographic position of the detected structure.

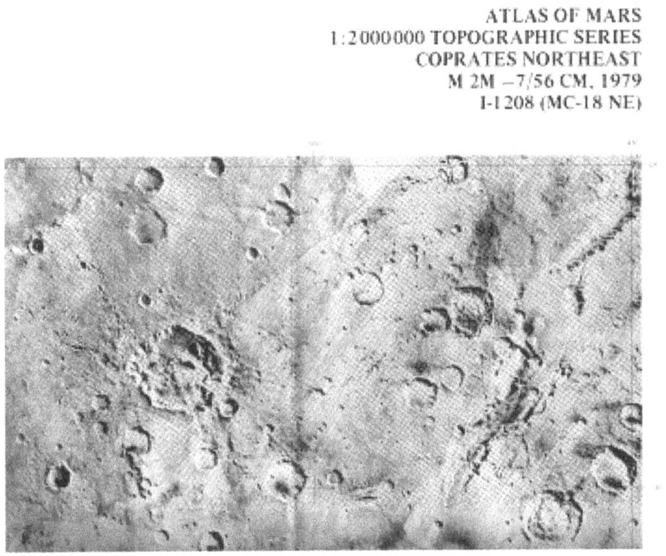

Tab. 105A	Ophir Planum – Map MC 18 1208 NE (portion), where the analyzed structure falls.

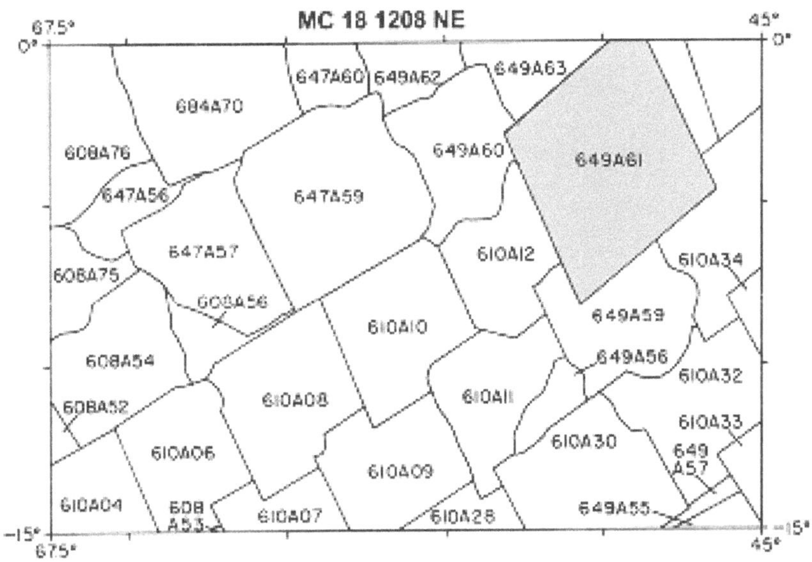

Tab. 105B	Ophir Planum – Map MC 18 1208 NE – Index of Viking images.

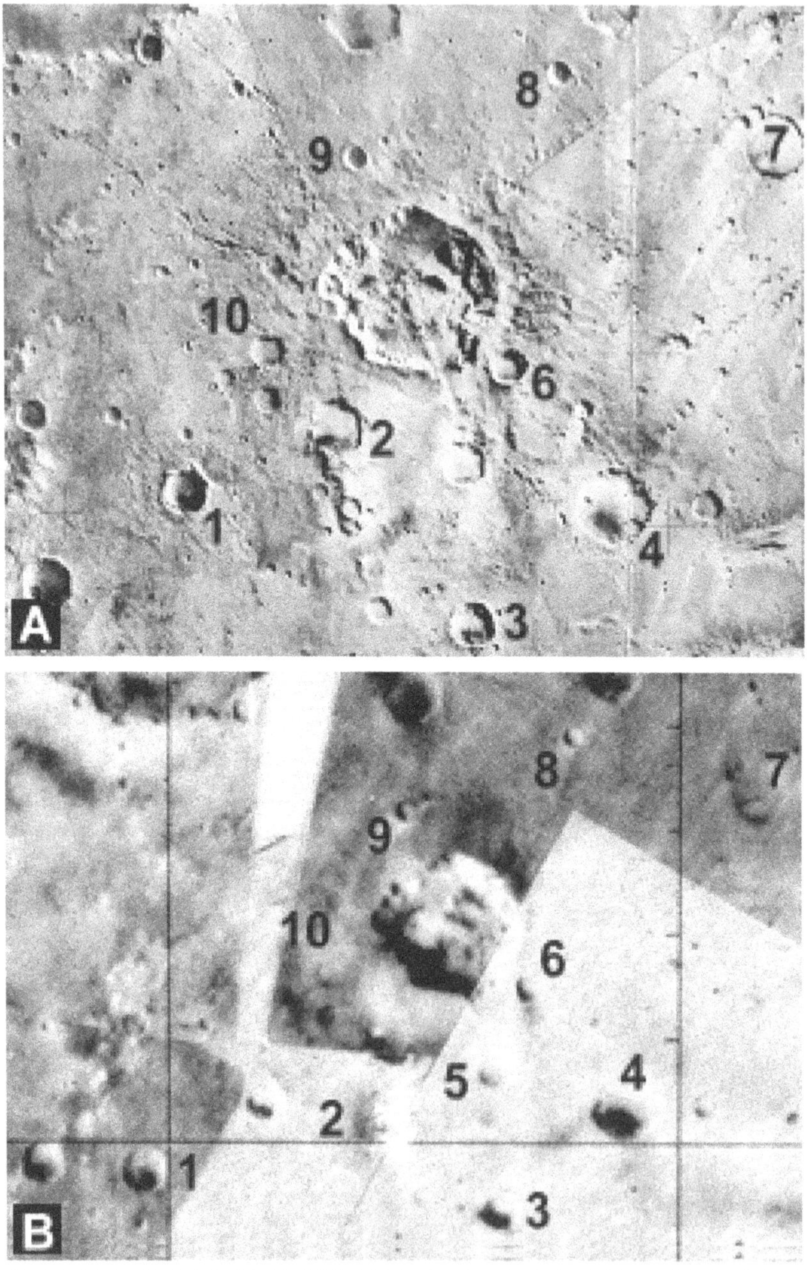

Tab. 106	Ophir Planum: B) Original photo with reference points; A) Map MC 18 1208 NE (portion) with the structure deleted and the finding of reference points.

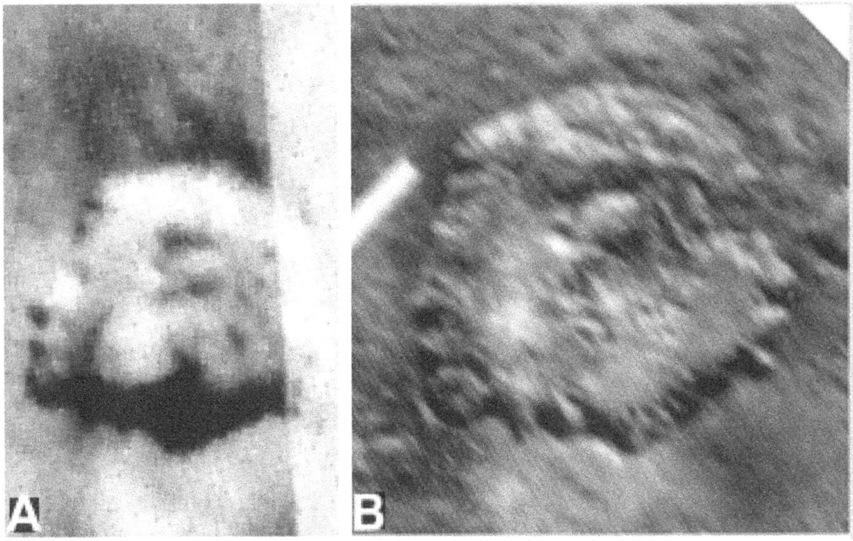

Tab. 107	Ophir Planum: A) Enlarged structure; B) Satellite survey of the Mars Odyssey (2004): in this case the structure appears two-faced.

Mariner 9 Das 7614843

INST	TYPE	TIME (GMT) D	H	M	S	PERI TIME H	M	S	SPACECRAFT LAT	SPACECRAFT LONG-W	HGT	PLATFORM CONE	PLATFORM CLOCK	INTERCEPTING LAT	INTERCEPTING LONG-W	INTERCEPTING RANGE	VIEW ANGLE	LIGHT ANGLE	PHASE ANGLE	DAS REFERENCE TIME
TETRAD NO 1																				
1	B	36	23	5	3	-0	34	10	-63.19	141.41	3314	115.93	115.98	-86.78	69.62	3945	48.03	78.75	64.06	7,612,428
2	A	36	23	5	45	-0	33	28	-63.41	139.40	3259	115.93	115.98	-84.81	51.85	3933	49.69	77.44	64.14	7,612,463
3	B	36	23	6	27	-0	32	46	-63.60	137.32	3205	119.15	114.42	-85.13	10.59	4009	54.15	80.43	60.84	7,612,498
4	B	36	23	7	51	-0	31	22	-63.88	132.97	3098	121.33	115.45	-84.44	22.54	3862	53.22	79.20	58.65	7,612,568
SINGLE B NO 1																				
5	B	36	23	19	3	-0	20	10	-59.40	95.15	2323	132.27	161.83	-61.67	100.58	2338	8.57	54.65	47.72	7,613,128
SINGLE B NO 2																				
6	B	36	23	24	39	-0	14	34	-52.27	80.06	2018	134.02	171.48	-51.55	81.24	2020	2.75	43.73	45.97	7,613,408
MAPPING NO 1																				
7	A	36	23	40	45	0	1	31	-20.40	55.38	1655	139.96	245.87	-24.65	57.97	1692	14.70	35.71	40.11	7,614,213
8	B	36	23	41	27	0	2	13	-18.84	54.68	1660	139.85	245.94	-22.57	56.79	1688	12.72	36.16	40.14	7,614,248
9	A	36	23	42	9	0	2	55	-17.28	54.00	1666	139.75	245.99	-20.05	55.72	1682	9.68	36.58	40.32	7,614,283
10	B	36	23	42	51	0	3	37	-15.73	53.34	1675	139.67	246.02	-17.96	54.62	1685	7.66	37.32	40.32	7,614,318
11	A	36	23	43	33	0	4	19	-14.17	52.69	1685	139.63	245.93	-15.40	53.60	1688	4.53	38.06	40.44	7,614,353
12	B	36	23	44	15	0	5	1	-12.62	52.06	1697	139.58	245.85	-13.27	52.55	1698	2.40	39.06	40.41	7,614,388
13	A	36	23	44	57	0	5	43	-11.08	51.44	1710	139.69	245.86	-10.74	51.51	1710	1.03	40.16	40.37	7,614,423
14	B	36	23	45	39	0	6	25	-9.54	50.83	1725	139.84	245.88	-8.64	50.40	1727	2.95	41.47	40.15	7,614,458
15	A	36	23	46	21	0	7	7	-8.01	50.24	1742	140.03	245.93	-6.10	49.36	1749	6.15	42.83	40.07	7,614,493
16	B	36	23	47	3	0	7	49	-6.49	49.65	1761	140.12	245.94	-3.99	48.28	1773	8.29	44.31	39.87	7,614,528
MAPPING NO 2																				
17	B	36	23	52	39	0	13	25	5.21	45.29	1964	117.90	280.84	-3.52	49.86	2096	25.91	44.16	62.09	7,614,808
■ 18	A	36	23	53	21	0	14	7	6.60	44.78	1996	118.09	280.77	-.98	49.00	2098	22.79	45.58	61.98	7,614,843 ■
19	B	36	23	54	3	0	14	49	7.98	44.26	2029	118.29	280.80	.99	47.84	2112	20.49	47.26	61.70	7,614,878
20	A	36	23	54	45	0	15	31	9.34	43.76	2064	118.48	280.81	3.42	46.99	2124	17.50	48.81	61.58	7,614,913
21	B	36	23	55	27	0	16	13	10.68	43.25	2099	118.54	280.72	5.38	45.96	2146	15.34	50.46	61.45	7,614,948
22	A	36	23	56	9	0	16	55	12.00	42.75	2136	118.56	280.72	7.73	45.23	2168	12.61	51.96	61.51	7,614,983
23	B	36	23	56	51	0	17	37	13.30	42.25	2174	118.53	280.74	9.56	44.24	2197	10.73	53.60	61.49	7,615,018
24	A	36	23	57	33	0	18	19	14.58	41.75	2213	118.44	280.80	11.84	43.54	2227	8.20	55.12	61.63	7,615,053
25	B	36	23	58	15	0	19	1	15.85	41.26	2253	118.40	280.83	13.63	42.51	2261	6.31	56.81	61.59	7,615,088
26	A	36	23	58	57	0	19	43	17.09	40.76	2294	118.36	280.92	15.88	41.77	2297	3.83	58.40	61.71	7,615,123
TETRAD NO 2																				
27	A	37	0	3	9	0	23	55	24.11	37.80	2560	101.09	304.02	5.23	41.68	2968	41.34	56.36	78.98	7,615,333
28	A	37	0	4	33	0	25	18	26.29	36.81	2655	101.29	297.71	14.75	42.06	2833	27.56	59.00	78.78	7,615,403
29	B	37	0	5	15	0	26	0	27.35	36.32	2704	101.46	297.48	16.59	40.92	2855	25.34	60.79	78.53	7,615,438
30	A	37	0	5	57	0	26	42	28.39	35.82	2753	101.56	291.56	23.31	42.38	2822	17.13	62.15	78.51	7,615,473
DYAD																				
31	A	37	0	8	45	0	29	31	32.35	33.79	2957	112.69	283.50	40.61	30.51	3039	18.34	78.65	67.38	7,615,613
32	B	37	0	9	27	0	30	13	33.29	33.28	3009	112.83	283.54	42.20	28.61	3109	20.15	80.67	67.16	7,615,648
33	A	37	0	10	9	0	30	55	34.21	32.76	3061	112.89	283.52	44.58	27.14	3195	23.14	82.59	67.17	7,615,683

JPL TECHNICAL MEMORANDUM 33-585, VOL. II

6-367

Tab. 108 Ophir Planum: Photo Das 7614843 (48) – Detailed satellite survey information.

5. Region of Aurorae Planum

Map MC 18 1208 NE

9.58 S – 51.39 W (alt. 10.0 km)

In quadrant MC- 18 Coprates (**Table 102A**) we take into consideration picture No. 30 of the attached list (**Table 102B**), with the following identification number (Das 7614423).

In **table 109A** we observe the pattern of detection and in **table 109B** the original frame.

In **table 110A** the detected structure is inserted in the context of Equatorial Photomosaic 211-4294; the asterisk (*) indicates the detected structure.

It could be described as a female human face, placed in profile, with an apparent headdress, shaped like a helmet, having the following dimensions: km 75x79x88 (**Table 111A**).

Located in the region of Aurorae Planum (**Table 110B**), where its cartographic position is indicated, the original structure has completely disappeared on map MC.18.1208.NE (**Table 112A**).

A cartographic analysis shows that, in the corresponding part occupied by the original structure, there is the 610A11 frame attributed to the Viking 1 mission (**Table 112B**).

Given that the structure found has considerable photographic evidence, it has been heavily modified on map MC 18 1208 NE (**Table 113A**) to such an extent that it is no longer recognizable from the original photo (**Table 114B**).

Reconfirmation process

The structure in question was detected in 1972 (and the released frame was not modified – see **Table 114A**), then a modified frame was presented (**Table 114B**), finally it was reconfirmed by a survey carried out by the Mars Odyssey probe in 2004 (**Tables 113B, 114C**).

In the updated topographic cartography (**Table 111B**) the structure was named “Morella”, and was classified as “crater”.

In **table 115** we have inserted the detailed data of the frame Das7614423.

Stable elements of experimentation/photographic points of reference: (9.0-13.0 somebody S – 50.0-55.0 person W); 1) Crater Ds – Morella (main structure); 2) Ganges Cavus; 3) Sommerset (point); 4) Johnstown (point); 5) Elaver Valles (river structure connecting with the Ganges Chasma); 6) Contours Craters: W Fs, Es, S Dy.

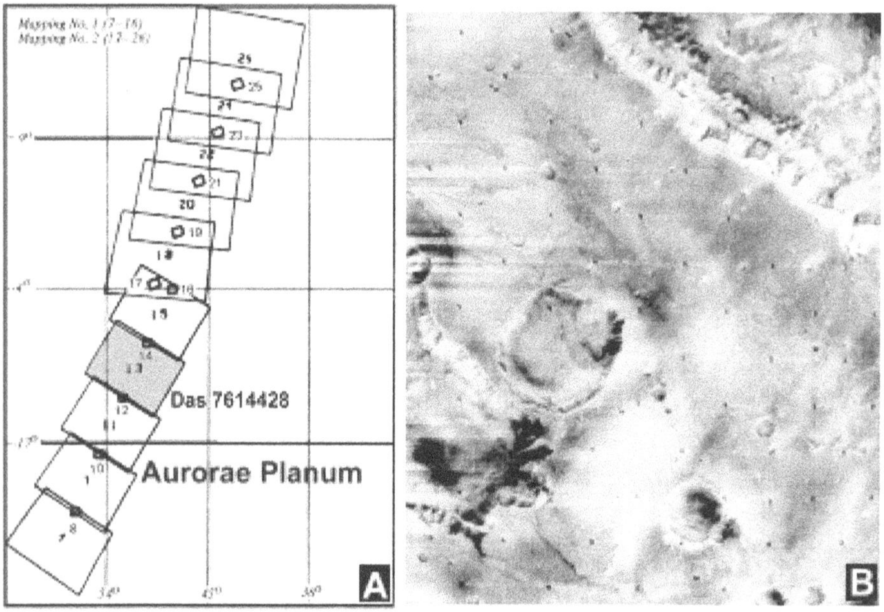

Tab. 109	Aurorae Planum – Photo Das 7614428: A) Detection scheme; B) Original frame.

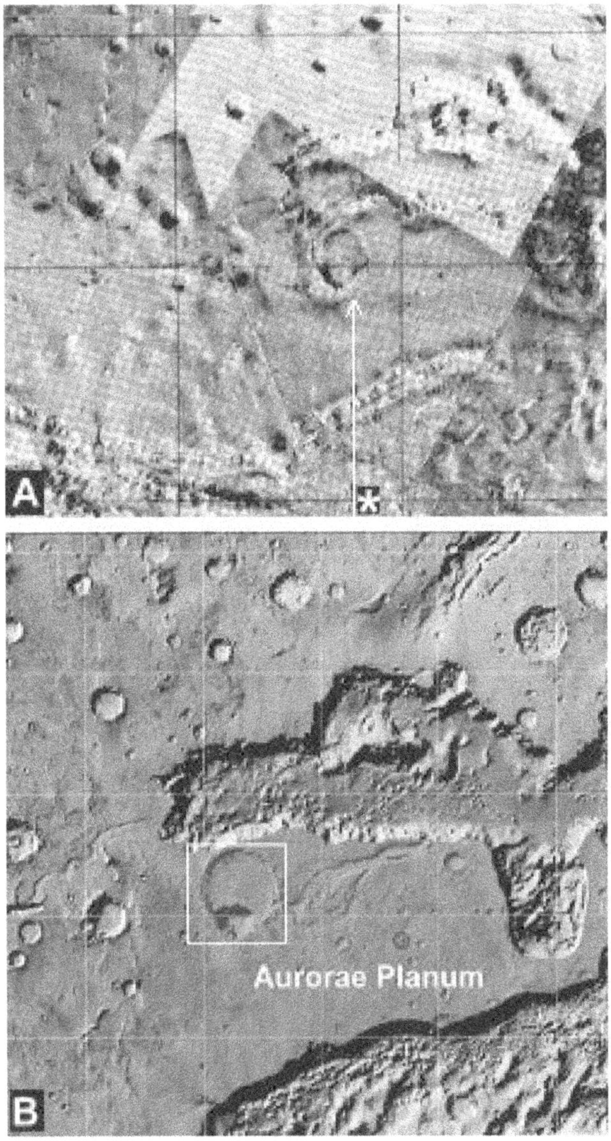

Tab. 110	Aurorae Planum: A) Photomosaic 211-4294 (portion) with indication (*) of the detected structure; B) Cartographic position of the detected structure.

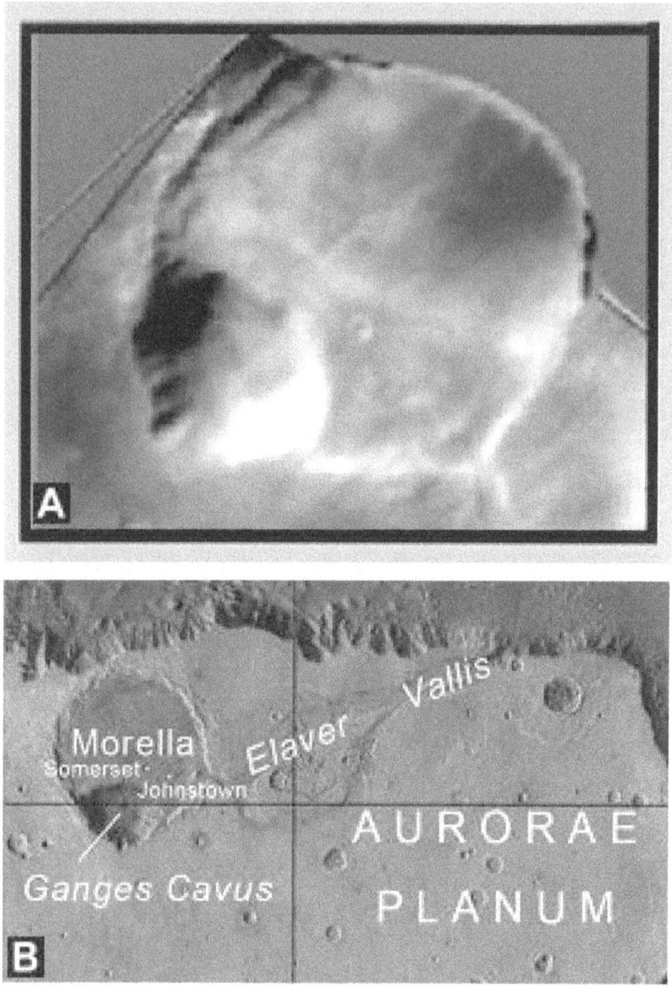

Tab. 111	Aurorae Planum: A) Expanded structure; B) Cartography with updated nomenclature. The structure analyzed is topographically called “Morella”.

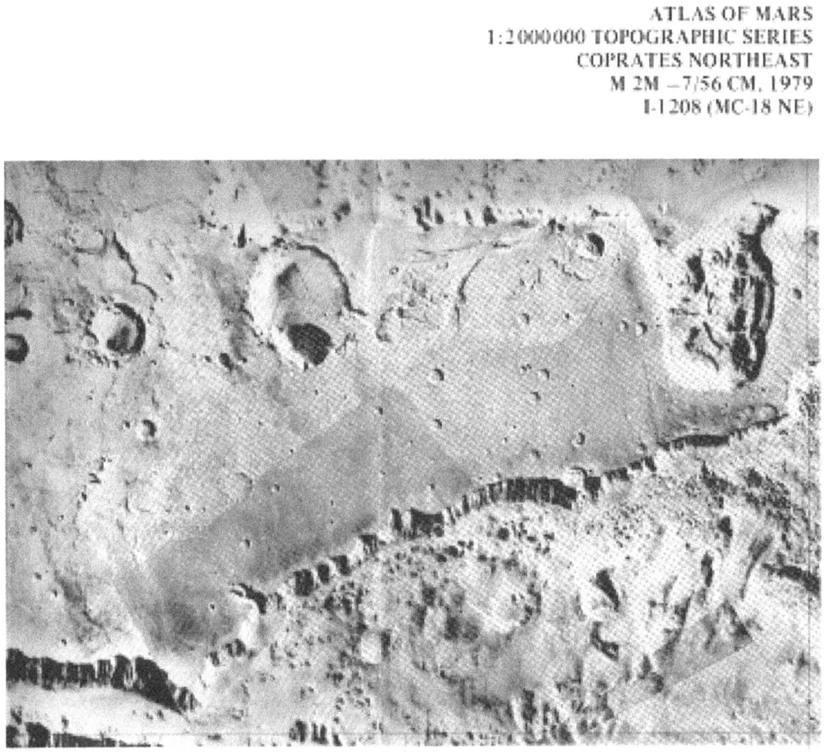

Tab. 112A	Aurorae Planum – Map MC 18 1208 NE (portion), where the structure analyzed falls.

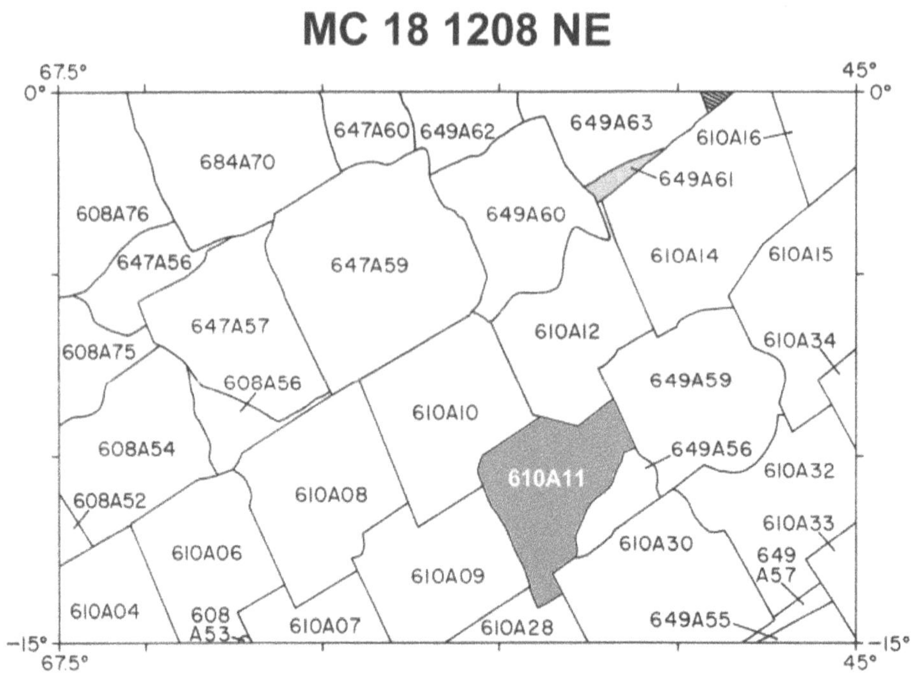

Tab. 112B	Aurorae Planum – Mpa MC 18 1208 NE – Index of Viking images.

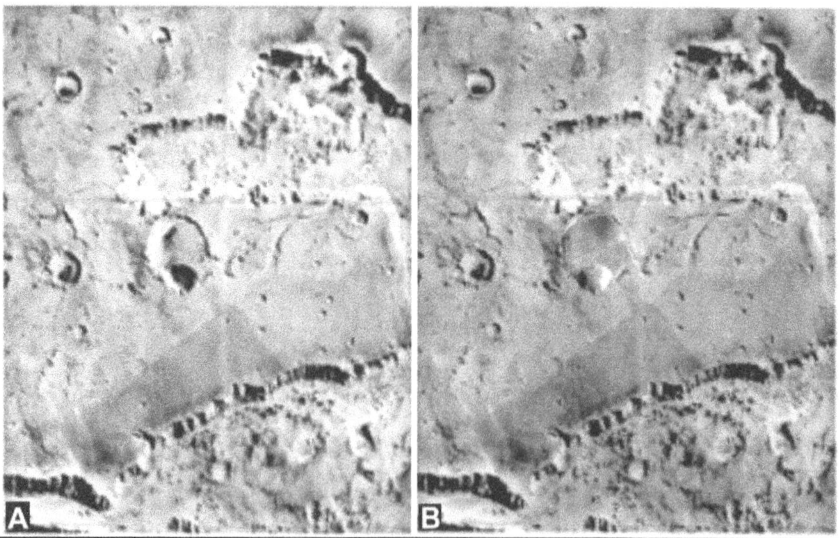

Tab. 113	Aurorae Planum – Map MC 18 1208 NE: A) Photo rigged; B: Relocation of the authentic structure (taken over by Mars Global Surveyor in 2004), on map MC 18 1208 NE.

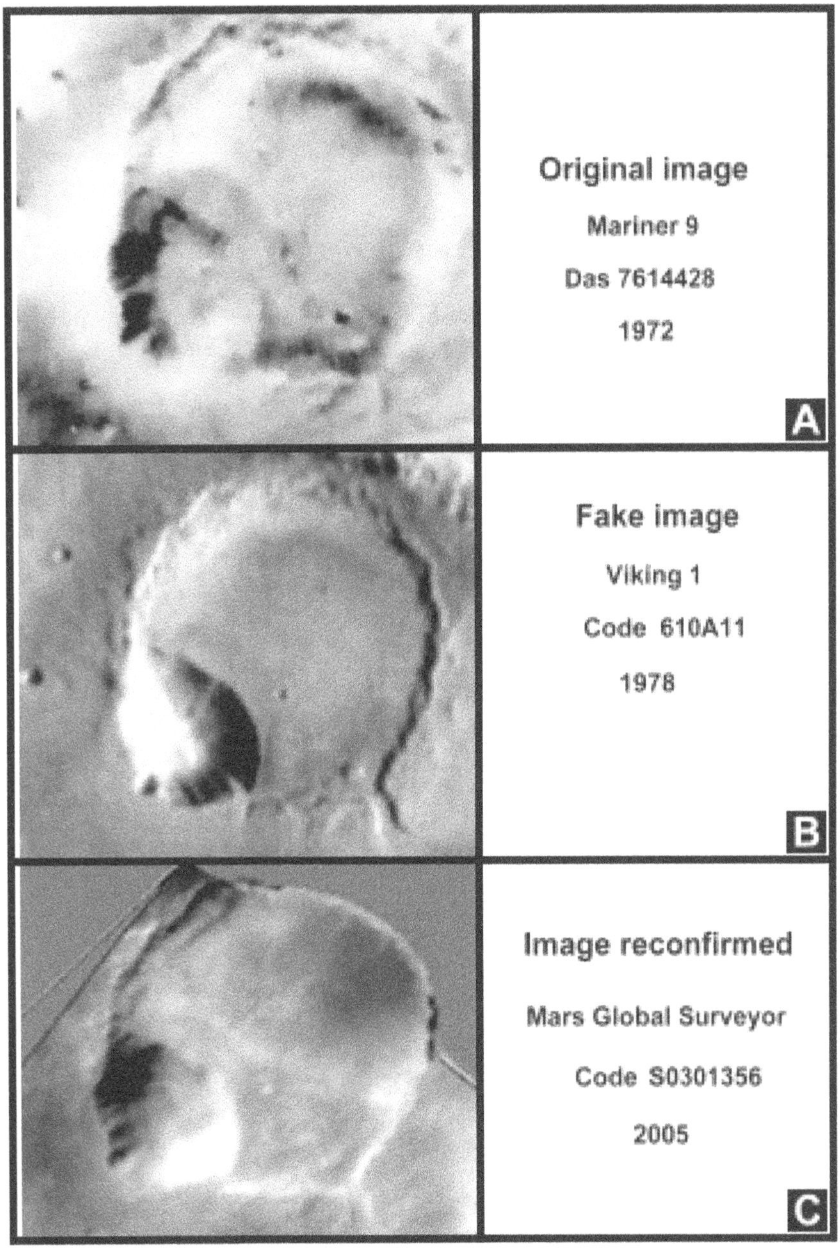

Tab. 114	Aurorae Planum – Process of falsification and reconfirmation; A: original image; B: fake image; C: image reconfirmed.

Mariner 9 Das 7614423

INST TYPE	TIME (GMT) D H M S	PERI TIME H M S	SPACECRAFT LAT	SPACECRAFT LONG-W	SPACECRAFT HGT	PLATFORM CONE	PLATFORM CLOCK	INTERCEPTING LAT	INTERCEPTING LONG-W	INTERCEPTING RANGE	VIEW ANGLE	LIGHT ANGLE	PHASE ANGLE	DAS REFERENCE TIME
TETRAD NO 1														
1 B	36 23 5 3	-0 34 10	-63.19	141.41	3314	115.93	115.98	-86.78	69.62	3945	48.03	78.75	64.06	7,612,428
2 A	36 23 5 45	-0 33 28	-63.41	139.40	3259	115.93	115.98	-84.81	51.85	3933	49.69	77.44	64.14	7,612,463
3 B	36 23 6 27	-0 32 46	-63.60	137.32	3205	119.15	114.42	-85.13	10.59	4009	54.15	80.43	60.84	7,612,498
4 B	36 23 7 51	-0 31 22	-63.88	132.97	3098	121.33	115.45	-84.44	22.54	3862	53.22	79.20	58.65	7,612,568
SINGLE B NO 1														
5 B	36 23 19 3	-0 20 10	-59.40	95.15	2323	132.27	161.83	-61.67	100.58	2338	8.57	54.65	47.72	7,613,128
SINGLE B NO 2														
6 B	36 23 24 39	-0 14 34	-52.27	80.06	2018	134.02	171.48	-51.55	81.24	2020	2.75	43.73	45.97	7,613,408
MAPPING NO 1														
7 A	36 23 40 45	0 1 31	-20.40	55.38	1655	139.96	245.87	-24.65	57.97	1692	14.70	35.71	40.11	7,614,213
8 B	36 23 41 27	0 2 13	-18.84	54.68	1660	139.85	245.94	-22.57	56.79	1688	12.72	36.16	40.14	7,614,248
9 A	36 23 42 9	0 2 55	-17.28	54.00	1666	139.75	245.99	-20.05	55.72	1682	9.68	36.58	40.32	7,614,283
10 B	36 23 42 51	0 3 37	-15.73	53.34	1675	139.67	246.02	-17.96	54.62	1685	7.66	37.32	40.32	7,614,318
11 A	36 23 43 33	0 4 19	-14.17	52.69	1685	139.63	245.93	-15.40	53.60	1688	4.53	38.06	40.44	7,614,353
12 B	36 23 44 15	0 5 1	-12.62	52.06	1697	139.58	245.85	-13.27	52.55	1698	2.40	39.06	40.41	7,614,388
■ 13 A	36 23 44 57	0 5 43	-11.08	51.44	1710	139.69	245.86	-10.74	51.51	1710	1.03	40.16	40.37	7,614,423 ■
14 B	36 23 45 39	0 6 25	-9.54	50.83	1725	139.84	245.88	-8.64	50.40	1727	2.95	41.47	40.15	7,614,458
15 A	36 23 46 21	0 7 7	-8.01	50.24	1742	140.03	245.93	-6.10	49.36	1749	6.15	42.83	40.07	7,614,493
16 B	36 23 47 3	0 7 49	-6.49	49.65	1761	140.12	245.94	-3.99	48.28	1773	8.29	44.31	39.87	7,614,528
MAPPING NO 2														
17 B	36 23 52 39	0 13 25	5.21	45.29	1964	117.90	280.84	-3.52	49.86	2096	25.91	44.16	62.09	7,614,808
18 A	36 23 53 21	0 14 7	6.60	44.78	1996	118.09	280.77	-.98	49.00	2098	22.79	45.58	61.98	7,614,843
19 B	36 23 54 3	0 14 49	7.98	44.26	2029	118.29	280.80	.99	47.84	2112	20.49	47.26	61.70	7,614,878
20 A	36 23 54 45	0 15 31	9.34	43.76	2064	118.48	280.81	3.42	46.99	2124	17.50	48.81	61.58	7,614,913
21 B	36 23 55 27	0 16 13	10.68	43.25	2099	118.54	280.72	5.38	45.96	2146	15.34	50.46	61.45	7,614,948
22 A	36 23 56 9	0 16 55	12.00	42.75	2136	118.56	280.72	7.73	45.23	2168	12.61	51.96	61.51	7,614,983
23 B	36 23 56 51	0 17 37	13.30	42.25	2174	118.53	280.74	9.56	44.24	2197	10.73	53.60	61.49	7,615,018
24 A	36 23 57 33	0 18 19	14.58	41.75	2213	118.44	280.80	11.84	43.54	2227	8.20	55.12	61.63	7,615,053
25 B	36 23 58 15	0 19 1	15.85	41.26	2253	118.40	280.83	13.63	42.51	2261	6.31	56.81	61.59	7,615,088
26 A	36 23 58 57	0 19 43	17.09	40.76	2294	118.36	280.92	15.88	41.77	2297	3.83	58.40	61.71	7,615,123
TETRAD NO 2														
27 A	37 0 3 9	0 23 55	24.11	37.80	2560	101.09	304.02	5.23	41.68	2968	41.34	56.36	78.98	7,615,333
28 A	37 0 4 33	0 25 18	26.29	36.81	2655	101.29	297.71	14.75	42.06	2833	27.56	59.00	78.78	7,615,403
29 B	37 0 5 15	0 26 0	27.35	36.32	2704	101.46	297.48	16.59	40.92	2855	25.34	60.79	78.53	7,615,438
30 A	37 0 5 57	0 26 42	28.39	35.82	2753	101.56	291.56	23.31	42.38	2822	17.13	62.15	78.51	7,615,473
DYAD														
31 A	37 0 8 45	0 29 31	32.35	33.79	2957	112.69	283.50	40.61	30.51	3039	18.34	78.65	67.38	7,615,613
32 B	37 0 9 27	0 30 13	33.29	33.28	3009	112.83	283.54	42.20	28.61	3109	20.15	80.67	67.16	7,615,648
33 A	37 0 10 9	0 30 55	34.21	32.76	3061	112.89	283.52	44.58	27.14	3195	23.14	82.59	67.17	7,615,683

JPL TECHNICAL MEMORANDUM 33-585, VOL. II

6-367

Tab. 115 Aurore Planum – Photo Das 7614423 (28). Detailed satellite survey information.

6. Region of Capri Chasma

Map MC 19 1381 NW

6.90° S – 39.15° W (alt. km 1.0)

In quadrant MC-19 Margaritifer Sinus (**Table 116A**) we take into consideration picture No. 71 of the attached list (**Table 116B**) with the following identification number (Das 7686383).

In **table 117A** we observe the pattern of detection and in table **117B** the original frame.

In **table 118A** the detected structure is inserted in the context of Equatorial Photomosaic 211-4294; the asterisk (*) indicates the detected structure.

It could be described as a female human face, placed in three quarters, having the following dimensions: 80x104 km (**Tables 119,120**).

Located in the region of Capri Chasma (**Table 118B**), where the cartographic position is indicated, the original structure is completely disappeared on map MC.19.1381.NW (**Table 122A**).

A cartographic analysis shows that, in the corresponding part occupied by the original structure, there is the 610A36 frame attributed to the Viking 1 mission (**Table 122B**).

In any case, before being destroyed on mpa (MC.19. 1381.NW), the structure detected had undergone a first transformation within the original frame (**Table 121A**), giving rise to a false frame (**Table 121B**).

To superimpose the original image on the "false" one, taking into account the original photo (**Table 122C**), we fix some photographic control points (**Tables 123A/B**), then we find them on map MC 19 1381 NW (**Tables 123C/D**) and proceed to the relocation of the real structure, on the portion of map MC 19 1381 NW (**Tables 123D-124A, B, C, D**).

In **table 125** we have the area of Capri Chasma with the updated nomenclature.

In **table 126** we have inserted the detailed data of the frame Das7686383.

Stable elements of experimentation/photographic points of reference: (3.5-7.0°S – 37.0-41.0° W); 1) NW: Crater Azusa Xv; 2) SW: Crater Innsbruch Xu; 3) NE: Crater Berseba Vw; 4) SE: Crater Timbuktu Vv.

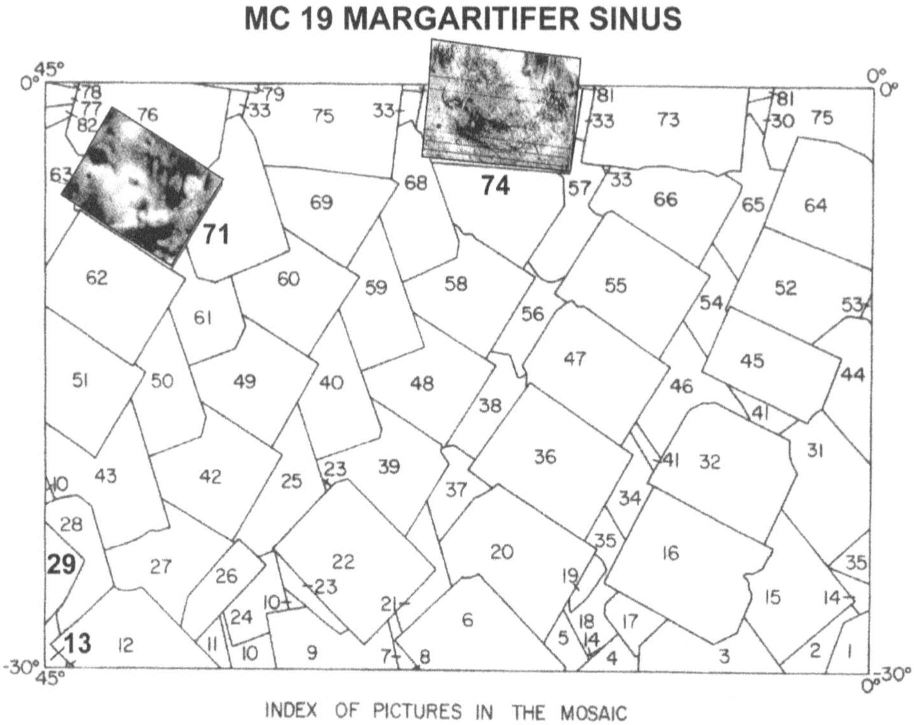

Tab. 116A	MC 19 Margaritifer Sinus – Index of images Mariner 9.

IDENTIFICATION NUMBERS OF PICTURES IN THE MOSAIC

Index No.	DAS No.	Index No.	DAS No.	Index No.	DAS No.
1	8117303	■ 29	6283023	57	9304859
2	5239393	30	5884793	58	7830093
3	8045413	31	8045553	59	9232899
4	5167433	32	6570793	60	7758203
5	7973453	33	5812763	61	9161009
6	6498623	34	7973593	62	7686313
7	5740803	35	6642683	63	9089189
8	6498553	36	7901843	64	6571003
9	9304509	47	9304649	65	7973803
10	5668773	38	9304719	66	7902053
11	9232619	39	7829953	67	7830163
12	6354843	40	9232829	68	9232969
■13	9160729	41	5956753	69	7758273
14	5239463	42	7758063	70	9161079
15	8045483	43	9160869	■ 71	7686383
16	6570723	44	8045623	72	6571353
17	5167503	45	6570863	73	7902403
18	7973523	46	7973663	■ 74	7830513
19	6498693	47	7901913	75	7758623
20	7901773	48	7830023	76	7686733
21	9304579	49	7758133	77	7615333
22	6426803	50	9160939	78	7614913
23	7829883	51	7686243	79	7686803
24	9232689	52	6570933	80	7758693
25	9232759	53	8045693	81	7830583
26	6354913	54	7973733	82	10492689
27	7757993	55	7901983		
28	9160799	56	9304789		

Tab. 116B MC 19 Margaritifer Sinus – Image identification number.

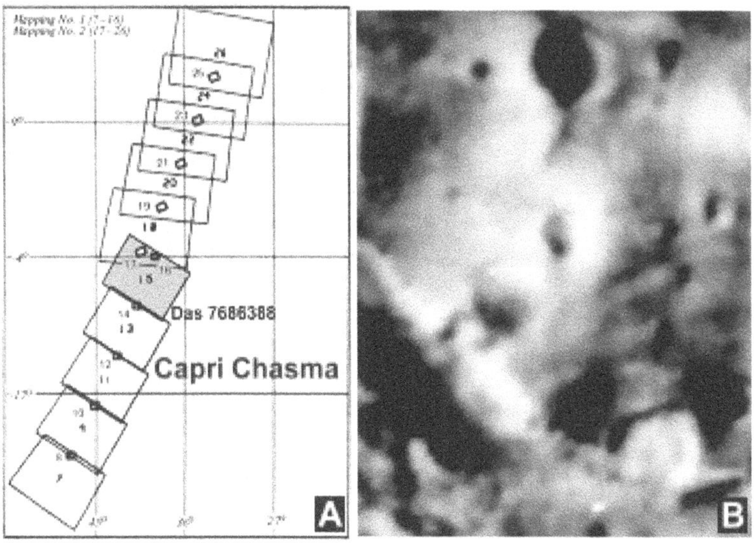

Tab. 117 Capri Chasma. Photo Das 7686388: A) Detection scheme; B) Original frame.

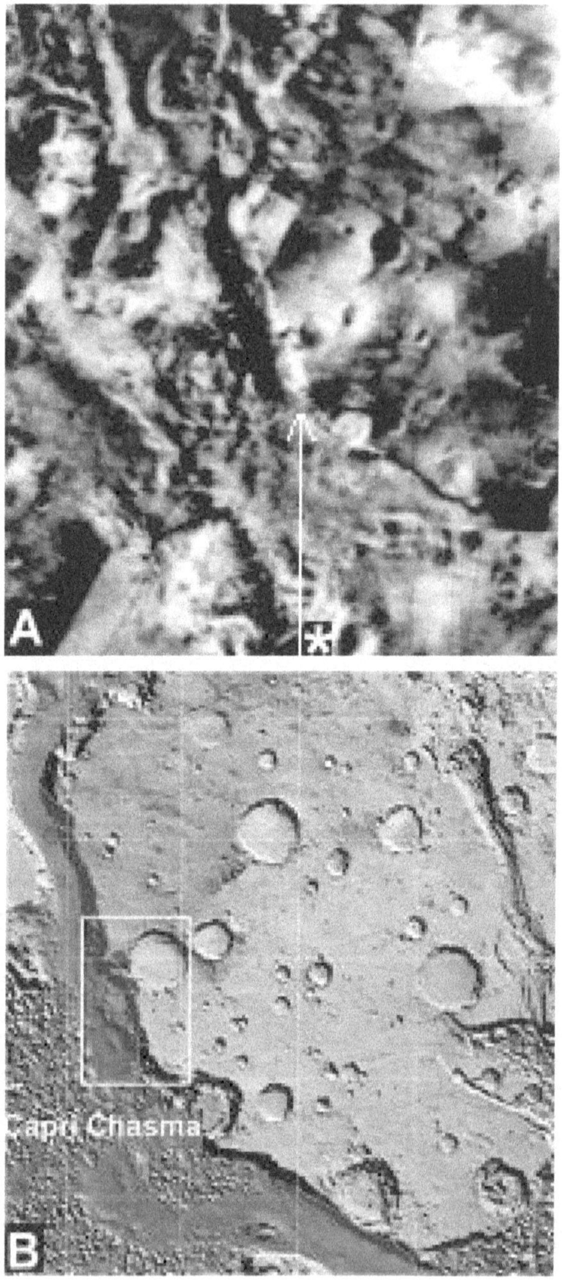

Tab. 118	Capri Chasma: A) Photomosaic 211- 4294 (portion), with indication (*) of the structure detected; B) Cartographic position of the structure detected.

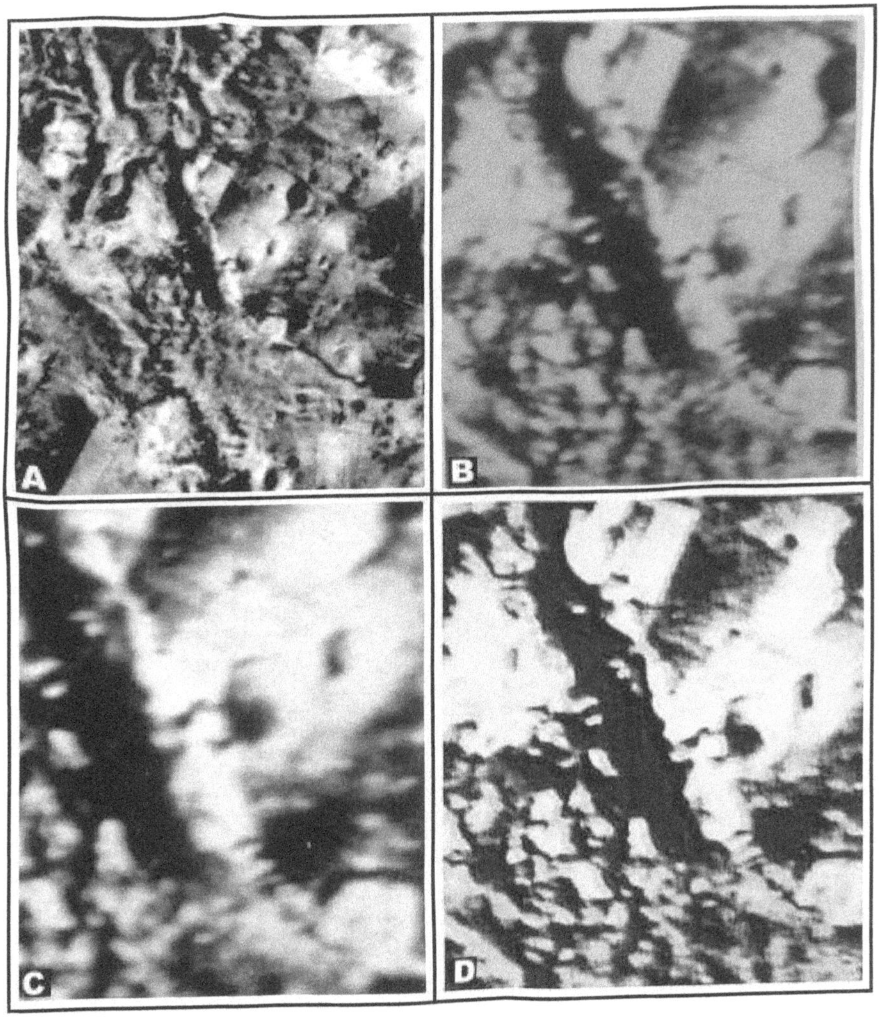

Tab. 119	Capri Chasma: A-B-C) Structure detected; D) Bas-relief effect.

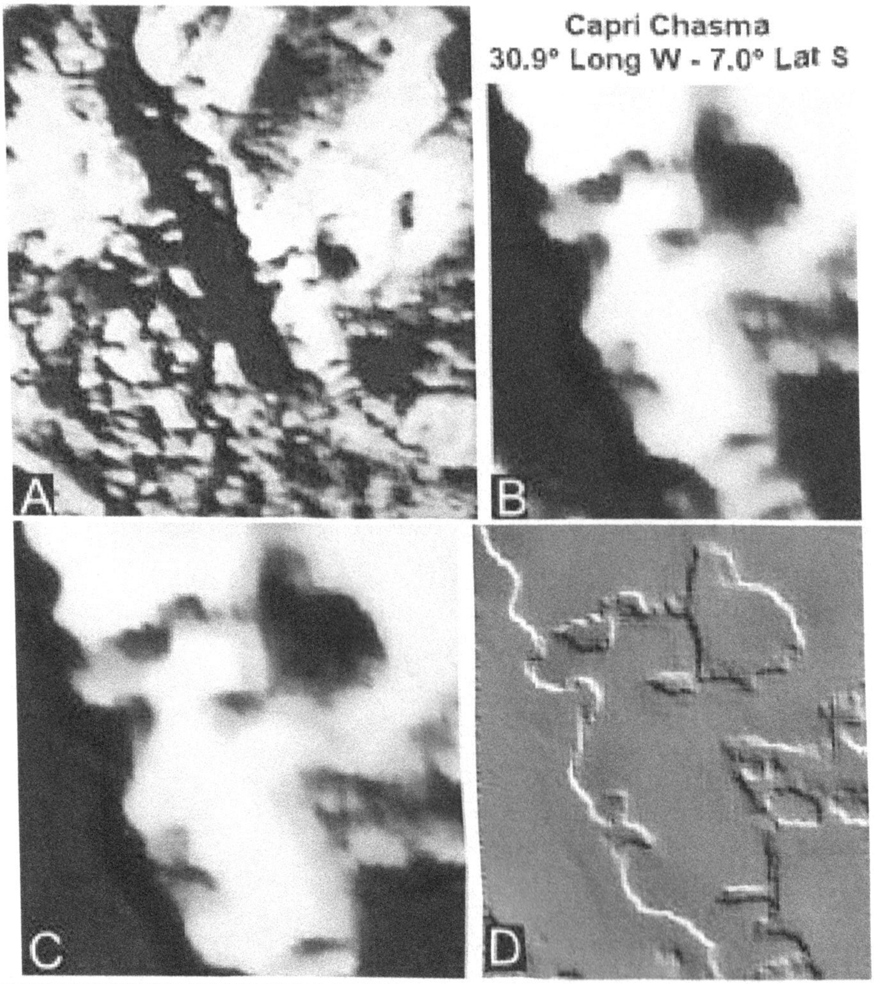

Tab. 120	Capri Chasma: A-B-C) Structure detected; D) Bas-relief effect.

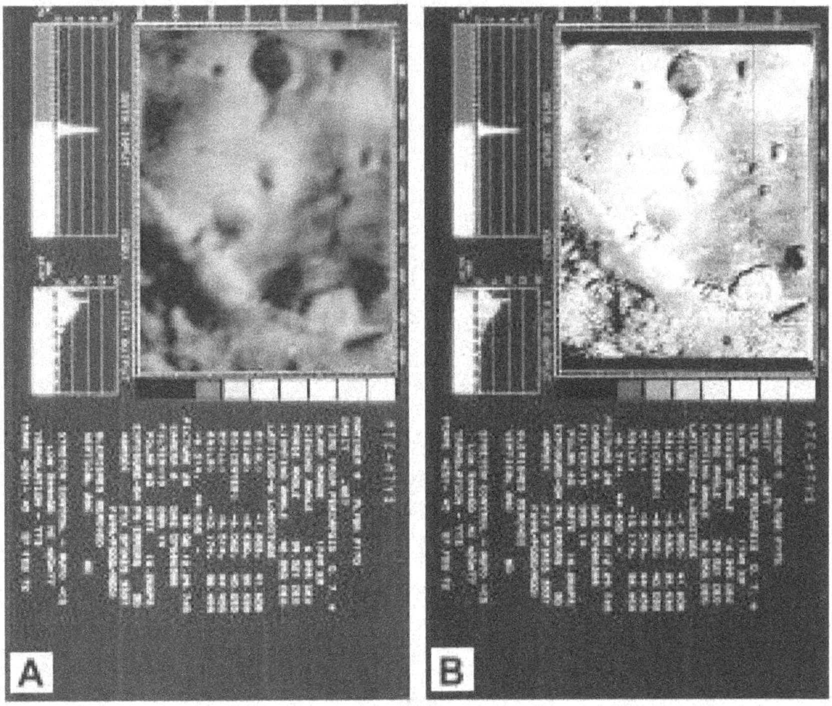

Tab. 121	Capri Chasma: A) Frame 7686388 (authentic); B) Modified frame, in the photographic presentation through microfiche.

Tab. 122A	Capri Chasma – Map MC 19 1381 NW (portion), where falls the structure analyzed.

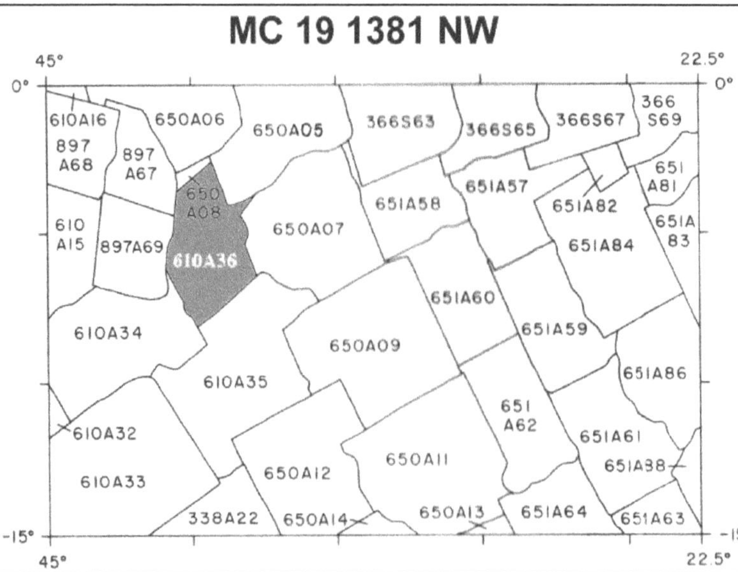

Tab. 122B	Map MC 19 1381 NW – Index of Viking images.

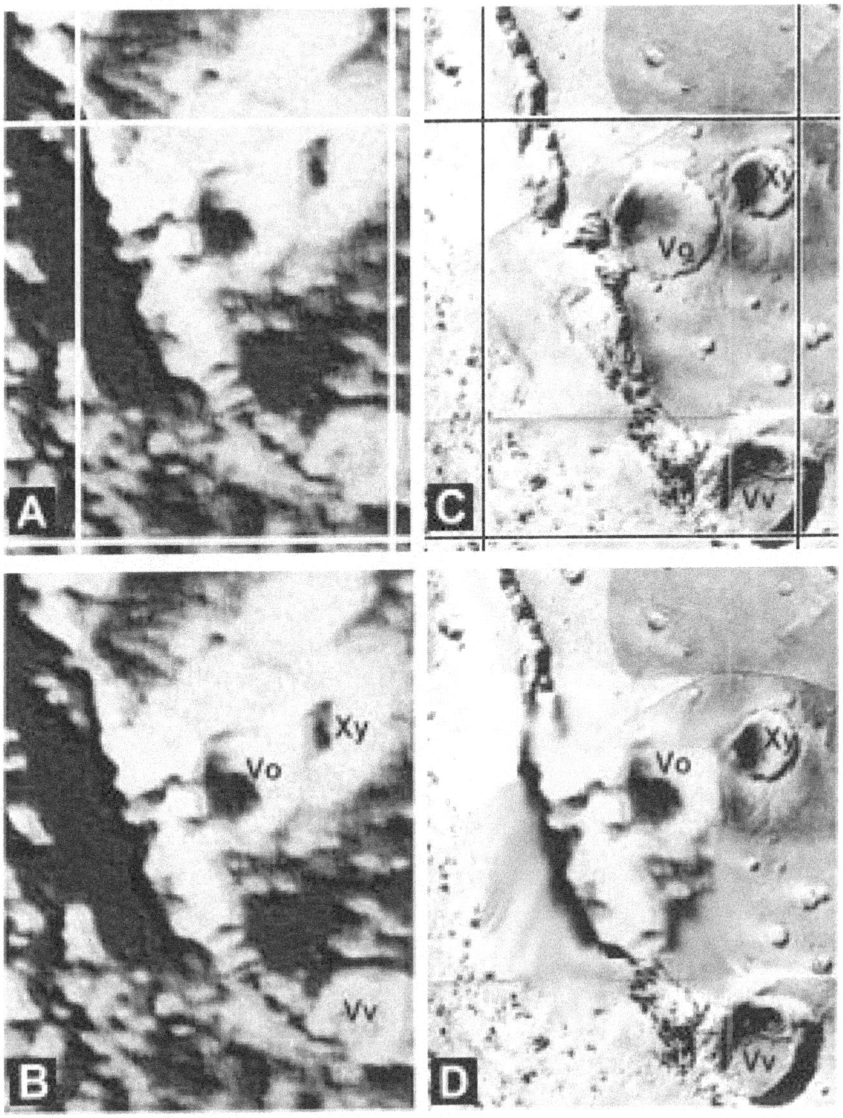

Tab. 123	Capri Chasma: A) Delimitation of the detected structure (Mariner 9 – Das 7686388); B) Indication of three photographic control points; C) Cancellation of the frame on the corresponding part of the Map MC 19 1381 NW (portion) and finding of the three photographic control points; D) Artistic relocation of the deleted structure on the map, with indication of the three photographic control points.

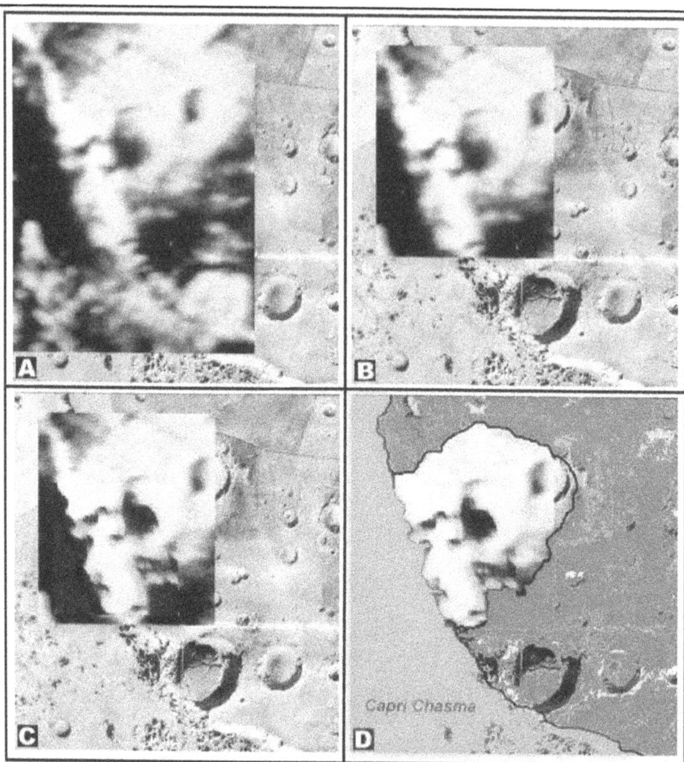

Tab. 124 Capri Chasma: A,B,C) Reallocation of the structure on Map MC 19 1381 NW; D) Artistic reconstruction.

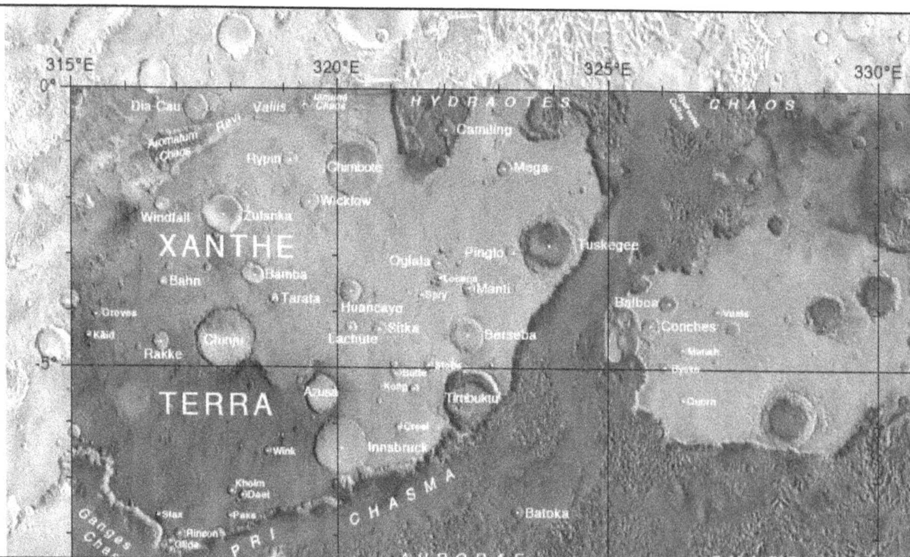

Tab. 125 Capri Chasma: Cartography with updated nomenclature.

Mariner 9 Das 7686383

INST	TYPE	TIME (GMT) D H M S	PERI TIME H M S	SPACECRAFT LAT	SPACECRAFT LONG-W	HGT	PLATFORM CONE	PLATFORM CLOCK	INTERCEPTING LAT	INTERCEPTING LONG-W	INTERCEPTING RANGE	VIEW ANGLE	LIGHT ANGLE	PHASE ANGLE	DAS REFERENCE TIME
TETRAD NO 1															
1	B	37 23 2 51	-0 34 13	-63.19	132.10	3317	109.64	146.55	-56.55	136.37	3367	13.98	69.30	70.35	7,684,318
2	A	37 23 3 33	-0 33 31	-63.41	130.09	3263	110.02	146.47	-57.81	133.08	3298	11.73	67.95	70.05	7,684,353
3	B	37 23 4 15	-0 32 49	-63.60	128.02	3208	112.37	149.28	-56.73	135.95	3274	16.08	68.95	67.62	7,684,388
4	B	37 23 5 39	-0 31-25	-63.88	123.67	3101	115.21	152.19	-57.01	135.45	3186	18.49	68.61	64.78	7,684,458
SINGLE B NO 1															
5	B	37 23 16 51	-0 20 13	-59.46	85.83	2326	125.19	155.22	-57.90	83.04	2331	5.22	50.33	54.79	7,685,018
SINGLE B NO 2															
6	B	37 23 22 27	-0 14 37	-52.36	70.70	2021	132.43	165.43	-50.47	68.50	2028	6.23	43.17	47.56	7,685,298
MAPPING NO 1															
7	A	37 23 38 33	0 1 28	-20.52	45.96	1656	140.54	245.04	-24.71	48.60	1692	14.59	35.20	39.54	7,686,103
8	B	37 23 39 15	0 2 10	-18.96	45.26	1661	140.43	245.12	-22.63	47.42	1688	12.61	35.61	39.56	7,686,138
9	A	37 23 39 57	0 2 52	-17.41	44.58	1667	140.52	245.35	-20.17	46.25	1683	9.59	36.09	39.55	7,686,173
10	B	37 23 40 39	0 3 34	-15.85	43.92	1675	140.60	245.60	-18.15	45.07	1685	7.67	36.88	39.39	7,686,208
11	A	37 23 41 21	0 4 16	-14.30	43.27	1685	140.73	245.58	-15.62	43.96	1688	4.43	37.67	39.34	7,686,243
12	B	37 23 42 3	0 4 58	-12.75	42.64	1697	140.91	245.45	-13.48	42.79	1698	2.22	38.75	39.08	7,686,278
13	A	37 23 42 45	0 5 40	-11.20	42.02	1710	140.92	245.22	-10.87	41.80	1710	1.19	39.77	39.15	7,686,313
14	B	37 23 43 27	0 6 22	-9.67	41.41	1725	140.87	244.97	-8.68	40.78	1727	3.44	40.96	39.11	7,686,348
■ 15	A	37 23 44 9	0 7 4	-8.14	40.82	1742	140.73	244.77	-6.06	39.91	1750	6.65	42.16	39.35	7,686,383 ■
16	B	37 23 44 51	0 7 46	-6.62	40.23	1760	140.56	244.69	-3.91	38.98	1774	8.68	43.49	39.43	7,686,418
MAPPING NO 2															
17	B	37 23 50 27	0 13 22	5.09	35.87	1963	118.88	280.51	-3.46	40.36	2090	25.43	43.53	61.11	7,686,698
18	A	37 23 51 9	0 14 4	6.48	35.35	1995	119.06	280.65	-1.04	39.46	2095	22.52	44.96	61.01	7,686,733
19	B	37 23 51 51	0 14 46	7.86	34.84	2028	119.24	280.72	.90	38.30	2109	20.29	46.63	60.74	7,686,768
20	A	37 23 52 33	0 15 28	9.22	34.33	2062	119.44	280.86	3.27	37.43	2122	17.41	48.17	60.62	7,686,803
21	B	37 23 53 15	0 16 10	10.56	33.83	2098	119.53	280.91	5.16	36.35	2145	15.36	49.85	60.46	7,686,838
22	A	37 23 53 57	0 16 52	11.88	33.33	2134	119.56	280.91	7.52	35.60	2166	12.59	51.36	60.51	7,686,873
23	B	37 23 54 39	0 17 34	13.19	32.83	2172	119.53	280.92	9.36	34.60	2196	10.68	53.01	60.46	7,686,908
24	A	37 23 55 21	0 18 16	14.47	32.33	2211	119.46	280.90	11.67	33.91	2225	8.05	54.52	60.61	7,686,943
25	B	37 23 56 3	0 18 58	15.73	31.84	2251	119.41	280.88	13.49	32.90	2259	6.16	56.20	60.58	7,686,978
26	A	37 23 56 45	0 19 40	16.98	31.34	2292	119.34	280.85	15.80	32.21	2295	3.56	57.77	60.72	7,687,013
TETRAD NO 2															
27	A	38 0 0 57	0 23 52	24.01	28.39	2558	104.71	303.90	6.86	29.97	2889	37.51	58.22	75.36	7,687,223
28	A	38 0 2 21	0 25 16	26.19	27.40	2653	104.89	298.89	14.95	29.90	2803	25.39	60.83	75.18	7,687,293
29	B	38 0 3 3	0 25 58	27.25	26.90	2702	105.08	298.80	16.63	28.67	2832	23.61	62.64	74.91	7,687,328
30	A	38 0 3 45	0 26 40	28.29	26.41	2751	105.14	293.75	22.60	29.76	2798	14.25	63.87	74.93	7,687,363
DYAD															
31	A	38 0 6 33	0 29 28	32.25	24.39	2954	104.78	282.26	38.75	31.01	3032	17.87	70.30	75.28	7,687,503
32	B	38 0 7 15	0 30 10	33.20	23.87	3006	104.79	282.16	40.47	29.82	3088	18.22	72.03	75.19	7,687,538
~~33~~	A	38 0 7 57	0 30 52	34.12	23.36	3059	104.77	282.12	42.85	29.20	3162	20.39	73.67	75.30	~~7,687,573~~

JPL TECHNICAL MEMORANDUM 33-585, VOL. II

6-371

Tab. 126 Capri Chasma: Photo Das 7686383 (88) – Detailed satellite survey information.

7. Region of Iani Chaos

Map MC 19 1382 NE

3.66° S – 18.46° W (alt. km 0.5)

In quadrant MC- 19 Margaritifer Sinus (**Table 116A**) we take into consideration picture No. 74 of the attached list (**Table 116B**), with the following identification number (Das 7830513).

In **table 127A** we observe the pattern of detection and in **table 127B** the original frame.

In **table 128A** the detected structure is inserted in the context of Equatorial Photomosaic 211-4294; the arrows indicate the detected structures.

They could be described as: I) Linear formations delimited by terraces with great variations in altitude. This area has a total width of 150x200 km (**Table 129-I**) Human face placed frontally, with Egyptian traits and headgear. This property has a total width of 20x50 km (**Table 130-II**).

Located in the region of Iani Chaos (**Table 128B**), where the cartographic positions are indicated, the original structures have completely disappeared on map MC.19.1382.NE (**Table 131**).

A cartographic analysis shows that, in the corresponding part occupied by the original structures, there are the frames 651A83, 653A52 and 277A66 (**Table 132A**), attributed to the Viking 1 mission.

In **table 132B** we marked the frames used on the paper and those that should have been used.

In **tables 133A-134A,B,C** we have the images with the anthropomorphic structure and the linear structures taken in 1972 (Fotomosaico Mariner 9 – 211-4294); they were then reconfirmed altogether (in 1977) by the Viking mission (v. photomosaic Viking 1 – 211-5571 – **Tables 133B-134D**).

In **table 135** we present the three satellite surveys carried out by Viking 1 (**Tables 134A, B, C**), which confirm unequivocally the anthropomorphic structure detected by the Mariner 9 mission in 1972 (**Tables 127B-128A-133A-134A, B, C**).

We observe the photomosaic "Phobos Transit 211-5571" (**Table 136A**) with the detail of the various satellite surveys carried out by the Viking 1 mission (**Table 136B**).

In **table 137 (A1)** we extrapolated a portion of the photomosaic where the regions of Ares Vallis, Aram Chaos and Iani Chaos are indicated. Finally, the Iani Chaos region (where the structures analysed fall) is presented in detail in **table 137 (A2).**

Now we will analyze the photomosaic of the Viking 1 code 211-5571, in relation to paper MC 19 1382 NE (**Table 131**). As a first operation, we apply a cartographic grid (**Table 138**), then we report the same grid on the photomosaic 211-5571 (**Table 139**) and, in the next step, we highlight the part that is made up on map MC 19 1382 NE (**Table 140**).

To superimpose the original image (**Table 137-A2**) on the false one (**Table 131**), we fix some photographic control points on the photomosaic (**Table 141A**), then found on paper MC 19 1382 NE (**Table 141B**).

As a last operation we take into consideration the paper MC.19.1382.NE and on the latter we mark the tricked portion (**Table 142A**), then we apply the original image of the photomosaic (**Table 142B**). The effect is impressive, since you can see very clearly what has been the level of manipulation, make-up and concealment of some original structures, totally hidden in cartography.

Reconfirmation process

As mentioned above, the images of the Iani Chaos region, taken in 1972 by the Mariner 9 mission (**Tables 133A-134A,B,C**), were later reconfirmed in 1977 by the filming by Viking 1 (**Tables 133B-134D-135**).

Nevertheless, twenty-nine years later, the Viking images were imaginatively modified, using some images allegedly attributed to the Mars Express mission (**Tables 143,144,145**). Of course, Mars Express images don't really exist.

In **table 146** we have inserted the detailed data of the frame Das 7830513.

Stable elements of experimentation/photographic points of reference: (0.0-7.0"S -16.0-24.0" W); 1) S: Crater Kv; 2) SE: Crater Jv; 3) Contours craters: SW Hwb, NW Nz, W Mx, N Lz.

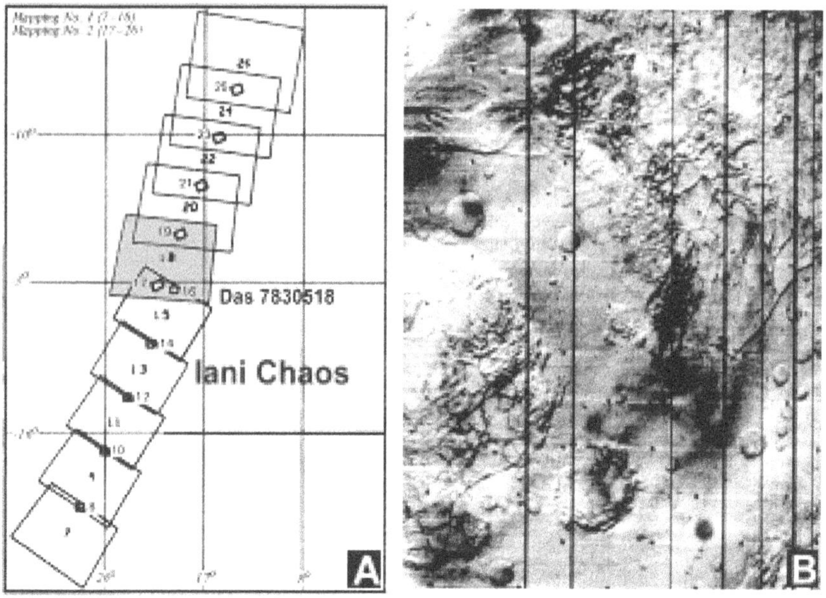

Tab. 127	Iani Chaos – Photo Das 7830518: A) Detection scheme; B) Original frame.

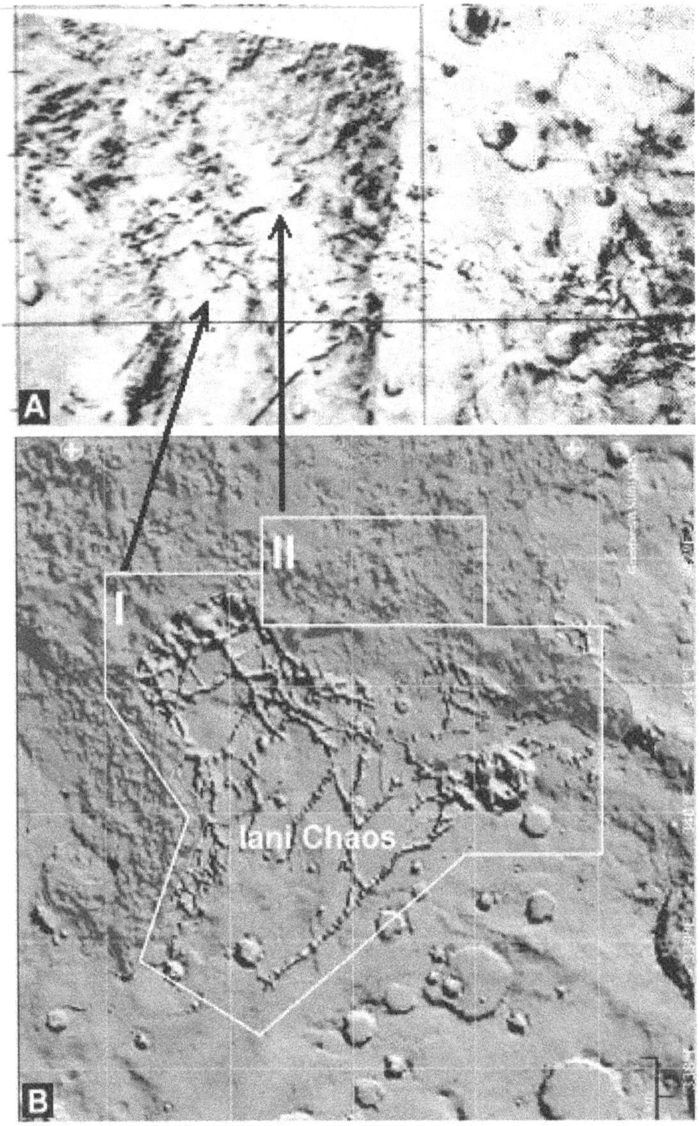

Tab. 128	Iani Chaos – A) Portion of the photomosaic 211-4294 with indication of the detected structures; B) Cartographic position of the detected structures and localization of the Iani Chaos region. I) miscellaneous structures; II) anthropomorphic structure.

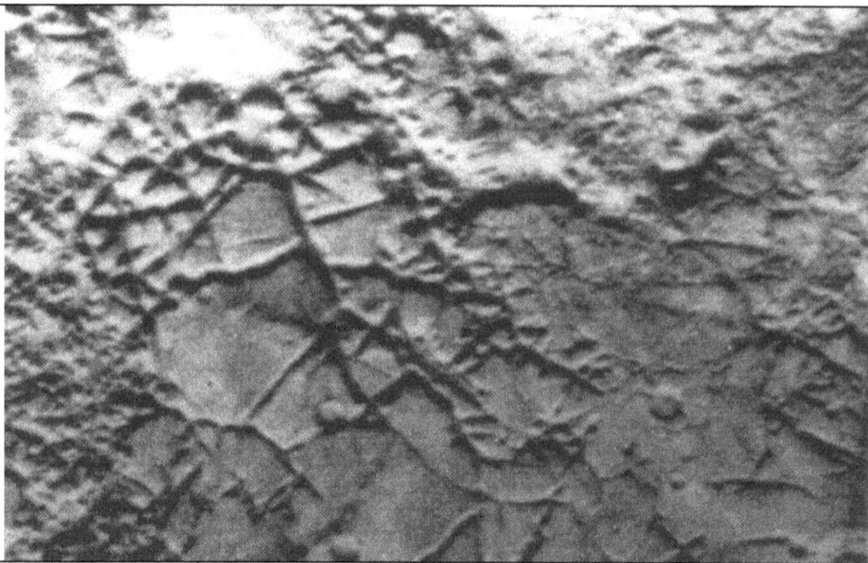

Tab. 129	Iani Chaos. Photo Das 451A06: western part, with structures supposedly roads, eastern part, with a supposedly anthropomorphic structure (I).

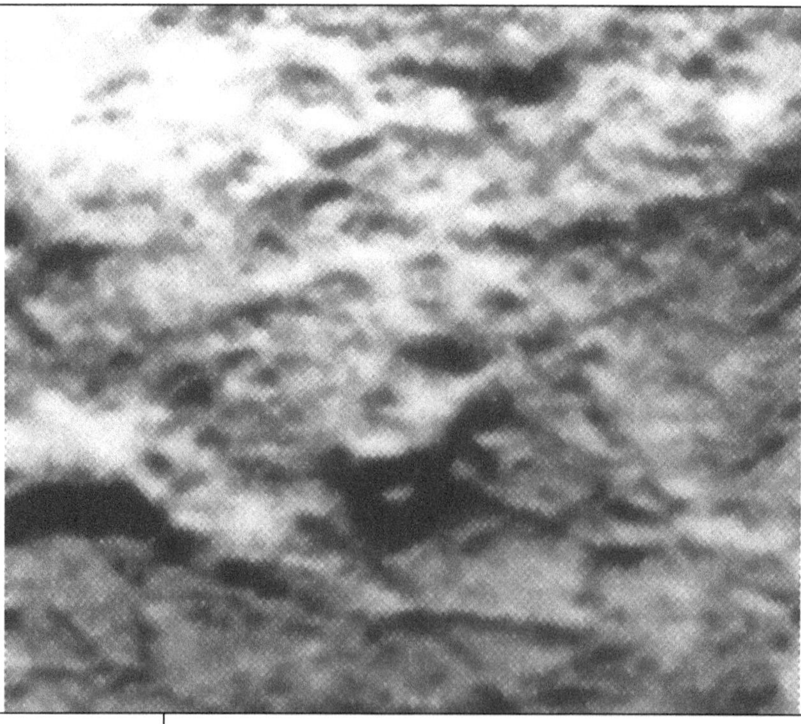

Tab. 130	Iani Chaos. Photo Das 451A06: anthropomorphic structure, female human face placed frontally, with Egyptian headgear (II).

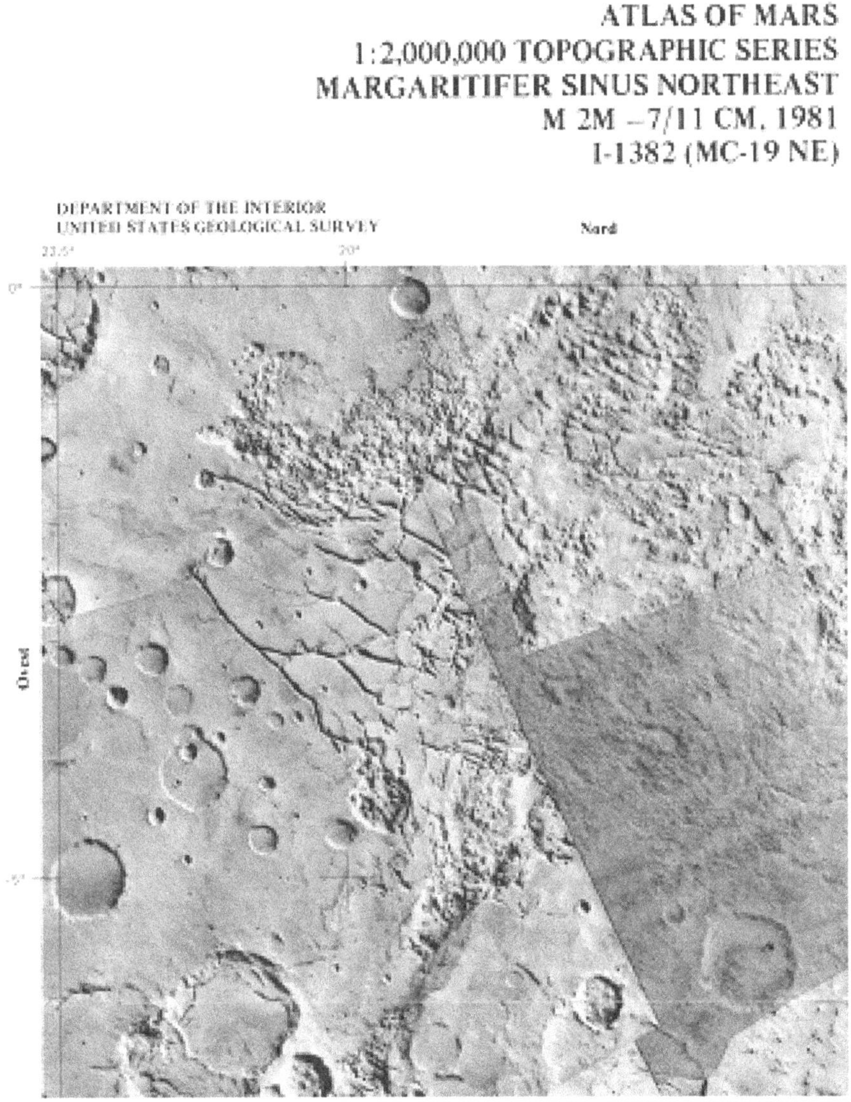

Tab. 131	**Iani Chaos:** Map MC 19 1382 NE (portion), where the structure analyzed falls. The dark part is the one made up.

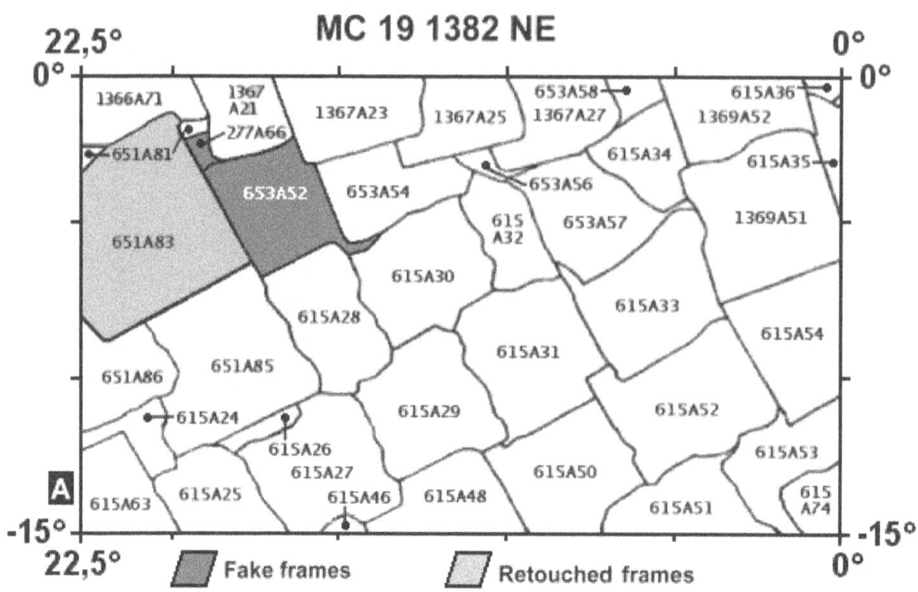

FRAMES USED

Code	Resolution	Probe	Year	Note
277A66	168,00*	Viking 1	1977	Fake frame
653A52	247,00*	Viking 1	1977	Fake frame
651A83	262,00*	Viking 1	1977	Retouched frame

* The value of geometric resolution is invented.

FRAMES TO USE

Code	Resolution	Probe	Year	Note
451A06	343,05	Viking 1	1977	Authentic frame
451A08	343,09	Viking 1	1977	Authentic frame
451A09	343,99	Viking 1	1977	Authentic frame
451A11	343,00	Viking 1	1977	Authentic frame

B

Tab. 132	Iani Chaos: A) Map MC 191382 NE; index of Viking images, with the highlighting of fixed frames; B) The frames used and those to be used.

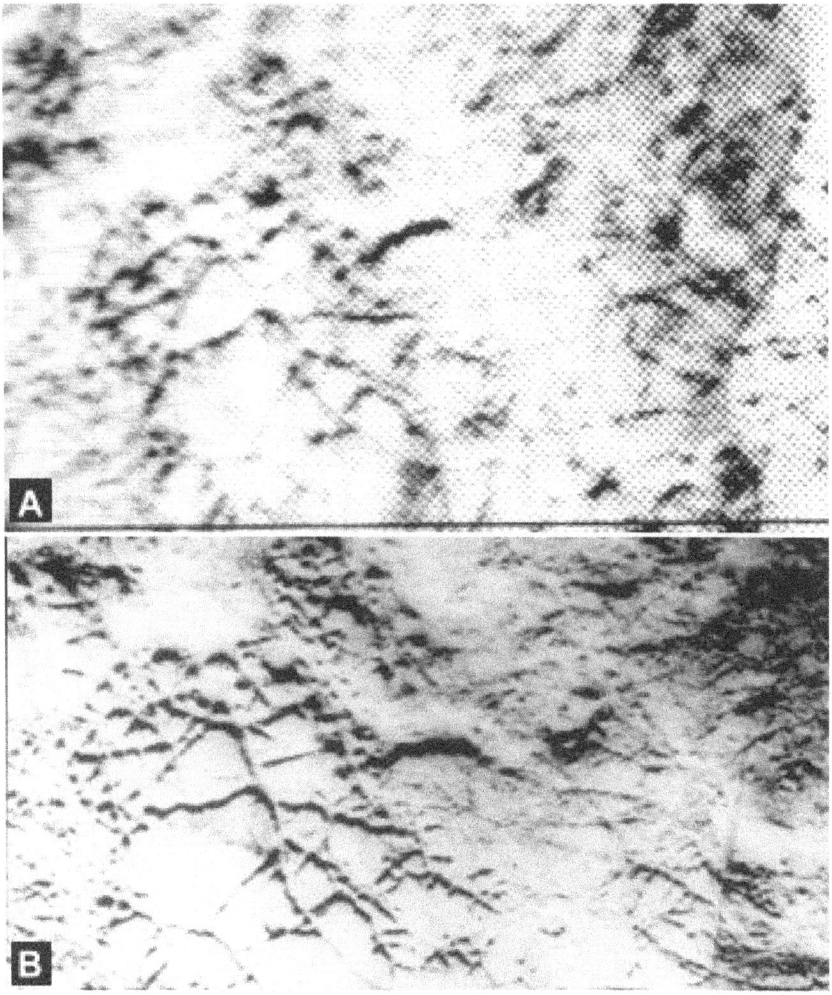

Tab. 133	Iani Chaos – Reconfirmation process: A) Photomosaic 211-4294 (portion); B) Photomosaic 211-5571 (portion).

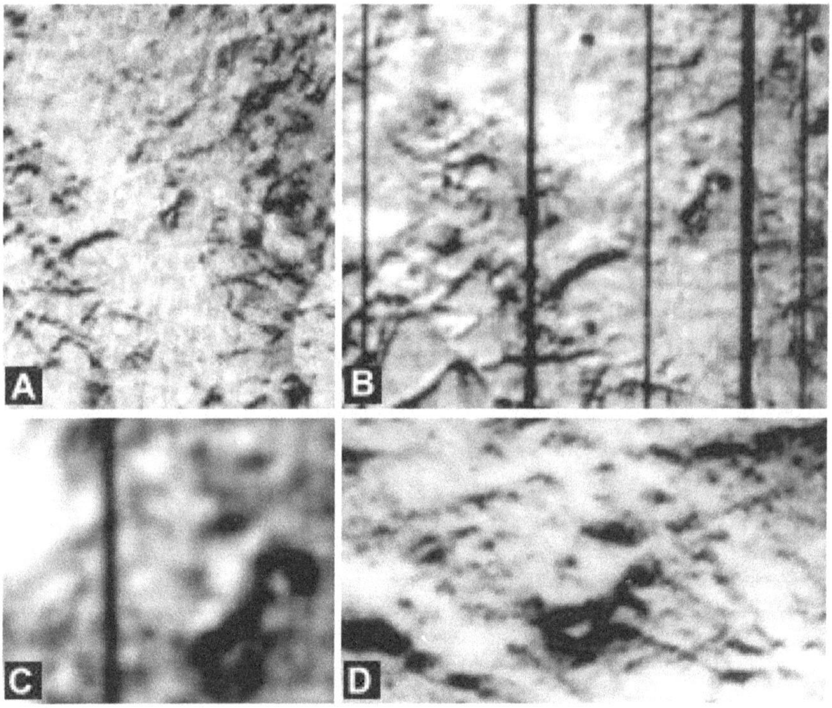

Tab. 134	Iani Chaos – Anthropomorphic structures – Reconfirming process; A,B,C) Photomosaic 211-4294 (portion); D) Photomosaic 211-5571 (portion).

Viking Orbiter 1 Mission
Iani Chaos - 1977

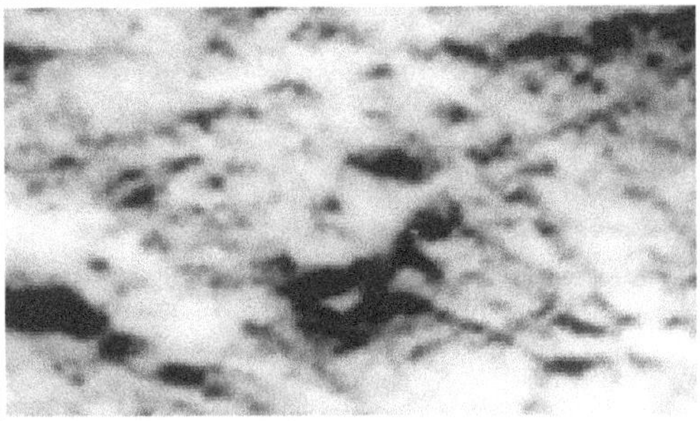	451A06
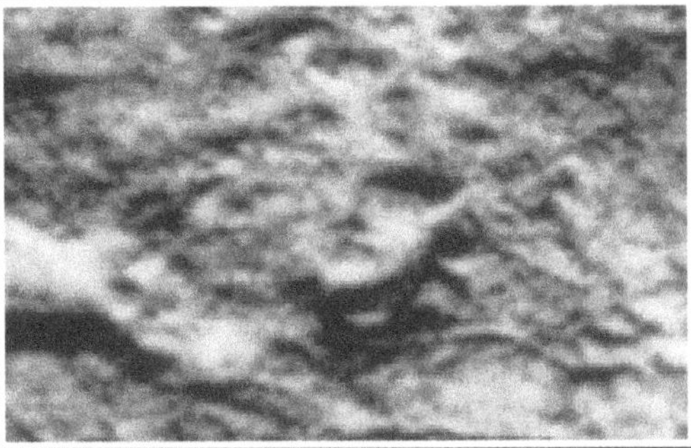	451A08
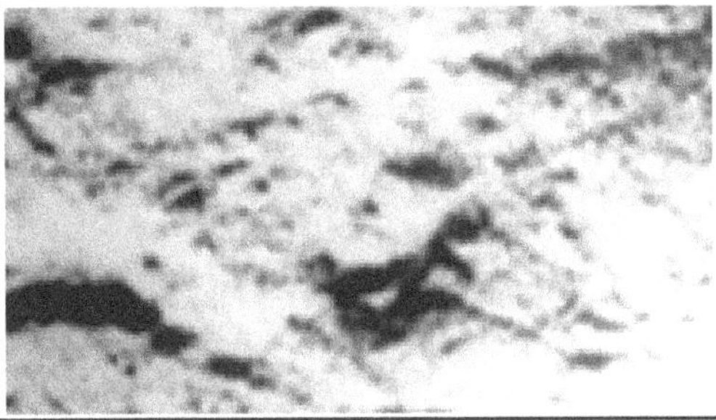	451A09

Tab. 135	Iani Chaos. The anthropomorphic structure, with Viking 1 codes.

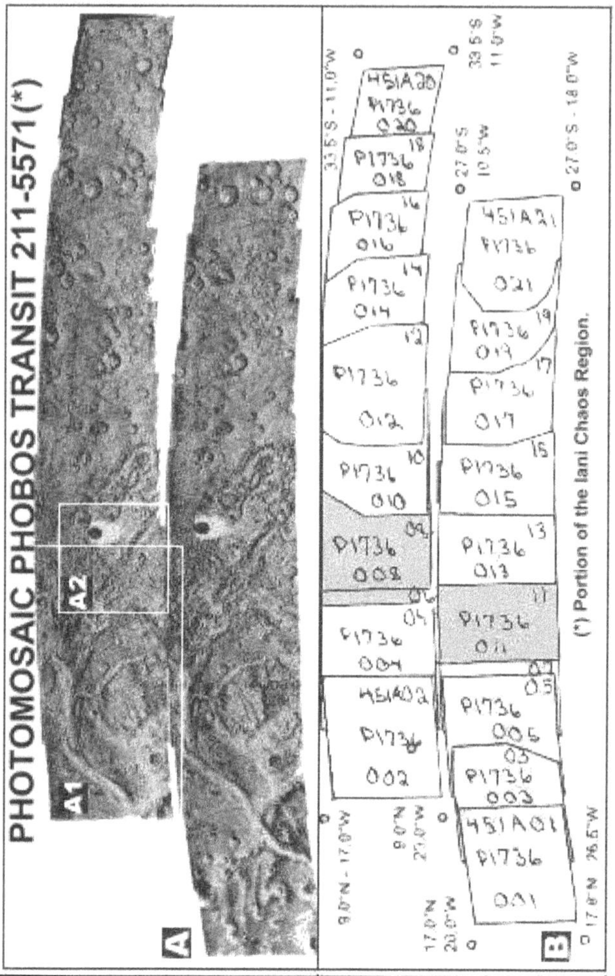

Tab. 136	Iani Chaos: A) Photomosaic Phobos Transit 211-5571 (portion); B) detection patterns.

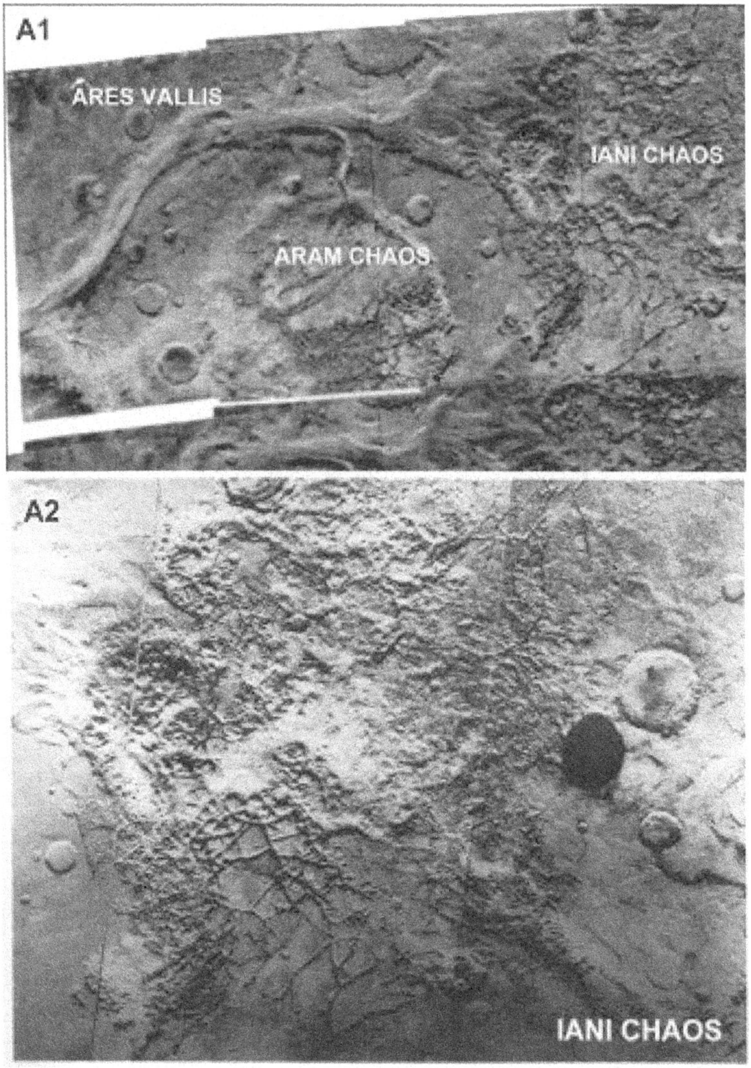

Tab. 137 Iani Chaos – Photomosaic 211-5571 (portion): A1) Ares Vallis, Aram Chaos and Iani Chaos regions; A2) Iani Chaos region, enlargement.

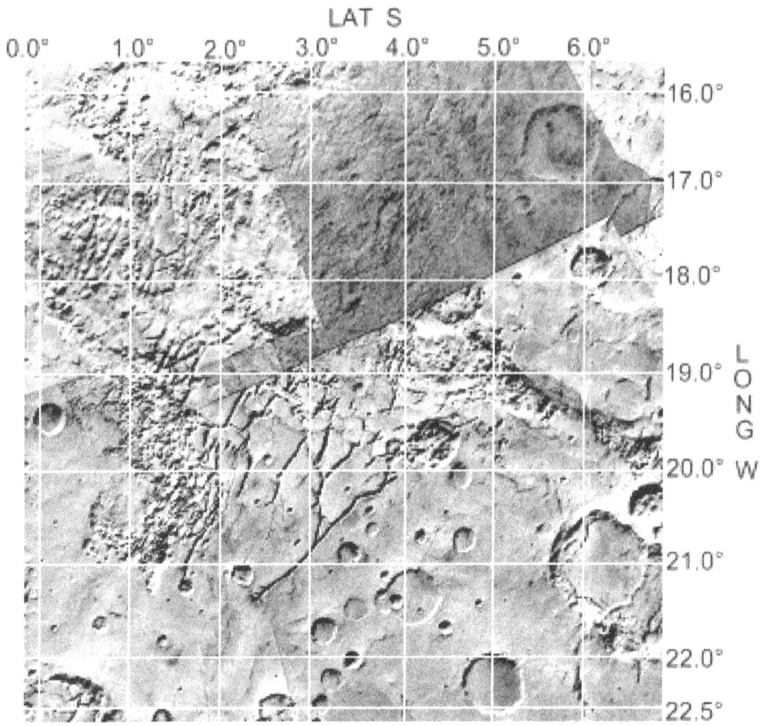

Tab. 138	Iani Chaos. Map MC 19 1382 NE (portion) – with cartographic grid; the anthropomorphic structure at 3.5 S and 18.0 W.

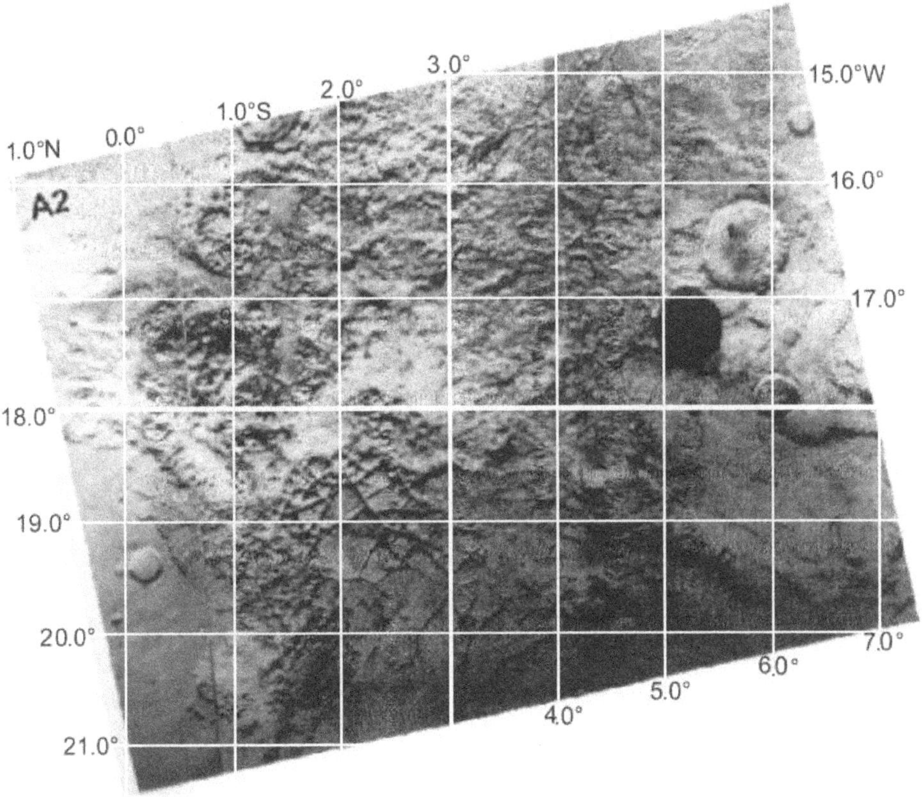

Tab. 139	Iani Chaos. Photomosaic 211-5571 (portion) with a cartographic grid; the anthropomorphic structure at 3.5 S and 18.0 W.

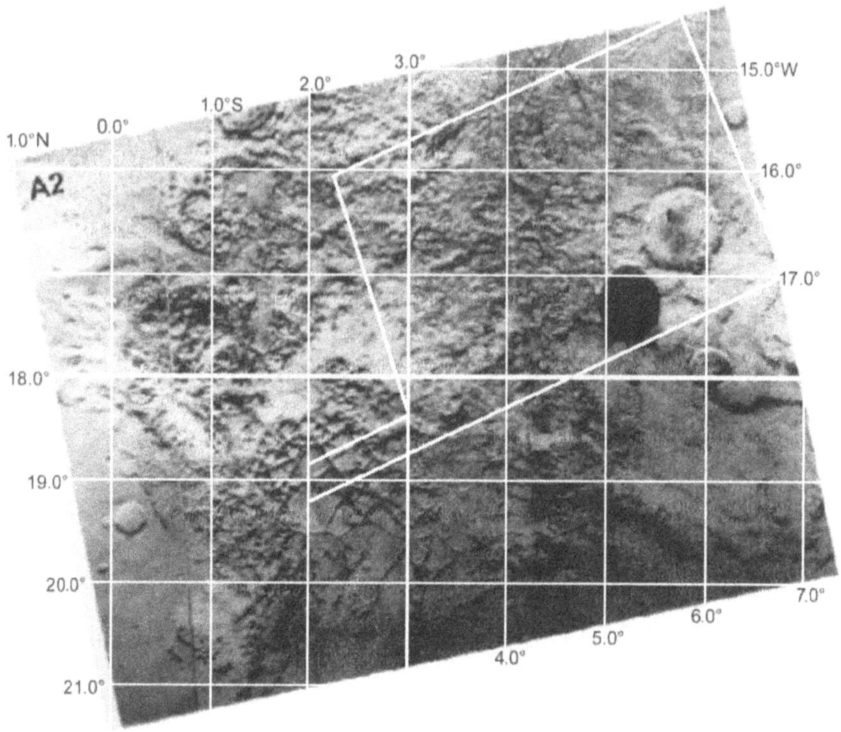

Tab. 140	Iani Chaos. Photomosaic 211-5571 (portion) with cartographic grid; the anthropomorphic structure at 3.5 S and 18.0 W; delimitation of the rigged area on the map.

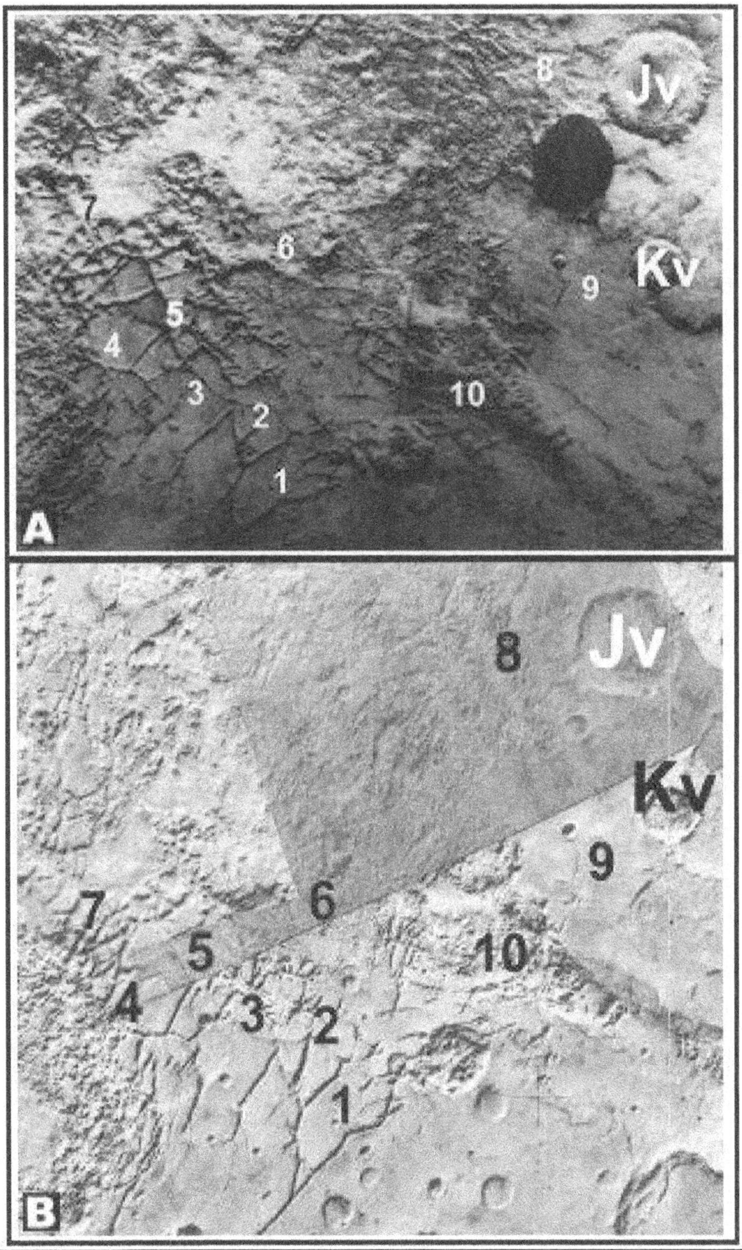

Tab. 141	Iani Chaos – Photographic control points; A) Photomosaic 211-5571 (portion).; B) Map MC.19.1382-NE (portion).

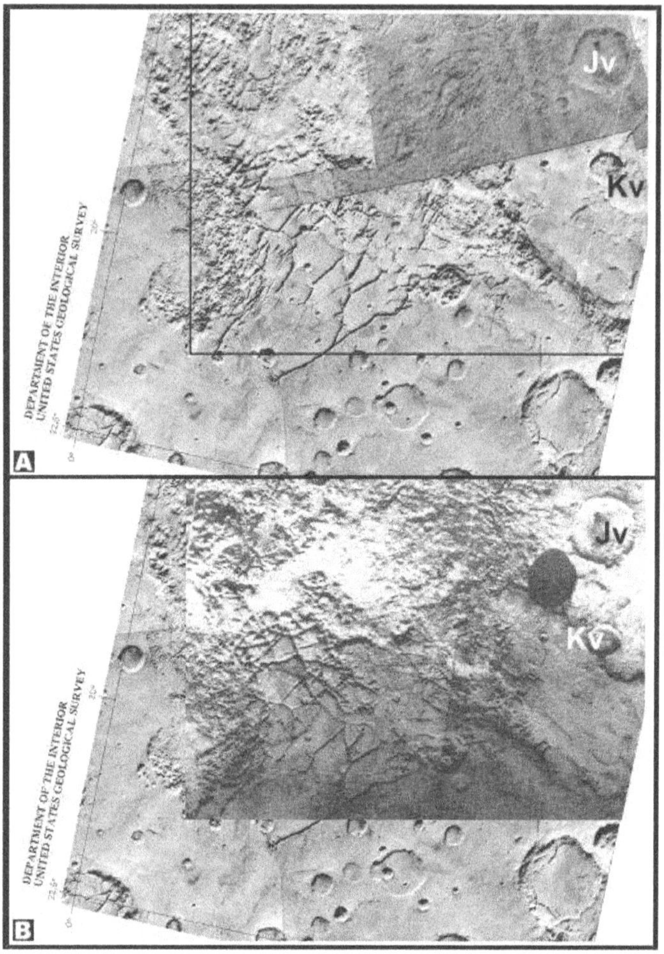

Tab. 142

Iani Chaos: A) Map MC 19 1382 NE (portion) with delimitation of the rigged part; B) relocation, on the rigged part of the Map MC 19 1382 NE, of the authentic photo (photomosaic 211-5571).

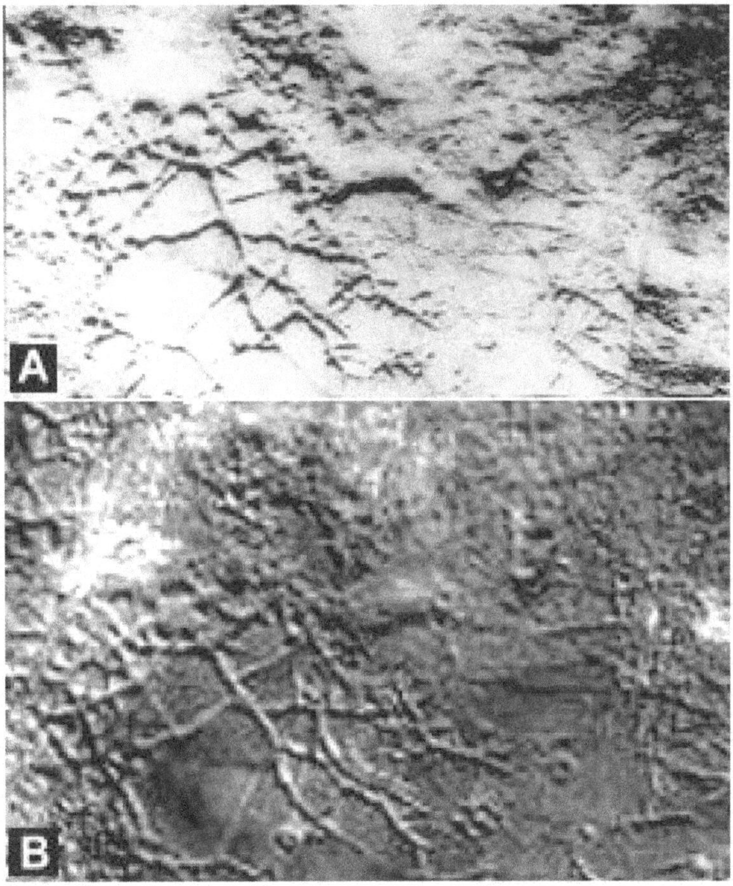

Tab. 143	Iani Chaos: A) Photo Das 451A06; B) Tricks attributed to Mars Express.

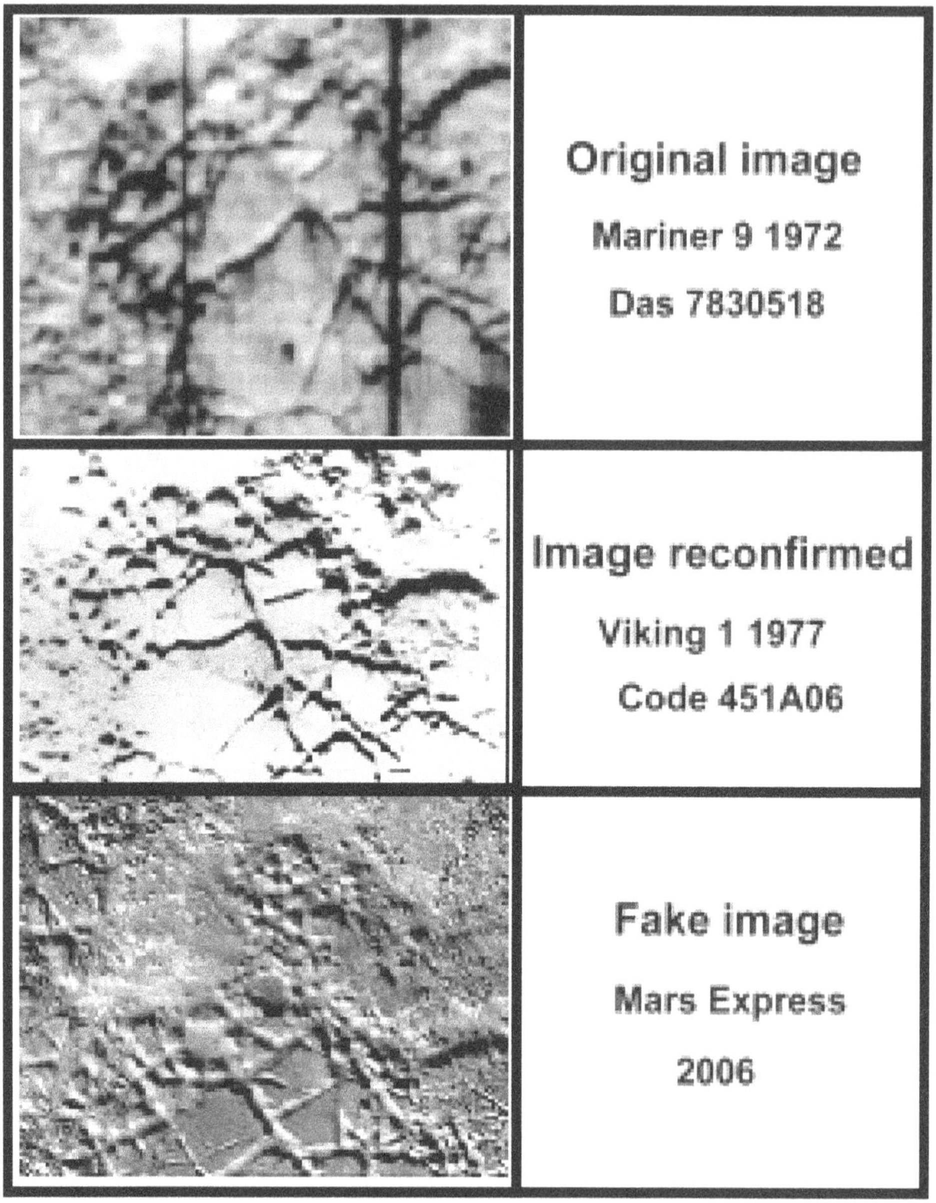

Tab. 144	Iani Chaos. Falsification process attributed to Mars Express.

Tab. 145 Iani Chaos. Falsification process attributed to Mars Express.

Mariner 9 Das 7830513

INST	TYPE	TIME (GMT) D H M S	PERI TIME H M S	SPACECRAFT LAT	SPACECRAFT LONG-W	HGT	PLATFORM CONE	PLATFORM CLOCK	INTERCEPTING LAT	INTERCEPTING LONG-W	RANGE	VIEW ANGLE	LIGHT ANGLE	PHASE ANGLE	DAS REFERENCE TIME
TETRAD NO 1															
1	B	39 22 58 26	-0 33 59	-63.26	112.57	3300	113.31	128.91	-73.21	112.63	3402	19.85	74.62	66.68	7,828,098
2	A	39 22 59 8	-0 33 17	-63.47	110.54	3246	113.14	128.86	-73.76	104.51	3360	21.08	72.87	66.92	7,828,133
3	B	39 22 59 50	-0 32 35	-63.66	108.44	3192	114.15	124.33	-77.72	91.67	3421	29.81	73.13	65.84	7,828,168
4	B	39 23 1 14	-0 31 11	-63.92	104.06	3085	116.44	125.69	-77.20	86.41	3300	29.14	71.97	63.55	7,828,238
SINGLE B NO 1															
5	B	39 23 12 26	-0 19 59	-59.27	66.22	2313	124.39	153.32	-57.54	62.09	2323	6.80	50.34	55.60	7,828,798
SINGLE B NO 2															
6	B	39 23 18 2	-0 14 23	-52.06	51.23	2011	109.06	141.16	-27.11	29.41	2982	64.24	32.56	70.93	7,829,078
MAPPING NO 1															
7	A	39 23 34 8	0 1 42	-20.09	26.76	1659	141.62	244.89	-24.31	29.35	1695	14.59	34.42	38.45	7,829,883
8	B	39 23 34 50	0 2 24	-18.53	26.07	1664	141.47	245.09	-22.27	28.21	1692	12.75	34.77	38.51	7,829,918
9	A	39 23 35 32	0 3 6	-16.98	25.39	1671	141.34	245.17	-19.75	27.16	1687	9.76	35.08	38.73	7,829,953
10	B	39 23 36 14	0 3 48	-15.42	24.73	1680	141.35	245.21	-17.67	26.02	1690	7.70	35.78	38.64	7,829,988
11	A	39 23 36 56	0 4 30	-13.87	24.09	1690	141.37	245.23	-15.14	24.97	1694	4.59	36.49	38.70	7,830,023
12	B	39 23 37 38	0 5 12	-12.33	23.46	1702	141.41	245.28	-13.07	23.88	1703	2.53	37.48	38.58	7,830,058
13	A	39 23 38 20	0 5 54	-10.78	22.85	1716	141.45	245.34	-10.55	22.88	1716	.71	38.49	38.62	7,830,093
14	B	39 23 39 2	0 6 36	-9.25	22.24	1732	141.52	245.37	-8.46	21.82	1733	2.65	39.71	38.47	7,830,128
15	A	39 23 39 44	0 7 18	-7.72	21.65	1749	141.62	245.23	-5.86	20.82	1755	5.96	41.01	38.45	7,830,163
16	B	39 23 40 26	0 8 0	-6.21	21 07	1768	141.72	245.11	-3.71	19.75	1780	8.19	42.46	38.26	7,830,198
MAPPING NO 2															
17	B	39 23 46 2	0 13 36	5.46	16.73	1974	119.81	281.00	-3.28	21.27	2106	25.83	42.36	60.18	7,830,478
18	A	39 23 46 44	0 14 18	6.85	16.22	2007	120.04	280.99	-.78	20.38	2108	22.73	43.79	60.03	7,830,513
19	B	39 23 47 26	0 15 0	8.22	15.71	2040	120.27	280.96	1.23	19.22	2122	20.34	45.47	59.72	7,830,548
20	A	39 23 48 8	0 15 42	9.57	15.20	2075	120.49	280.96	3.67	18.36	2134	17.32	47.01	59.58	7,830,583
21	B	39 23 48 50	0 16 24	10.91	14.70	2111	120.50	280.98	5.56	17.33	2158	15.31	48.64	59.49	7,830,618
22	A	39 23 49 32	0 17 6	12.22	14.20	2148	120.43	280.91	7.91	16.67	2180	12.65	50.08	59.64	7,830,653
23	B	39 23 50 14	0 17 48	13.52	13.70	2186	120.35	280.85	9.78	15.71	2209	10.71	51.70	59.64	7,830,688
24	A	39 23 50 56	0 18 30	14.80	13.21	2225	120.29	280.78	12.11	15.04	2239	8.09	53.22	59.78	7,830,723
25	B	39 23 51 38	0 19 12	16.05	12.71	2265	120.24	280.72	13.95	14.04	2273	6.12	54.90	59.75	7,830,758
26	A	39 23 52 20	0 19 54	17.29	12.22	2307	120.20	280.69	16.26	13.34	2310	3.66	56.49	59.87	7,830,793
TETRAD NO 2															
27	A	39 23 56 32	0 24 6	24.28	9.27	2574	104.82	304.08	6.62	11.52	2925	38.50	56.31	75.25	7,831,003
28	A	39 23 57 56	0 25 30	26.45	8.28	2670	105.01	298.85	14.94	11.58	2830	26.20	58.88	75.06	7,831,073
29	B	39 23 58 38	0 26 12	27.50	7.79	2718	105.11	298.64	16.70	10.46	2857	24.26	60.62	74.89	7,831,108
30	A	39 23 59 20	0 26 54	28.54	7.29	2768	105.20	293 53	22.73	11.63	2823	15.40	61.88	74.87	7,831,143
DYAD															
31	A	40 0 2 8	0 29 42	32.48	5.27	2971	108.40	270.92	50.64	15.58	3369	39.49	73.32	71.68	7,831,283
32	B	40 0 2 50	0 30 24	33.41	4.76	3024	106.91	272.85	49.94	15.09	3364	36.52	73.26	73.08	7,831,318
~~33~~	A	40 0 3 32	0 31 6	34.33	4.25	3076	106.76	272.89	52.72	15.03	3476	39.32	74.97	73.31	~~7,831,353~~

JPL TECHNICAL MEMORANDUM 33-585, VOL. II

6-379

Tab. 146	Iani Chaos – Photo Das 7830513 (18) – Detailed satellite survey information.

8. Region of Aeolis Mensae

Map MC 23 1213 NW

5.0° S – 222.0° W (alt. km 2.0)

In quadrant MC- 23 Aeolis (**Table 147A**) we consider picture No. 2 of the attached list (**Table 147B**) with the identification number (Das7650473).

In **table 148A** we observe the pattern of detection and in **table 148B** the original frame.

In **table 149A** the detected structure is inserted in the context of Equatorial Photomosaic 211-4294; the asterisk (*) indicates the detected structure.

It could be described as a human face, circumscribed in a crater, measuring 70x90 km (**Table 150B**).

Located in the region of Aeolis Mensae (**Table 149B**), where the cartographic position is indicated, the original structure is completely disappeared on map MC.23.1213.NW (**Table 151A**).

A cartographic analysis shows that, in the corresponding part occupied by the original structure, there is the frame 631A05 (**Table 151B**), attributed to the mission Viking 1.

To superimpose the original image on the "false" one, taking into account the original photo (**Table 148B**), we fix some photographic control points (**Table 152B**), then we find them on paper MC.23.1213.NW (**Tables 152A,C**). As a last operation we reconnect the original structure (**Table 152B**) on the corresponding portion of paper MC.23.1213.NW (**Table 152D**).

In **table 150A** we have updated topographical data.

In **table 153** we have inserted the detailed data of the frame Das7650473.

Stable elements of experimentation/photographic points of reference: (0.0-10.0°S – 215.0-225.0°W); 1) Crater Gale Yv; 2) Aeolis Mons (inside the Crater); 3) N: Aeolis Palus; 4) W: Dukce Vallis, Sakya Vallis; 5) NW: Peace Vallis, Gediz Vallis, Zy Crater; 6) SW: Farah Vallis.

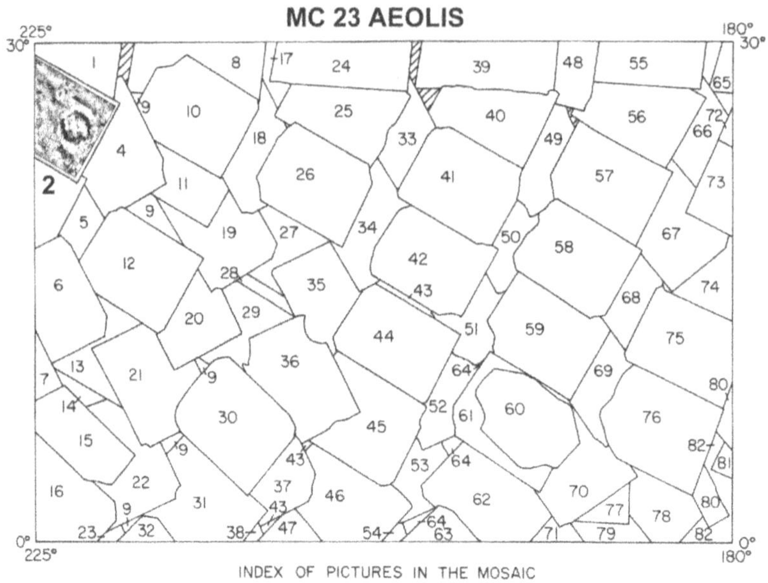

Tab. 147A	MC 23 Aeolis – Index of images Mariner 9.

IDENTIFICATION NUMBERS OF PICTURES IN THE MOSAIC

Index No.	DAS No.	Index No.	DAS No.	Index No.	DAS No.
1	7650823	29	7794043	57	7937963
■ 2	7650473	30	6390893	58	7937893
3	7650403	31	6390823	59	7937823
4	9125169	32	6390753	60	6534673
5	9125099	33	9268949	61	7937753
6	9125029	34	9268879	62	6534603
7	9124959	35	9268809	63	6534533
8	7722713	36	9268739	64	5920773
9	5632793	37	9268669	65	6607333
10	7722363	38	9268599	66	8009783
11	7722293	39	7866493	67	8009713
12	7722223	40	7866143	68	8009643
13	7722153	41	7866073	69	8009573
14	7722083	42	7866003	70	8009503
15	6319003	43	5848813	71	8009433
16	6318933	44	7865933	72	6606983
17	5704753	45	7865863	73	6606913
18	9197059	46	6462713	74	6606843
19	9196989	47	6462643	75	6606773
20	9196919	48	12326301	76	6606703
21	9196849	49	9340839	77	9484199
22	9196779	50	9340769	78	5203483
23	9196709	51	9340699	79	5203413
24	7794603	52	9340629	80	8081393
25	7794253	53	9340559	81	6678593
26	7794183	54	9340489	82	9412799
27	7794113	55	7938383		
28	5776783	56	7938033		

Tab. 147B	MC 23 Aeolis – Image identification number.

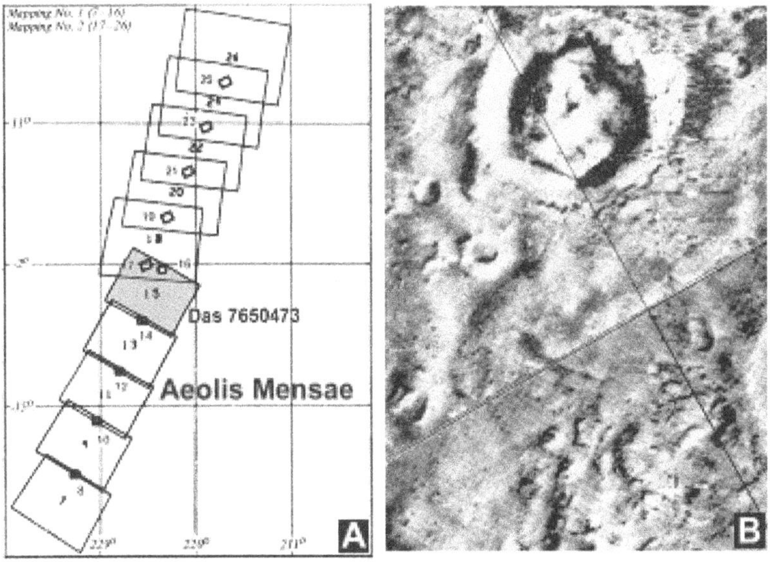

Tab. 148	Aeolis Mensae. Photo Das 7650473: A) Detection scheme; B) Original frame.

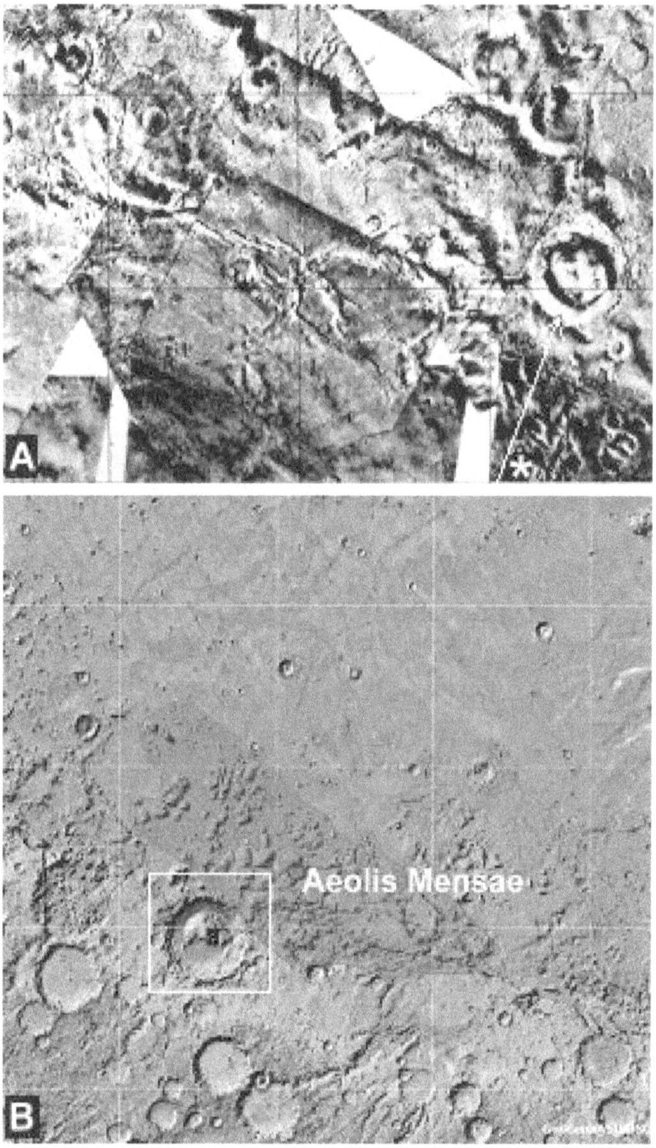

Tab. 149

Aeolis Mensae: A) Photomosaic Atlas of Mars 1979 MC 23 Aeolis (portion), with indication (*) of the detected structure; B) Cartographic position of the detected structure.

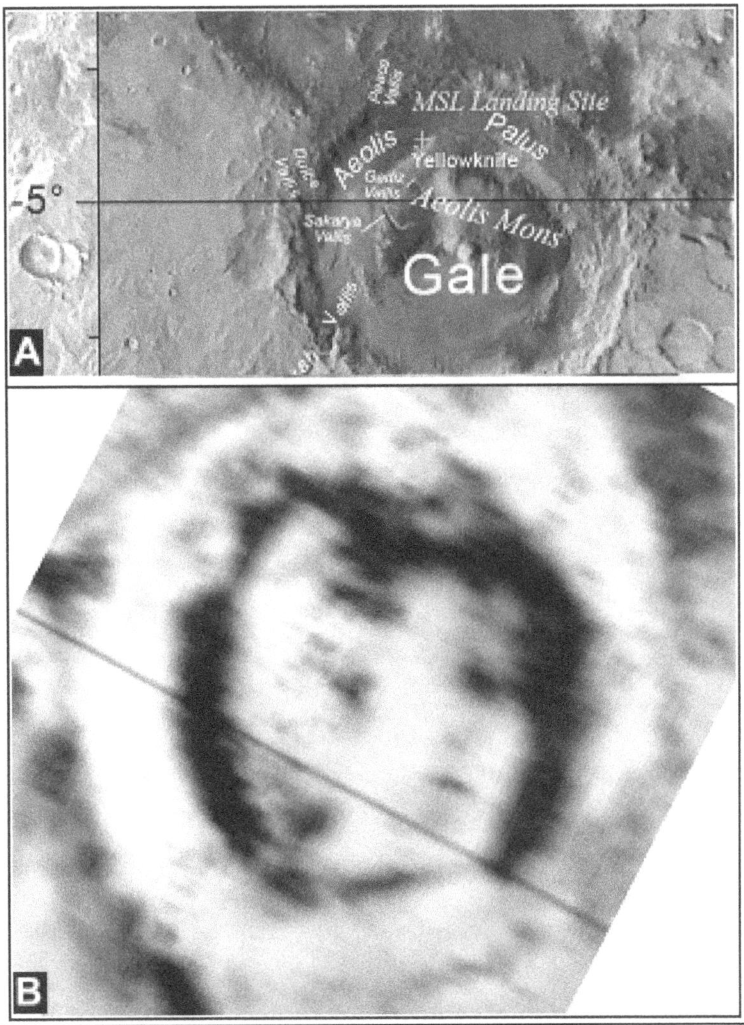

Tab. 150	Aeolis Mensae. A) Cartography with updated nomenclature; B) Enlarged structure.

Tab. 151A	Aeolis Mensae – Map MC 23 1213 NE (portion), where the structure analysed falls.

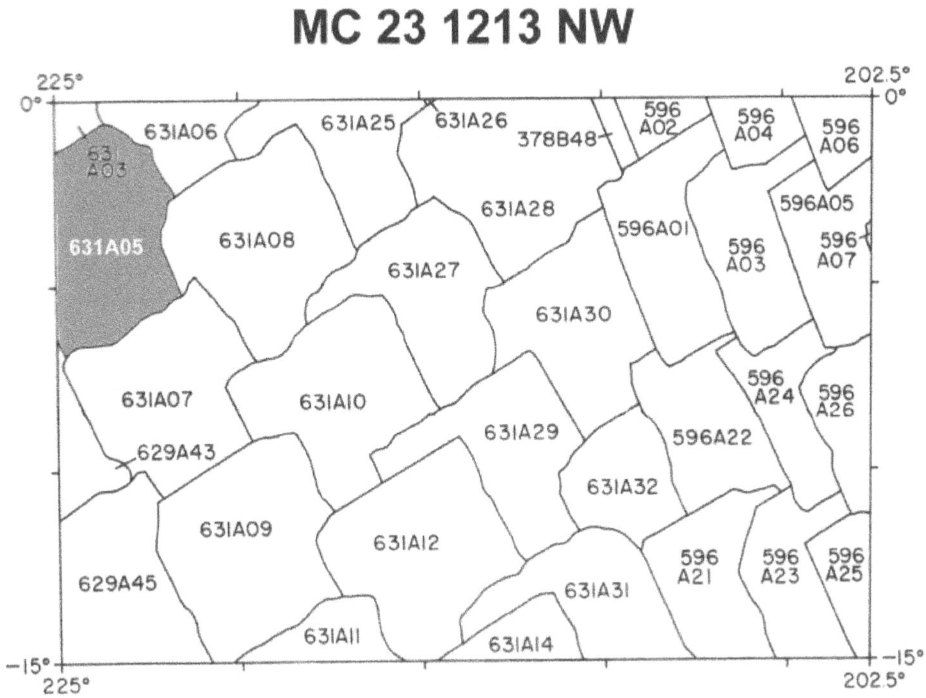

Tab. 151B	Aeolis Mensae – Viking image index.

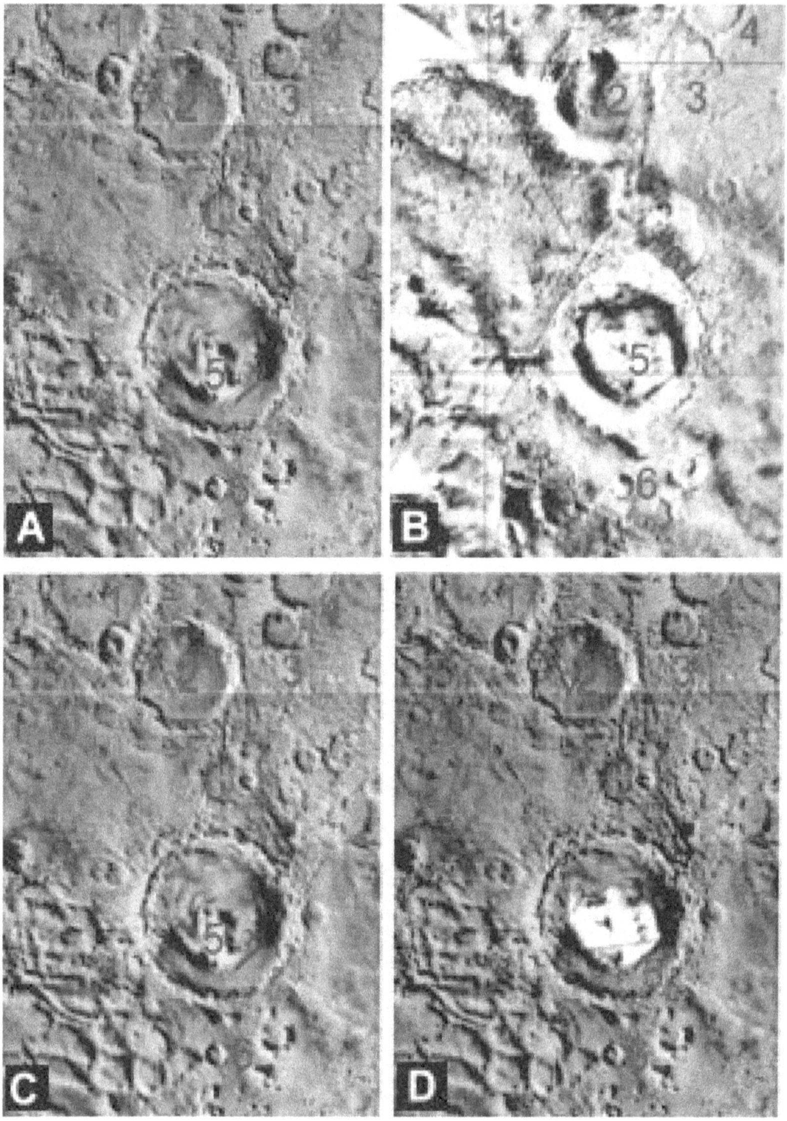

Tab. 152	Aeolis Mensae. Images with photographic control points; A,C) Map MC 23 1213 NE (portion); B) Photomosaic 211-4294 (portion); D) Relocation of the original structure, on Map MC 23 1231 NW (portion).

Mariner 9 Das 7650473

INST	TYPE	TIME (GMT) D H M S	PERI TIME H M S	SPACECRAFT LAT	SPACECRAFT LONG-W	HGT	PLATFORM CONE	PLATFORM CLOCK	INTERCEPTING LAT	INTERCEPTING LONG-W	INTERCEPTING RANGE	VIEW ANGLE	LIGHT ANGLE	PHASE ANGLE	DAS REFERENCE TIME
SINGLE B NO 1															
1	B	37 11 0 27	-0 37 43	-61.69	325.81	3595	107.77	130.92	-67.16	325.15	3625	10.59	74.43	72.22	7,648,198
TRIAD															
2	A	37 11 5 21	-0 32 49	-63.60	312.76	3208	118.01	141.69	-65.32	325.06	3241	11.38	73.39	62.06	7,648,443
3	B	37 11 6 3	-0 32 7	-63.75	310.62	3155	119.41	141.49	-66.22	323.85	3194	12.56	73.14	60.58	7,648,478
4	B	37 11 7 27	-0 30 43	-63.96	306.16	3048	122.08	145.62	-65.31	322.81	3105	15.13	72.30	57.91	7,648,548
SINGLE B NO 2															
5	B	37 11 18 39	-0 19 31	-58.74	268.46	2284	130.12	161.08	-59.32	271.04	2286	3.59	51.89	49.87	7,649,108
SINGLE B NO 3															
6	B	37 11 24 15	-0 13 55	-51.27	253.85	1988	134.58	170.27	-50.27	253.43	1989	2.79	42.92	45.40	7,649,388
MAPPING NO 1															
7	A	37 11 40 21	0 2 10	-18.96	230.00	1660	139.70	247.90	-23.24	232.48	1696	14.59	35.77	40.37	7,650,193
8	B	37 11 41 3	0 2 52	-17.40	229.32	1666	139.51	247.46	-20.99	231.36	1692	12.23	36.22	40.47	7,650,228
9	A	37 11 41 45	0 3 34	-15.85	228.66	1674	139.33	247.24	-18.38	230.37	1689	9.06	36.67	40.75	7,650,263
10	B	37 11 42 27	0 4 16	-14.29	228.01	1684	139.13	247.37	-16.32	229.35	1693	7.21	37.43	40.86	7,650,298
11	A	37 11 43 9	0 4 58	-12.74	227.38	1696	139.11	247.51	-13.84	228.35	1699	4.33	38.28	40.96	7,650,333
12	B	37 11 43 51	0 5 40	-11.20	226.76	1709	139.25	247.72	-11.81	227.21	1710	2.26	39.44	40.74	7,650,368
13	A	37 11 44 33	0 6 22	-9.66	226.15	1724	139.42	247.92	-9.34	226.17	1725	.93	40.62	40.65	7,650,403
14	B	37 11 45 15	0 7 4	-8.13	225.56	1741	139.59	247.90	-7.24	225.05	1743	3.02	42.00	40.40	7,650,438
■ 15	A	37 11 45 57	0 7 46	-6.61	224.97	1760	139.63	247.86	-4.67	224.09	1766	6.20	43.35	40.44	7,650,473 ■
16	B	37 11 46 39	0 8 28	-5.10	224.40	1780	139.48	247.75	-2.52	223.15	1792	8.26	44.74	40.51	7,650,508
MAPPING NO 2															
17	B	37 11 52 15	0 14 4	6.49	220.09	1994	117.38	281.50	-2.09	224.68	2122	25.39	44.64	62.61	7,650,788
18	A	37 11 52 57	0 14 46	7.87	219.58	2027	117.55	281.45	.41	223.84	2128	22.34	46.09	62.52	7,650,823
19	B	37 11 53 39	0 15 28	9.23	219.07	2061	117.75	281.54	2.34	222.68	2142	20.12	47.80	62.24	7,650,858
20	A	37 11 54 21	0 16 10	10.57	218.56	2097	117.96	281.60	4.73	221.82	2156	17.18	49.37	62.11	7,650,893
21	B	37 11 55 3	0 16 52	11.90	218.06	2134	118.10	281.63	6.63	220.71	2179	15.04	51.10	61.89	7,650,928
22	A	37 11 55 45	0 17 34	13.20	217.56	2172	118.07	281.69	8.92	220.00	2203	12.43	52.60	62.00	7,650,963
23	B	37 11 56 27	0 18 16	14.48	217.07	2210	118.04	281.73	10.73	218.98	2233	10.55	54.26	61.95	7,650,998
24	A	37 11 57 9	0 18 58	15.75	216.57	2250	118.03	281.76	13.02	218.26	2264	7.94	55.82	62.04	7,651,033
25	B	37 11 57 51	0 19 40	16.99	216.08	2292	118.01	281.82	14.78	217.20	2299	6.08	57.53	61.98	7,651,068
26	A	37 11 58 33	0 20 22	18.21	215.58	2334	118.01	281.86	17.05	216.45	2336	3.49	59.15	62.06	7,651,103
TETRAD															
27	A	37 12 2 45	0 24 34	25.12	212.63	2604	101.05	303.66	7.25	216.55	2971	39.19	57.05	79.02	7,651,313
28	A	37 12 4 9	0 25 58	27.27	211.63	2701	101.31	297.92	16.18	216.70	2864	26.31	59.85	78.76	7,651,383
29	B	37 12 4 51	0 26 40	28.31	211.13	2750	101.47	297.98	17.75	215.46	2893	24.52	61.64	78.52	7,651,418
30	A	37 12 5 33	0 27 22	29.33	210.63	2800	101.61	292.04	24.54	216.94	2861	16.12	63.04	78.46	7,651,453
DYAD															
31	A	37 12 8 21	0 30 10	33.21	208.60	3005	105.69	303.81	23.58	204.95	3116	21.19	73.56	74.38	7,651,593
32	B	37 12 9 3	0 30 52	34.13	208.08	3058	105.73	303.92	24.72	203.49	3169	21.15	75.32	74.26	7,651,628

JPL TECHNICAL MEMORANDUM 33-585, VOL. II — 6-369

Tab. 153 Aeolis Mensae. Photo Das 7650473 – Detailed satellite survey information.

9. Region of Nirgal Vallis

Map MC 26 1191 NW – MC 26 1192 NC

31° S – 46° W (alt. km 1.0)

In quadrant MC-26 Argyre (**Table 154A**) we take into consideration images nos. 4 and 6 of the attached list (**Table 154B**) with the following identification numbers (Das6282953- 6354773).

The image also includes other portions of the surface falling into MC.18.1184.SE (Das6283023) and MC.19.1209. SW (Das9160729). (v. **Tables 162,163,164**).

In **Tables 155A-156A-157A** we see the detection patterns and in **Tables 155B-156B-157B** the original frames.

In **table 162** we have the original photo, with the highlighting of the structure; in **table 163** we have the original photo with the highlighting of the frames that make up the complex of the structure; Finally, in **table 164** we have the picture of the union of the frames that make up the original image.

In **table 158A** the detected structure is inserted in the context of Equatorial Photomosaic 211-4294, with the asterisk (*) indicating the structure detected.

It could be described as a female human face placed in front, with a probable hexagonal structure headdress, having the following dimensions: 48x68 km.

Located in the region of Nirgal Vallis (**Table 158B**), where the map position is indicated, the original structure is completely disappeared in the picture of union of the maps MC.18.1184.SE; MC.19.1209.SW; MC.26.1191-NW; MC.26.1192.NC (**Table 159**).

A cartographic analysis shows that, in the corresponding part occupied by the original structure, there are the frames 611A03, 611A05 and 611A24 (**Table 160**), attributed to the Viking 1 mission.

To superimpose the original image on the "false" one, we fix some photographic control points (**Table 161**), then we find them on the planetary map (**Table 158B**).

Finally, we have relocated the original structure (**Table 166**) within the planetary chart (**Table 159**) and this is the result (**Table 165**). As a last elaboration, in three steps, we redefined the structure (**Table 166**).

In **tables 167,168,169** we have inserted the detailed data of the frames Das6282953,6354773 and 6283023.

Stable elements of experimentation/photographic points of reference:: (28.0-32.0 somebody S – 44.0-48.0 person W); 1) Nirgal Vallis (MC 18 1184 SE, MC 19 1209 SW); 2) Ladon Valles (MC 19 1209 SW); 3) Uzboi Vallis (MC 19 1209 SW); 4) MC 18 1184 SE: Crater Aa (28.8°S-46.0°W), Crater Aad (28.9°S-46.20°W); 5) MC 19 1209 SW: Crater Za (28.8°S-44.7°W); 6) MC 26 1191 NW: Crater Uz (30.8°S-46.5°W), Crater Uv (32.5°S-44.8°W); 7) MC 26 1192 NC: Crater Tz (30.8°S-44.8°W).

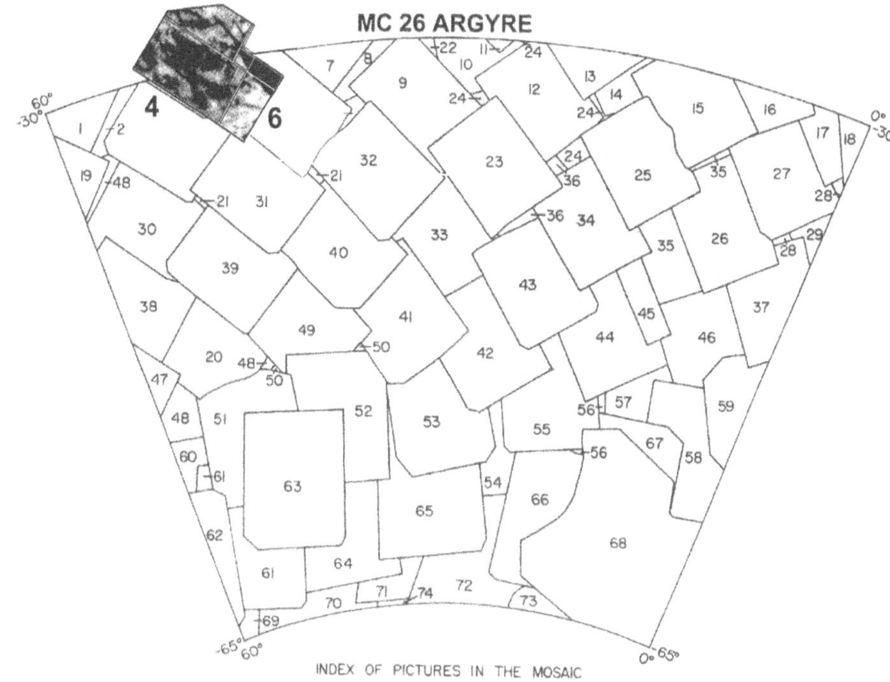

Tab. 154A	MC 26 Argyre – Index of images Mariner 9.

IDENTIFICATION NUMBERS OF PICTURES IN THE MOSAIC

Index No.	DAS No.	Index No.	DAS No.	Index No.	DAS No.
1	6211063	26	5239253	51	6426103
2	9088839	27	5239323	52	6497993
3	6282883	28	6028433	53	5166943
■ 4	6282953	29	5311213	54	5238833
5	9160729	30	6282813	55	5238903
■ 6	6354773	31	6354703	56	5956613
7	6354843	32	6426593	57	5310863
8	9232619	33	6498413	58	5382753
9	6426663	34	5167293	59	5382963
10	9304509	35	6028573	60	6426033
11	6426733	36	5812623	61	6497853
12	6498553	37	5311143	62	6425963
13	6498623	38	6282743	63	6497923
14	7973453	39	6354633	64	5166803
15	5167433	40	6426523	65	5238763
16	8045413	41	6498343	66	5382613
17	5239393	42	5167153	67	5382683
18	8117303	43	5167223	68	7432703
19	6210993	44	5239113	69	5166733
20	6354563	45	5239183	70	5094703
21	5596673	46	5311073	71	5238693
22	5957873	47	6282673	72	5382543
23	6498483	48	5596603	73	5454503
24	5812693	49	6426453	74	5740523
25	5167363	50	6856743		

Tab. 154B	MC 26 Argyre Image identification number.

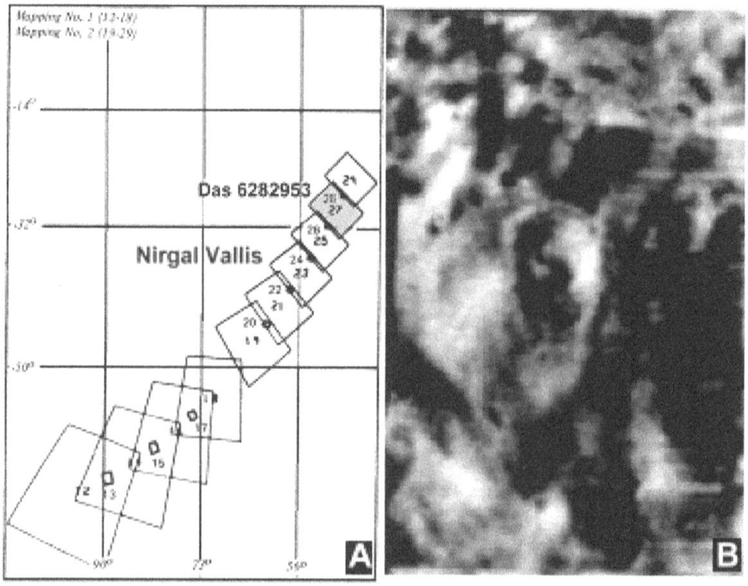

Tab. 155	Nirgal Vallis. Photo Das 6282953: A) Detection scheme; B) Original frame.

Tab. 156	Nirgal Vallis. Photo Das 6354773: A) Detection scheme; B) Original frame.

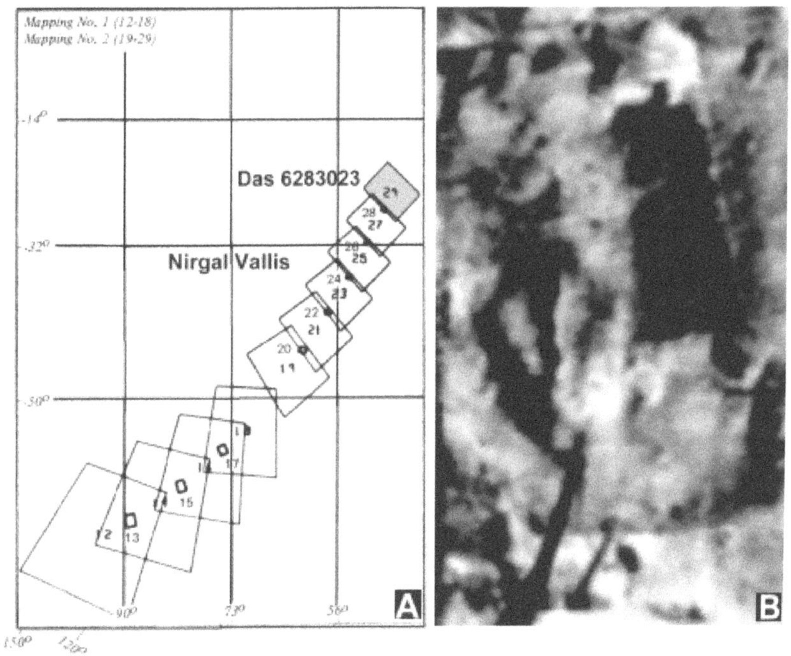

Tab. 157	Nirgal Vallis. Photo Das 6283023: A) Detection scheme; B) Original frame.

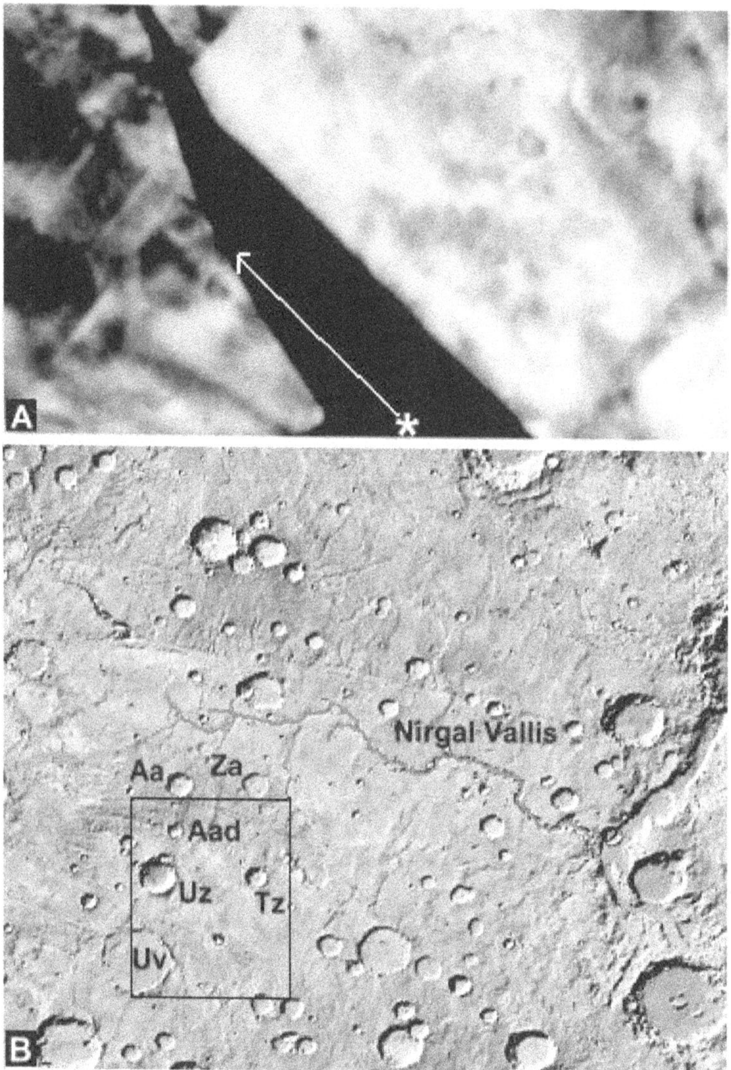

Tab. 158	Nirgal Vallis: A) Photomosaic 211-4294 (portion), with indication (*) of the detected structure; B) Cartographic position of the detected structure.

ATLAS OF MARS
1:2000000 TOPOGRAPHIC SERIES
COPRATES SOUTHEAST
M 2M −22/56 CM, 1979
I-1184 (MC-18 SE)

ATLAS OF MARS
1:2000000 TOPOGRAPHIC SERIES
MARGARITIFER SINUS SOUTHWEST
M 2M −22/34 CM, 1979
I-1209 (MC-19 SW)

ATLAS OF MARS
1:2000000 TOPOGRAPHIC SERIES
ARGYRE NORTHWEST
M 2M −39/56 CM, 1979
I-1191 (MC-26 NW)

ATLAS OF MARS
1:2000000 TOPOGRAPHIC SERIES
ARGYRE NORTH-CENTRAL
M 2M −39/34 CM, 1979
I-1192 (MC-26 N-C)

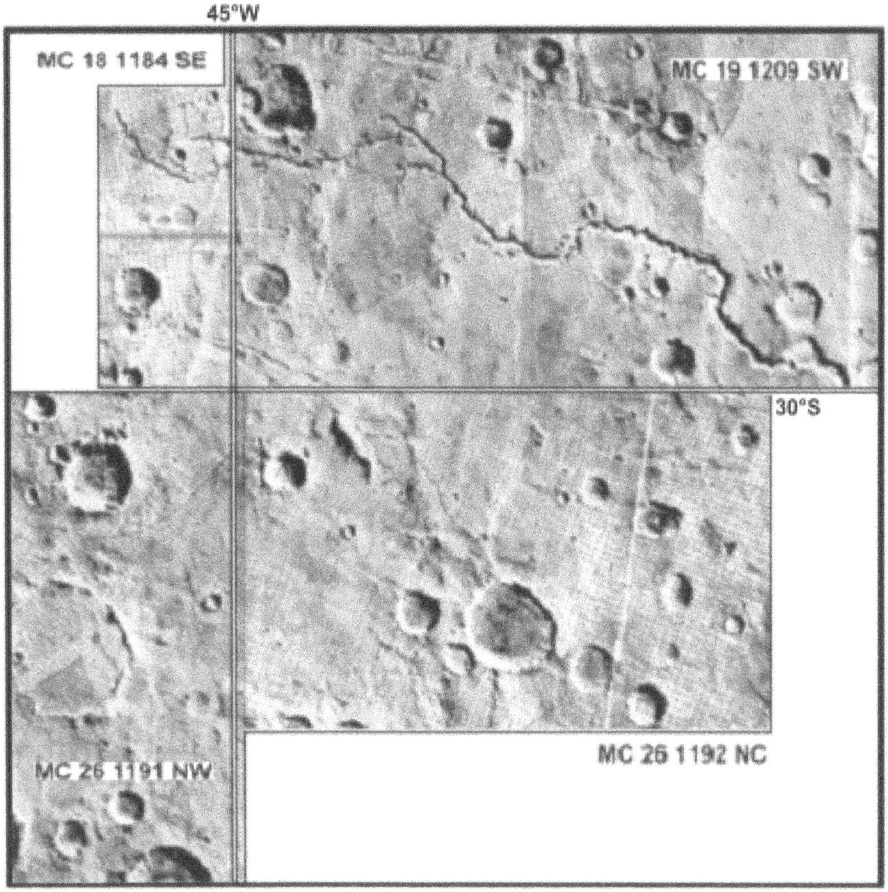

Tab. 159	Nirgal Vallis – Map Union MC 18 1184 SE, MC 19 1209 SW, MC 26 1191 NW, MC 26 1192 NC (portions). Indicated coordinates: 45.0W – 30.0 S.

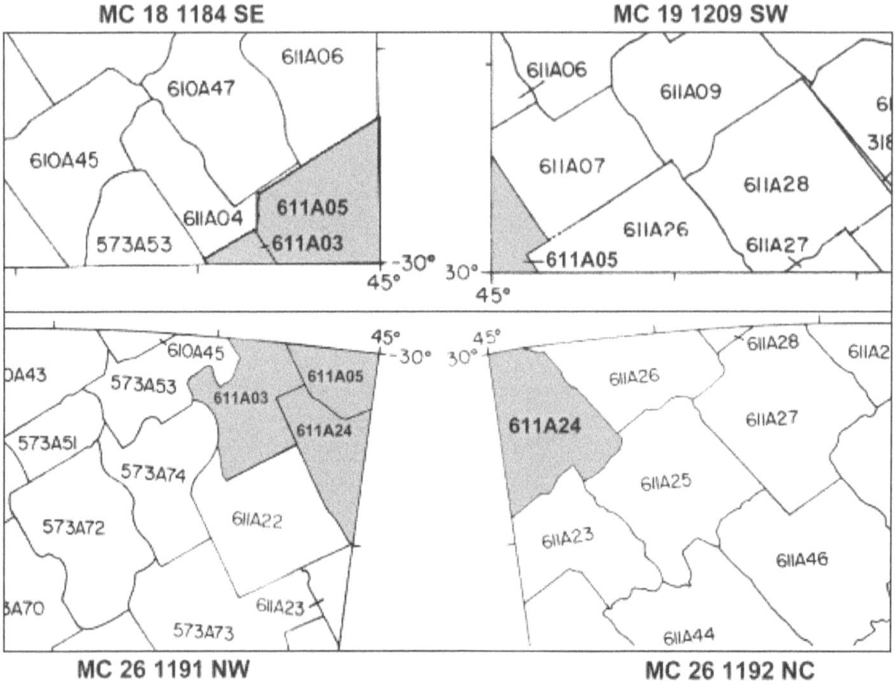

Tab. 160	Nirgal Vallis – Maps MC 18 1184 SE; MC 19 1209 SW; MC 26 1191 NW; MC 26 1192 NC – Viking image index. Highlighted frames: 611A03 – 611A05 – 611A24.

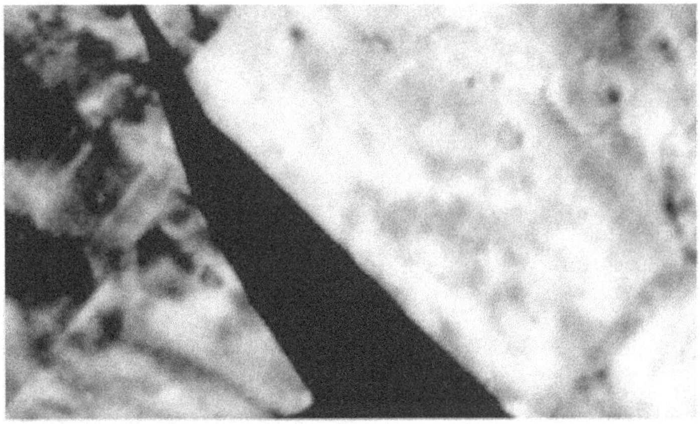

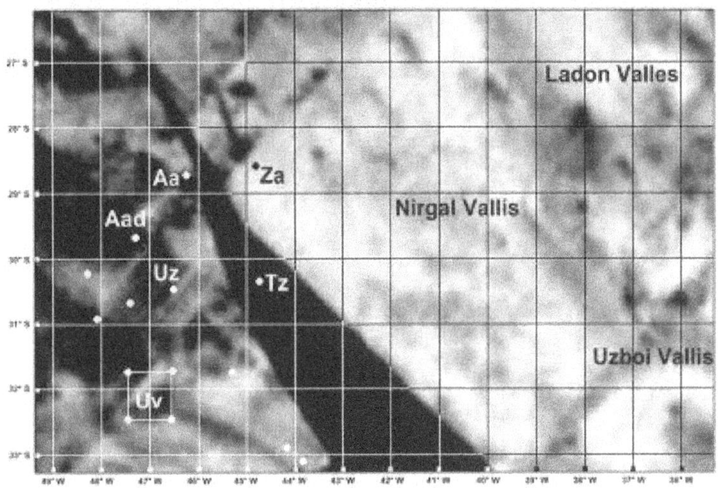

Tab. 161	Nirgal Vallis. Original photo and photo with cartographic grid, with photographic control points.

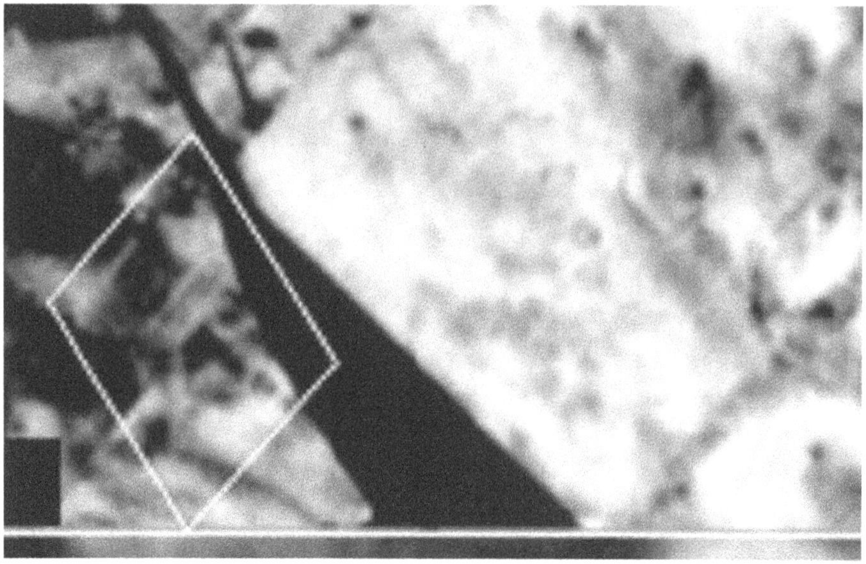

Tab. 162	Nirgal Vallis. Original photo showing the structure.

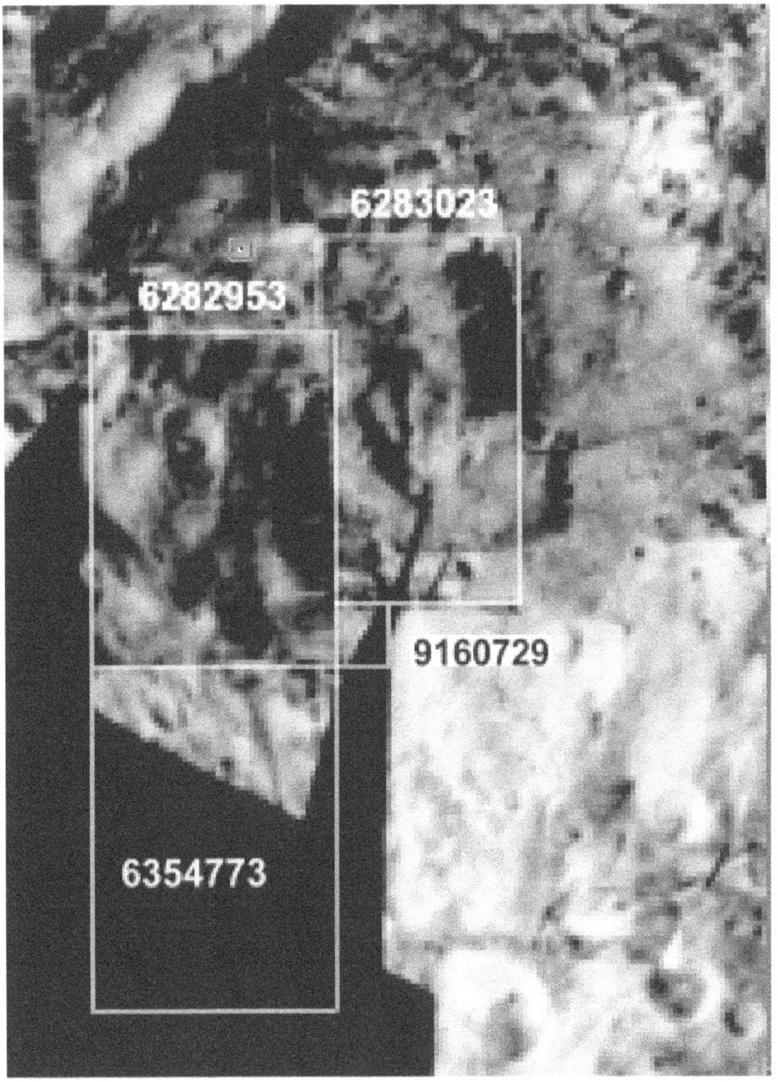

Tab. 163	Nirgal Vallis – Original photo with the highlighting of the frames that make up the complex of the structure.

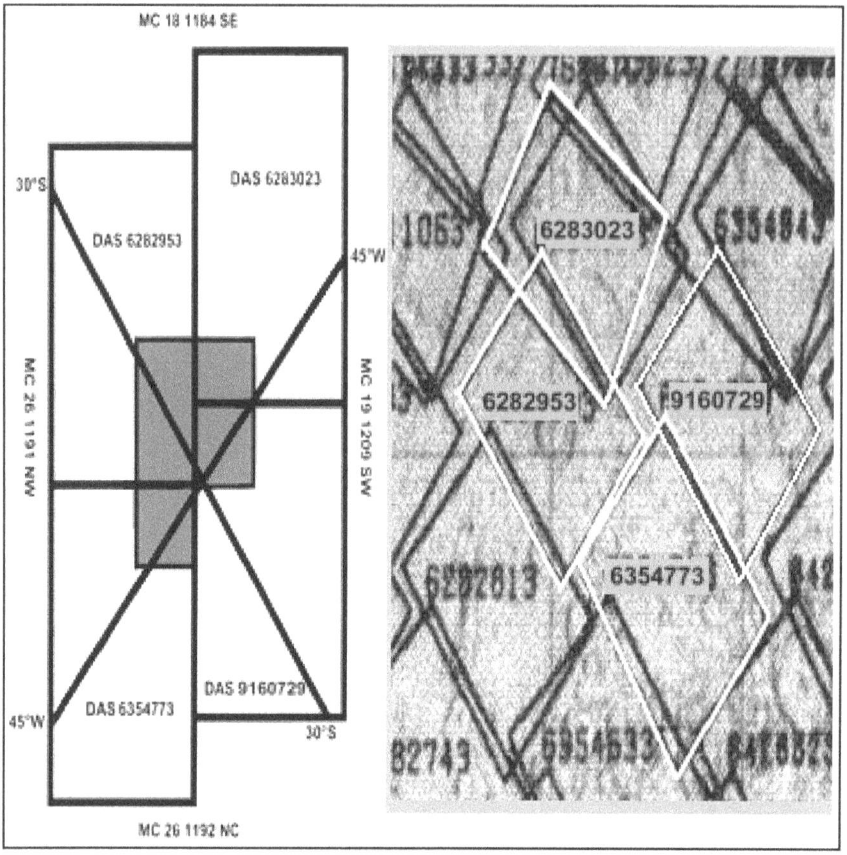

Tab. 164	Nirgal Vallis – Picture of the union of the frames that make up the original image.

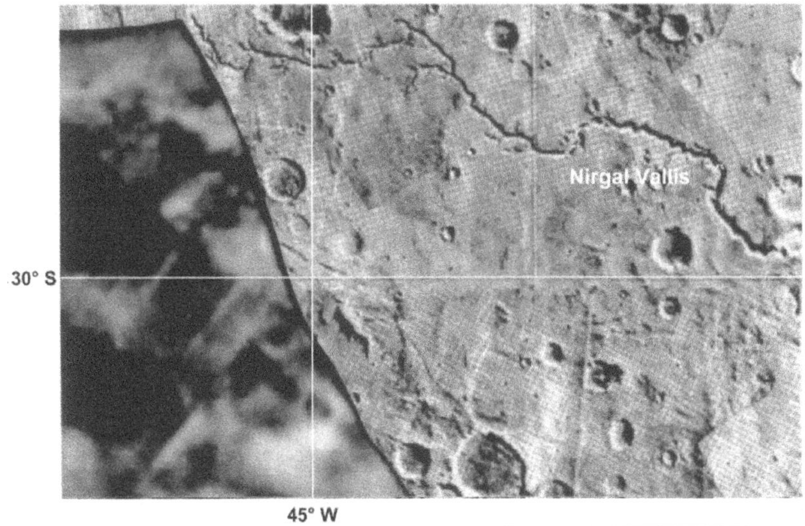

Tab. 165	Nirgal Vallis – Relocation of the structure (original photo) placed adjacent with planetary charts (Tav. 159); coordinates indicated: 45.0 W – 30.0 S.

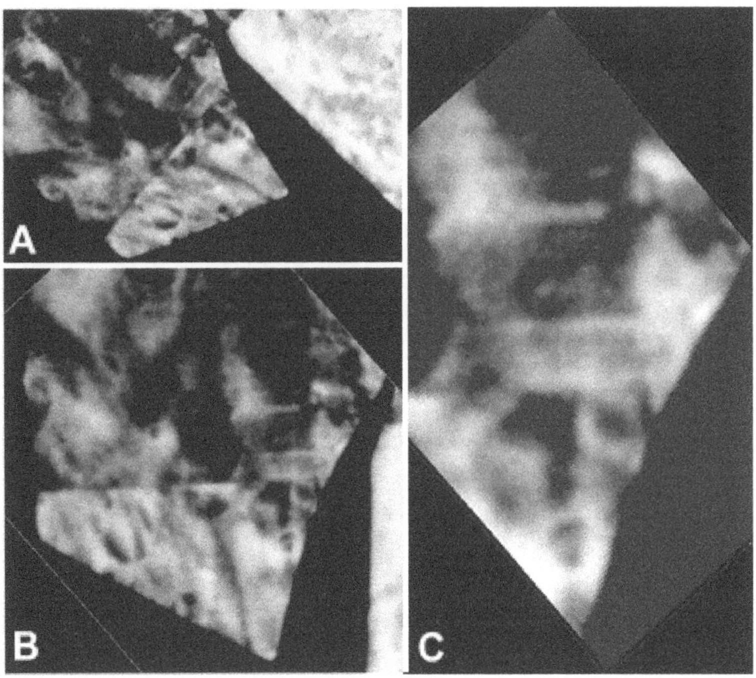

Tab. 166	Nirgal Vallis – Progressive photographic analysis of the original structure. A) Structure taken in the photomosaic portion; B) Straightening of the previous photographic portion; C) Enlargement of the structure.

Mariner 9 Das 6282953

INST	TYPE	TIME (GMT) D H M S	PERI TIME H M S	SPACECRAFT LAT	SPACECRAFT LONG-W	SPACECRAFT HGT	PLATFORM CONE	PLATFORM CLOCK	INTERCEPTING LAT	INTERCEPTING LONG-W	INTERCEPTING RANGE	VIEW ANGLE	LIGHT ANGLE	PHASE ANGLE	DAS REFERENCE TIME
PHOBOS															
1	B	18 10 25 20	-1 33 2	-30.63	175.25	8007	134.70	115.25	******	******	*****	123.71	134.93	.00	6,278,438
GLOBAL															
2	A	18 10 55 14	-1 0 8	-47.66	167.70	5444	96.45	109.82	-74.78	161.72	6022	42.15	76.63	83.62	6,280,083
3	A	18 10 59 38	-0 58 44	-48.54	166.90	5328	96.49	117.95	-60.79	156.99	5479	21.78	73.71	83.58	6,280,153
4	A	18 11 1 2	-0 57 20	-49.42	166.03	5213	96.37	126.83	-48.52	154.35	5263	12.66	71.15	83.70	6,280,223
5	A	18 11 2 26	-0 55 56	-50.32	165.07	5097	96.18	134.95	-37.73	152.23	5300	25.44	68.95	83.88	6,280,293
6	A	18 11 3 50	-0 54 32	-51.23	164.03	4981	96.03	143.83	-23.29	150.38	5687	47.12	67.40	84.07	6,280,363
TLR															
7	B	18 11 19 56	-0 38 26	-61.40	141.91	3653	107.19	121.12	-70.19	81 68	4218	44.75	58.24	72.79	6,281,168
TETRAD															
8	B	18 11 25 32	-0 32 50	-63.62	127.23	3210	131.11	110.93	-60.31	325.03	5437	86.13	93.67	48.88	6,281,448
9	A	18 11 26 14	-0 32 8	-63.77	125.08	3156	130.95	110.86	******	******	*****	90.45	94.17	84.07	6,281,483
10	B	18 11 26 56	-0 31 26	-63.89	122.88	3103	130.80	110.81	******	******	*****	90.18	92.99	48.88	6,281,518
11	A	18 11 27 38	-0 30 44	-63.97	120.62	3050	130.75	110.75	******	******	*****	90.82	91.11	84.07	6,281,553
MAPPING NO 1															
12	A	18 11 38 50	-0 19 32	-58.63	83.07	2285	136.56	182.84	-61.56	93.33	2329	14.45	49.71	43.51	6,282,113
13	B	18 11 39 32	-0 18 50	-57.86	81.02	2245	136.57	182.63	-60.57	89.28	2277	12.48	48.61	43.42	6,282,148
14	A	18 11 40 14	-0 18 8	-57.03	79.04	2205	136.50	182.48	-59.18	84.53	2222	9.10	47.33	43.57	6,282,183
15	B	18 11 40 56	-0 17 26	-56.16	77.13	2166	136.41	182.43	-57.90	81.27	2177	7.27	46.32	43.58	6,282,218
16	A	18 11 41 38	-0 16 44	-55.24	75.28	2128	136.32	182.39	-56.27	77.39	2132	4.04	45.24	43.75	6,282,253
17	B	18 11 42 20	-0 16 2	-54.28	73.51	2092	136.27	182.36	-54.84	74.64	2093	2.26	44.42	43.72	6,282,288
18	A	18 11 43 2	-0 15 20	-53.28	71.79	2057	136.23	182.33	-53.05	71.37	2057	.90	43.59	43.84	6,282,323
MAPPING NO 2															
19	A	18 11 50 2	-0 8 20	-41.19	57.95	1777	136.97	225.85	-46.74	64.06	1851	20.09	41.65	43.10	6,282,673
20	B	18 11 50 44	-0 7 38	-39.81	56.84	1758	137.02	225.88	-44.88	61.83	1816	17.90	41.50	42.97	6,282,708
21	A	18 11 51 26	-0 6 56	-38.41	55.78	1740	137.07	225.89	-42.51	59.64	1778	14.67	41.25	43.00	6,282,743
22	B	18 11 52 8	-0 6 14	-36.99	54.76	1723	137.14	225.96	-40.62	57.75	1751	12.66	41.38	42.85	6,282,778
23	A	18 11 52 50	-0 5 32	-35.55	53.77	1709	137.22	225.99	-38.23	55.88	1724	9.41	41.46	42.84	6,282,813
24	B	18 11 53 32	-0 4 50	-34.09	52.82	1695	137.30	225.89	-36.23	54.22	1705	7.26	41.82	42.69	6,282,848
25	A	18 11 54 14	-0 4 8	-32.61	51.90	1684	137.36	225.78	-33.77	52.60	1687	3.90	42.18	42.71	6,282,883
26	B	18 11 54 56	-0 3 26	-31.12	51.01	1675	137.31	225.77	-31.74	51.22	1675	1.95	42.72	42.68	6,282,918
■ 27	A	18 11 55 38	-0 2 44	-29.61	50.15	1667	137.25	225.76	-29 25	49.86	1667	1.30	43.30	42.82	6,282,953 ■
28	B	18 11 56 20	-0 2 2	-28.09	49.32	1661	137.19	225.76	-27.20	48.60	1663	3.32	44.06	42.80	6,282,988
29	A	18 11 57 2	-0 1 20	-26.56	48.52	1656	137.15	225.74	-24.69	47.36	1664	6.51	44.89	42.92	6,283,023
DYAD NO 1															
30	A	18 12 6 50	0 8 27	-4.97	39.24	1781	101.97	246.20	.69	53.68	2108	40.92	43.27	78.10	6,283,513
31	B	18 12 7 32	0 9 9	-3.47	38.67	1803	102.02	246.25	2.73	52.70	2122	40.30	45.02	77.96	6,283,548
DYAD NO 2															
32	A	18 12 16 38	0 18 15	14.60	31.90	2211	106.86	242.49	45.69	45.56	3325	66.97	74.30	73.21	6,284,003

JPL TECHNICAL MEMORANDUM 33-585, VOL. II

6-285

Tab. 167 Nirgal Vallis; Photo Das 6282953 – Detailed satellite survey information.

Mariner 9 Das 6354773

INST	TYPE	TIME (GMT) D H M S	PERI TIME H M S	SPACECRAFT LAT	SPACECRAFT LONG-W	HGT	PLATFORM CONE	PLATFORM CLOCK	INTERCEPTING LAT	INTERCEPTING LONG-W	INTERCEPTING RANGE	VIEW ANGLE	LIGHT ANGLE	PHASE ANGLE	DAS REFERENCE TIME
PHOBOS															
1	A	19 10 25 56	-1 30 8	-31.86	165.60	7794	109 32	164.81	******	******	*****	130.72	108.70	.00	6,350,468
GLOBAL															
2	A	19 10 56 2	-1 0 2	-47.71	158.17	5435	95 19	109 86	-74.51	154.61	5999	41.52	77.63	83.89	6,351,973
3	A	19 10 57 26	-0 58 38	-48.58	157.37	5320	96 30	117.82	-60.95	148.84	5466	21 46	74.88	83 77	6,352,043
4	A	19 10 58 50	-0 57 14	-49.47	156.49	5204	96.43	126.85	-48.51	146 30	5243	11.11	72.70	83 63	6,352,113
5	A	19 11 0 14	-0 55 50	-50.37	155.53	5089	96 44	135.23	-37 43	144.26	5283	24.95	70.81	83.63	6,352,183
6	A	19 11 1 38	-0 54 26	-51.28	154.49	4972	96.35	144.37	-22.39	142.35	5701	47.87	69.50	83 71	6,352,253
TLR															
7	B	19 11 17 44	-0 38 20	-61.45	132.24	3645	120.42	118.12	-85.95	110 41	4244	46.31	75.80	59.57	6,353,058
TETRAD															
8	B	19 11 23 20	-0 32 44	-63.65	117.48	3202	134.63	126.73	-85.80	164.97	3744	45.02	78.25	45.36	6,353,338
9	A	19 11 24 2	-0 32 2	-63.80	115.33	3149	134.69	126 71	-87.49	136 97	3702	45.62	77.05	45.38	6,353,373
10	B	19 11 24 44	-0 31 20	-63.91	113.11	3095	134.76	126.71	-86.96	101.59	3620	44.66	75.51	45 23	6,353,408
11	A	19 11 25 26	-0 30 38	-63.98	110.85	3042	134 84	126.66	-86.14	64.69	3587	45.60	74.45	45.23	6,353,443
MAPPING NO 1															
12	A	19 11 36 38	-0 19 26	-58.54	73.30	2280	136.49	181.85	-61.40	83 06	2321	13 89	49.75	43.58	6,354,003
13	B	19 11 37 20	-0 18 44	-57 76	71.27	2239	136.53	181.64	-60.39	79 06	2269	11.93	48 63	43.49	6,354,038
14	A	19 11 38 2	-0 18 2	-56.93	69.29	2200	136 45	181.55	-58.99	74 42	2215	8.62	47.34	43.63	6,354,073
15	B	19 11 38 44	-0 17 20	-56.05	67.39	2161	136.43	181.56	-57 74	71.23	2171	6 88	46 35	43 50	6,354,108
16	A	19 11 39 26	-0 16 38	-55.13	65.55	2124	136.37	181.60	-56 13	67 44	2127	3 78	45.28	43 70	6,354,143
17	B	19 11 40 8	-0 15 56	-54.16	63.78	2087	136.36	181.65	-54 73	64.75	2088	2.10	44.46	43.63	6,354,178
18	A	19 11 40 50	-0 15 14	-53.15	62.08	2052	136.35	181.70	-52 95	61 55	2052	.99	43.60	43.73	6,354,213
MAPPING NO 2															
19	A	19 11 47 50	-0 8 14	-41.02	48.31	1775	137.53	224.97	-46.40	54.14	1844	19 44	41.46	42.59	6,354,563
20	B	19 11 49 32	-0 7 32	-39.64	47.21	1756	137 41	225.06	-44.53	52 07	1811	17.48	41.23	42.58	6,354,598
21	A	19 11 49 14	-0 6 50	-38.24	46.15	1738	137.33	225.26	-42.21	50.02	1775	14.41	40.91	42 74	6,354,633
22	B	19 11 49 56	-0 6 9	-36.81	45.13	1722	137 26	225.49	-40.34	48.26	1750	12.62	40.95	42 73	6,354,668
23	A	19 11 50 38	-0 5 26	-35.37	44.15	1707	137.23	225.74	-38.00	46.52	1724	9.64	40 94	42 87	6,354,703
24	B	19 11 51 20	-0 4 44	-33.91	43.20	1695	137 13	225 80	-36.04	44.97	1705	7.69	41.21	42.86	6,354,738
■ 25	A	19 11 52 2	-0 4 2	-32.43	42.28	1683	137.07	225.78	-33.58	43 42	1687	4.51	41.47	43.00	6,354,773 ■
26	B	19 11 52 44	-0 3 20	-30.93	41.40	1674	137.15	225.67	-31.54	41.94	1675	2 32	42 05	42 84	6,354,808
27	A	19 11 53 26	-0 2 38	-29.42	40.55	1667	137.26	225.57	-29.06	40.48	1667	1.10	42.68	42.81	6,354,843
28	B	19 11 54 8	-0 1 56	-27.90	39.72	1661	137.40	225.47	-27.00	39.11	1662	3.20	43.52	42 59	6,354,878
29	A	19 11 54 50	-0 1 14	-26 37	38.92	1657	137.53	225.36	-24.47	37.77	1664	6.57	44.42	42.54	6,354,913
DYAD NO 1															
30	A	19 12 4 38	0 8 33	-4.79	29 67	1785	96 47	251.61	-2.68	48.63	2250	48.31	37.33	83.60	6,355,403
31	B	19 12 5 20	0 9 15	-3.29	29.11	1807	96 41	251 58	-.68	47 57	2260	47 23	39.09	83 59	6,355,438
DYAD NO 2															
32	A	19 12 14 26	0 18 21	14.75	22.35	2218	109.78	245.71	40.31	30.46	2986	56.96	74.13	70.29	6,355,893

JPL TECHNICAL MEMORANDUM 33-585, VOL II — 6-289

Tab. 168 — Nirgal Vallis: Photo Das 6354773 – Detailed satellite survey information.

Mariner 9 Das 6283023

INST	TYPE	TIME (GMT) D H M S	PERI TIME H M S	SPACECRAFT LAT	SPACECRAFT LONG-W	HGT	PLATFORM CONE	PLATFORM CLOCK	INTERCEPTING LAT	INTERCEPTING LONG-W	INTERCEPTING RANGE	VIEW ANGLE	LIGHT ANGLE	PHASE ANGLE	DAS REFERENCE TIME
PHOBOS															
1	B	18 10 25 20	-1 33 2	-30.63	175.25	8007	134.70	115.25	******	******	*****	123.71	134.93	.00	6,278,438
GLOBAL															
2	A	18 10 59 14	-1 0 8	-47.66	167.70	5444	96.45	109.82	-74.78	161.72	6022	42.15	76.63	83.62	6,280,083
3	A	18 10 59 38	-0 58 44	-48.54	166.90	5328	96.49	117.95	-60.79	156.99	5479	21.78	73.71	83.58	6,280,153
4	A	18 11 1 2	-0 57 20	-49.42	166.03	5213	96.37	126.83	-48.52	154.35	5263	12.66	71.15	83.70	6,280,223
5	A	18 11 2 26	-0 55 56	-50.32	165.07	5097	96.18	134.95	-37.73	152.23	5300	25.44	68.95	83.88	6,280,293
6	A	18 11 3 50	-0 54 32	-51.23	164.03	4981	96.00	143.83	-23.29	150.38	5687	47.12	67.40	84.07	6,280,363
TLR															
7	B	18 11 19 56	-0 38 26	-61.40	141.91	3653	107.19	121.12	-70.19	81.68	4218	44.75	58.24	72.79	6,281,168
TETRAD															
8	B	18 11 25 32	-0 32 50	-63.62	127.23	3210	131.11	110.93	-60.31	325.03	5437	86.13	93.67	48.88	6,281,448
9	A	18 11 26 14	-0 32 8	-63.77	125.08	3156	130.95	110.86	******	******	*****	90.45	94.17	84.07	6,281,483
10	B	18 11 26 56	-0 31 26	-63.89	122.88	3103	130.80	110.81	******	******	*****	90.18	92.99	48.88	6,281,518
11	A	18 11 27 38	-0 30 44	-63.97	120.62	3050	130.75	110.75	******	******	*****	90.82	91.11	84.07	6,281,553
MAPPING NO. 1															
12	A	18 11 38 50	-0 19 32	-58.63	83.07	2285	136.56	182.84	-61.56	93.33	2329	14.45	49.71	43.51	6,282,113
13	B	18 11 39 32	-0 18 50	-57.86	81.02	2245	136.57	182.63	-60.57	89.28	2277	12.48	48.61	43.42	6,282,148
14	A	18 11 40 14	-0 18 8	-57.03	79.04	2205	136.50	182.48	-59.18	84.53	2222	9.10	47.33	43.57	6,282,183
15	B	18 11 40 56	-0 17 26	-56.16	77.13	2166	136.41	182.43	-57.90	81.27	2177	7.27	46.32	43.58	6,282,218
16	A	18 11 41 38	-0 16 44	-55.24	75.28	2128	136.32	182.39	-56.27	77.39	2132	4.04	45.24	43.75	6,282,253
17	B	18 11 42 20	-0 16 2	-54.28	73.51	2092	136.27	182.36	-54.84	74.64	2093	2.26	44.42	43.72	6,282,288
18	A	18 11 43 2	-0 15 20	-53.28	71.79	2057	136.23	182.33	-53.05	71.37	2057	.90	43.59	43.84	6,282,323
MAPPING NO. 2															
19	A	18 11 50 2	-0 8 20	-41.19	57.95	1777	136.97	225.85	-46.74	64.06	1851	20.09	41.65	43.10	6,282,673
20	B	18 11 50 44	-0 7 38	-39.81	56.84	1758	137.02	225.88	-44.88	51.83	1816	17.99	41.50	42.97	6,282,708
21	A	18 11 51 26	-0 6 56	-38.41	55.78	1740	137.07	225.89	-42.51	59.64	1778	14.67	41.25	43.00	6,282,743
22	B	18 11 52 8	-0 6 14	-36.99	54.76	1723	137.14	225.96	-40.62	57.75	1751	12.66	41.38	42.85	6,282,778
23	A	18 11 52 50	-0 5 32	-35.55	53.77	1709	137.22	225.99	-38.23	55.88	1724	9.41	41.46	42.84	6,282,813
24	B	18 11 53 32	-0 4 50	-34.09	52.82	1695	137.30	225.89	-36.23	54.22	1705	7.26	41.82	42.69	6,282,848
25	A	18 11 54 14	-0 4 8	-32.61	51.90	1684	137.36	225.78	-33.77	52.60	1687	3.90	42.18	42.71	6,282,883
26	B	18 11 54 56	-0 3 26	-31.12	51.01	1675	137.31	225.77	-31.74	51.22	1675	1.95	42.72	42.68	6,282,918
27	A	18 11 55 38	-0 2 44	-29.61	50.15	1667	137.25	225.76	-29.25	49.86	1667	1.30	43.30	42.82	6,282,953
28	B	18 11 56 20	-0 2 2	-28.09	49.32	1661	137.19	225.76	-27.20	48.60	1663	3.32	44.06	42.80	6,282,988
■ 29	A	18 11 57 2	-0 1 20	-26.56	48.52	1656	137.15	225.74	-24.69	47.36	1664	6.51	44.89	42.92	6,283,023 ■
DYAD NO. 1															
30	A	18 12 6 50	0 8 27	-4.97	39.24	1781	101.97	246.20	.69	53.68	2108	40.92	43.27	78.10	6,283,513
31	B	18 12 7 32	0 9 9	-3.47	38.67	1803	102.02	246.25	2.73	52.70	2122	40.30	45.02	77.96	6,283,548
DYAD NO. 2															
32	A	18 12 16 38	0 18 15	14.60	31.90	2211	106.86	242.49	45.69	45.54	3325	66.97	74.30	73.21	6,284,003

JPL TECHNICAL MEMORANDUM 33-585, VOL. II — 6-285

Tab. 169 Nirgal Vallis: Photo Das 6283023 – Detailed satellite survey information.

Mariner 9 Mission (1971–1972)

DETAILED EXPLANATIONS OF THE TABLE HEADINGS

Frame Number: The first column of the tables contains the frame number. The frame number varies from 1, the first TV picture taken on a given orbit, to approximately 33, the tape recorder capacity. TV frames are numbered consecutively regardless of whether they are A or B frames.

A. Instrument Type: The instrument type refers to either TV-A (wide angle) or TV-B (narrow angle), denoted as "A" or "B" in the table.
B. Frame Time: The frame time is the GMT when the individual frame was shuttered aboard the spacecraft, in days, hours, minutes and seconds. Days are measured consecutively from 1 January.
C. Periapsis Time: The periapsis time is the time that the individual frame was shuttered in hours, minutes and seconds, as referenced to the periapsis of the orbit during which the picture was taken. A negative sign indicates a time before the periapsis passage, a positive value indicates time after periapsis passage.
D. Spacecraft position: The point on the planet directly below to spacecraft as the picture is shuttered, specified by the areocentric latitude and west longitude. in addition, height above the surface of the planet in kilometres is given.
E. Scan Platform Position: The position of the scan platform for each TV picture is specified by the cone and clock angles.
F. Centre point Intercept Position: The position of the TV frame on the planet is specified in the table by the centre point intercepting latitude and west longitude. in addition, the range from the spacecraft to the centre point intercepting position is given in kilometres. For limb, satellite astronomy and star calibration pictures there may be no planetary intercepting position. in that case, asterisks are inserted into these fields.
G. View Angle: The viewing angle is defined as the angle between a normal vector at the intercepting point and the intercepting point-spacecraft vector. Thus, with a viewing angle of 0° the scan platform is looking directly toward the centre of the planet. A 90° viewing angle means the centre point is intercepting the limb of the planet. A viewing angle of greater than 90° means the scan platform is pointed away from the planet, i.e., the TV frame centre point is non-intercepting. As stated in the previous para-

graph, asterisks will be inserted in the centre point, intercepting latitude, longitude and range when the centre point is non-intercepting.

H. Lighting Angle: The lighting angle is defined as the angle between a normal vector at the intercepting point and the Mars-sun vector (i.e.: the zenith distance of the sun). A lighting angle of 0° means the sun is at the zenith at the intercepting point. A 90° lighting angle means the intercepting point is on the terminator (sun rising or setting).

I. Phase Angle: The phase angle is defined as the angle between the Mars-sun vector and the intercepting point-spacecraft vector.

J. Das Reference Time: The Data Automation Subsystem (Das) reference time is the integer number of pulses of the DAS clock from the time the clock was started (when science power was turned on) to the point at which the individual TV frame was shuttered.

Conclusions

(*Truth is not a Destination: Truth is a Path*)

As far as we know, no one on earth has already been contracted for the exclusive use of science and this "right-to-use" does not even only belongs to those who, in fact, spend their lives in practicing it.

In our opinion, the only basic (better yet: fundamental) condition required to practise the science correctly and worthily, can only be found by working in accordance with the principle of a strict and accurate adherence to the *Experimental Scientific Method*; so, therefore, we must (or should, at least) refer to the observation, then the making of one or even more hypotheses and finally work with the aim of verifying its/their validity.

In this context, only by acquiring data, analysing them, hypothesising a reasonable and linear explanation, making an experiment, verifying the results, repeating it and then continuing this process until our craving for perfection is duly satisfied: well, we believe that only this whole process can be rightly called "science".

Now, according to this rigorously *Scientific Perspective*, for example, Mars should not be considered something other than what is usually meant by the term Earth, therefore a planet.

This is what should righteously be called, in our opinion, a *True Scientific Approach.*

On the contrary, the process of collecting a certain number of information and data that have already been packaged, considered, analysed, tested, defined, et cet. And then assume them all as *true* only because they came from *Influential Entities* (consider, for instance, many public and private institutions, professors, scientists, space agencies and so forth, all known and considered like the true *source of scientific authority*), well, this is, instead, just "faith".[152]

This means, according to the *True Scientific Approach* that we have proposed hereabove, that Mars should be studied for what it is and not as it has been imagined and interpreted or is still being imagined and interpreted right now, such as a *mysterious astronomical object*.

As a matter of fact, Mars might have been considered as such (i.e.: *a mysterious astronomical object*) just until the beginning of the Space Era, when it was only possible to observe it telescopically, from Earth.

After that moment and thanks to the acquisition of a great number of photographic material (which arrived on Earth starting from November 1971), Mars has become a relatively common object of study, at least as much as the Earth is.

It seems superfluous to add that all the geophysical observations related to Mars so far, such as its gravity, meteorology, geomorphology, etc., indicate that Mars is a place where the laws of physics, jointly with the principles of geomorphology as we conceive them, can and must be applied with just a few minor *calibrations* which are basically due to its gravity, atmospheric composition and density, as well as its distance from the sun.

It's just plain *nonsense* to assume that there is only one (or just a few) small places on the surface of Mars where these very same principles of geomorphology are not and cannot be considered as valid.

And it should also be well remembered that the above-mentioned consideration is nothing more than the practical application of the first principle of the Copernican revolution, which extended the physical laws already tested on our home planet Earth to the whole universe.

In order to better understand this tough subject, let's first ask us what the results of the studies of Mars have shown (we would not say *proved* or *fully demonstrated*) so far and whether or not the presence of "life" (as we know it) has been ascertained (and, possibly, in what form/s).

At this moment, it can be said that, from the analysis of all the satellite and surface images relating to the Mariner 9 (1971–72) and Viking 1 and 2 (1975–83) missions, as well as from the study conducted in relation to the biological experiments carried out by the equipment located on the landers of the two Viking probes that reached the surface of Mars, it was possible to infer that:

1) Mars once (but remember: we cannot say for sure "how long ago"!) had a geomorphological structure with a hydrogeological cycle identical (or at least very similar) to the one of the earth;

[152] This does not imply an overestimation of science and a disavowal of faith, but concerns a functional distinction where faith has its own dignity, which is not the subject of discussion here from any point of view.

2) Biological experiments have shown that "life" (still, as we know it) on that planet has never fully ceased to exist;
3) On tens of thousands of both satellite/orbital and terrestrial[153] images, it was possible (and remember: the study is still ongoing!) to detect some (we say) "clear evidence" of both past and most likely, still present *intelligent life.*[154]

Now, even though this is the "simple" reality (we mean: this a "fact"!), on one side there is an *Official Science* that uses the institutions set up for this purpose and conducts research according to methodologies that do not apply the *True Scientific Approach* as we named it before and, likewise, on the other side there are individuals and, rarely, groups, which fully use and respect the true scientific canons by systematically using the *Principle of Experimentation.*

§§§

This book represents a synthesis and a first organic attempt to correctly carry out a deep study of the planet Mars, that has been conducted with the decisive support of many satellite (orbital) images, which can be analysed in technical terms and with solid photogrammetric values.

The study, still ongoing, has been so far conducted by using something like 61.340 orbital images (of which 57.357 taken from orbit and 3.983 from the surface by the two Viking landers).

On the basis of the early results, deduced from an accurate geomorphological analysis of the planet, the existence in the past of seas, lakes, rivers and continents (for instance: Icarus, Elysium, Boreum and so forth) was hypothesised and over 100 Regions of, let's say, "archaeological interest" have been identified on the last two (Elysium and Icarus), with about 200 surface structures "hypothetically" artificial (or simply "non-fully natural", if you prefer).

The complete results of this research will be systematically presented in another book that shall follow, written by the same authors of this one.

---

[153] The term terrestrial refers to an image taken by a module placed on the ground; in this case on the Surface of Mars.

[154] See 'Atlas of Mars' by Gianni Viola (forthcoming).

SCIENTIFIC GLOSSARY

ALBEDO FEATURE. The term refers both to the capacity of reflectance of a given surface, with regard to the solar rays (or any other *lighting incidental source*), as well as to a certain portion of a celestial body's surface having a major or minor reflectance—as observed through a telescope and, obviously, from a very long distance. in general, a surface that reflects all the *incidental light* has Albedo 1, while a surface that absorbs all the *incidental light* has Albedo 0 (zero).

ALTITUDE. Altitude level 0 (zero—or "Datum") is the estimated level of a water surface (or some other liquid whatsoever), possibly present on a celestial body at the present time (like on Earth and the Saturnian moon Titan) or in the past (even a very distant one).

ANALOGUE PHOTO. Photographic format obtained by the recording of the visible light that is the wavelengths of the electromagnetic spectrum combined with the film photosensitive emulsion. in the analogue format, all data and information are saved in a continuative way and they are present in the photograph's extension as a whole without interruptions or detachment between the portions of the image itself.

ANALYSIS OF IMAGES. It consists of a *qualitative phase* obtained by the *photo interpretation* (see) that takes into consideration the nature of detected images examining the natural or artificial character and a *quantity phase* determined by a *digital elaboration* (see) that takes into consideration the images to evaluate their fake or the true nature (*quality of images*, see) and also applying a number of resolution parameters (see *optical or geometrical resolution, radiometric r., topographic r.*).

AREOGRAPHY. From Ares (Greek: Mars), means the scientific study of the planet Mars' surface (general term: *Planetography*, see).

ASTRONOMICAL OBSERVATORIES. Before the space age, so till the sixties—If we include the first planetary missions that resulted partially successful—the astronomical research was conducted exclusively through the optical instruments located on the Earth's surface. These are the spyglasses and the telescopes that are found in the *astronomical observatories*. in the observatories, three types of astronomical research can be carried out: the *Astronomy of Position* (see), the *Celestial Mechanics* (see) and the *Astrophysics* (see).

ASTRONOMITE. Disease that causes the subject to apply the principles of astronomy inappropriately, i.e. in areas where astronomy has no relevance. For example, in the analysis of the surface of planets, the study of which is the sole competence of planetography. The latter corresponds, in planetary terms, to terrestrial geography and applies the same methodologies applied to the study of the Earth's surface. A check on the pernicious existence and persistence of astronomy reveals a series of false studies where planets are not studied through the analysis of satellite surveys (as is done in relation to the Earth), but rather through a few useless and inconsistent images taken by descent modules that detect planetary features that are completely decontextualised from the surface complex of a given planet. In particular, the pathology in question appears in many false studies referring to the planet Mars.

ASTROPHILY. Aficionados of one of the branches of the vast field of astronomy, in particular of position astronomy (or astrometry, v.), receive unqualified approval from institutional research, if only because, in the past, some of the greatest discoveries made in the field of astronomy were made by amateur astronomers. In particular, a specific term has been coined for the latter, namely 'astronomers', who, according to the Encyclopaedia of Astronomy and Cosmology, are defined as: 'Not so much astronomy enthusiasts as amateurs who practise astronomical observation (i.e. position astronomy)".[155]

ASTROPHYSICS. It studies the physical processes associated with celestial bodies; this discipline extends the physical laws experienced on Earth (in relation to a principle established by Copernicus) to any celestial body. It also studies the chemical composition of stars by detecting specific types of radiation (light and radio waves) from a given celestial body and then using measurements of their quantity (Photometry) and quality (Spectroscopy). In 'university' terms, the Faculties of Astronomy and Astrophysics now tend to be - in terms of content - essentially identical.

ASTROPHYSICAL TECHNIQUES. It studies the physical and chemical nature of celestial bodies and operates through the spectrographic analysis of their light.

BIOPIC ACTIVITY. Any activity linked/connected to the presence of (some kind of) life.

CANCELLATION OF A PHOTOGRAM. The *cancellation* of a photogram is a graphic elaboration to delete in full all the structures detected by previous missions and the topographic context where these structures are inserted. The result is the general modification of the ground morphology in hundreds of kilometres using the *overlapping* of an image or of another element in order to hide the

[155] John Gribbin, Enciclopedia di Astronomia e Cosmologia, Garzanti Editore, Milano, 1998; p. 25.

given image and make it even invisible (but only to non professional people) exchanging the original image for another *built* on purpose.

Another cancellation technique is the overlapping of several vertical bands so close one to another to prevent the identification of a certain structure put in the portion of a surface.

CELESTIAL MECHANICS. Branch of the astronomy that studies the movement of celestial bodies and the forces that determine them; it is basically based on Newton's law of universal gravitation and Einstein's relativistic mechanics. The sector of "celestial mechanics" united to the *Astrometry* (see) represents the most popular part of astronomy and it is known as *Classical Astronomy.*

COGNITIVE DEPRIVATION. Taking into account the cognitive deprivation recorded in various investigations carried out by the writer, a 'clinical' synthesis includes the presence of widespread psychosis, which is characterised as loss of contact with reality, alteration of inter-individual ties and habitual unawareness of disturbances on the part of the subjects themselves. In the various 'characters' interviewed, in particular, a form of chronic psychosis, known as paranoia, characterised by the slow and progressive construction of a delusional system, which in this case is expressed in the best-known form of the disorder, known as 'delusions of grandeur', was recognised. From this generally stems the prevailing, basic misconception that invades the entire affective (and relational) ideal system, as well as the ritual, where the subject repeats (with acts and words) a series of actions always in the same order.

COSMOGONY. It can be understood both as a myth and a doctrine, but also as a poem that gives an interpretation of the origins and formation of the universe and a branch of the *astronomy* (see) that studies both the formation of celestial bodies and their possible evolution.

COSMOGRAPHY. It can be considered as a branch of *astronomy* (see) because it describes the "objects in the sky" that is "astronomical geography" (see *Planetography*).

COSMOLOGY. It can be intended as the science that studies the formation, the structure and the evolution of the universe or as a mere philosophical and theosophical speculation related to the origins and finalities of the cosmos.

DEFINED RESOLUTION (RD). The *defined resolution* takes into account satellite surveys of the Earth since 1972, by the Landsat satellites. Given that the Earth structures surveyed are part of our direct cultural heritage (because they are directly experienced through visual observations on site or visualisations of photographs), it is possible to obtain an optimal topographic resolution level by analysing the photographic quality of the several surveyed structures. From this detailed examination we obtained the topographic resolution level of 0.26, to be used as a fixed number to relate to the dimensions of each surveyed structure. This will provide an optimal level of geometric resolution called 'defined resolution' (DR). The DR is therefore a "relative" (also called "favourable") ratio between the dimensions of the surveyed structures and the geometric resolution obtained. The formula to be applied is D: 0.26, where "D" represents the maximum measurable linear dimension on a given structure and "0.26" is the fixed divisor number.

DESCENT MODULE. Autonomous Module (see *lander*) that descends to the ground of a given Planet, after the detachment from the Orbiting Module (see *Orbiter*) or from any kind of automatic spacecraft or probe (see *Unmanned Spacecraft*). (Tab169A-330)

DETECTION THRESHOLD. Minimum distance to observe a satellite image according to terms photogrammetrically relevant and where it is possible to take precise measurements.

DIGITAL ELABORATION. It is, at the same time, the images' analysis as well as the photographic process of monitoring; their quantitative transformation is carried out in order to verify the quality of the picture (see *Images' Quality*), as well as to detect the possible transformations that can occur and compromise the correspondence between what the *Photographic Analyst* (such as the person whose work is to interpret the content of a picture) can see in a frame and the real/authentic scene that has been captured by the camera (see also *Cancellation of an Image* and *Modification of an Image*).

DIGITAL PHOTO. Photographic format derived from images and information regarding the non-visible bands of the visible spectrum. This format saves every portion of information separately with elements organised in lines and columns in order to form a bi-dimensional matrix and represents a structure of space data in the form of a numeric matrix called *raster.*

Using this method and magnifying the image, it is possible to distinguish several squares (*pixels*) of different colours and brightness (qualities indicated by a number representing the medium radiance of a small area called the elemental cell). The different rules that define the organisation of data create the so-called *format of digital images.*

DIMENSIONS OF DETECTED STRUCTURES. The longest side and the diagonal are taken into consideration when structures are detected on the ground.

EXOBIOLOGY (synonyms: astrobiology or bio-astronomy). Study of hypothetical extraterrestrial life forms still existing or existed.

EYE-PIECE. Lens or system of lenses that in optical instruments (*telescopes, microscopes,* etc.) allow the direct and unequivocal observation of the image given by them.

FOCAL LENGTH. It is the distance measured along the optical axis from the point at the back of the lens to the critical focal plane of a very distant object and that determines the capacity of magnifying the optics in use.

FOOTAGE DISTANCE. It is the distance between the camera sensor on board of a satellite and what is detected on the Earth, also named *slant range* or *range to surface* or *intercepting range.*

FRAME. Complete single element results of a photographic terrestrial or satellite footage; the modulated combination of several frames is called *photomosaic* (see). The sequential combination of more frames related to the same area and captured in a relatively short period of time can originate a GIF movie. The single images (the so-called pristine or "raw frames" and in the case of mission Viking *Individual Viking Orbiter Image*) are individual frames originally *raw* that means without any geometrical corrections (see *optical resolution*) and radiometric (see *radiometric resolution*).

GENERAL ASTRONOMY. Scientific study of universe and everything in existence beyond the Earth's atmosphere. It is the science that helps us to understand in a mere *descriptive* way the universe and the elements (we actually know) and the elements that compose it (i.e. planets, asteroids, comets, bolides, single stars, multiple stars, constellations, asterisms, nebulae—in all their variants—galaxies—idem, etc.); their behaviour (as seen/investigated on the Earth or using space telescopes and automatic probes), their shape, their colours, some physical characteristics, their movements and some of the laws that rule them. The instrument to investigate it is the *telescope* (see).

GEODETHIC. Line (properly "curve") of shorter length that unites two points of a given "space".

GEOMETRIC AND OPTICAL RESOLUTION. The *geometric resolution* represents the ability to detect separately and subsequently to distinguish very small details on the ground and it is represented by the pixels. It is measured in metres. Properly speaking, it is the ability of the camera to resolve into two or more surface objects of bodies that - only apparently - appear to form a unitary element. An example is the eyes of a subject observed at a certain distance, which can only be distinguished when observed from very close up, whereas from a distance they appear to be united.

LANDER. Descent Module.

LENS. A single lens or a group of lenses that, in an optical system, gives a real image of an observed object. The image can be directly observed (through a *spyglass* and *telescope,* see) or indirectly with the aid of cameras or filming tools and eventually projected on a screen or/and visualised on a computer.

MODIFICATION OF THE FRAME. The methodologies of modification of an image it is possible to distinguish in *intervention sectors* and *used typologies.* As regards the first, they include *analytical operations* that examine an isolated image (*pristine* images see) or synthetic operations that deal with a group of images (photomosaic, see). The *modification* of the frame can be obtained through a geometrical elaboration of the resolution (see) and also through the variation of dimensions, increasing or diminishing the width of the sides in relation the ones with the others.

MODIFIED LANDSCAPE (or ARTIFICIAL). The "artificial" landscape takes into consideration the transformations produced by an intelligent civilisation most likely of humankind. The artificial causes are said "human causes" because they are caused by humans and considering that there are some non-natural modifications observed on other celestial bodies (as in the case of planet Mars), we refer to "intelligent causes". As regards to the Earth, (and hypothetically to planet Mars), we refer to an intervention of mankind on the environment with the making of *anthropic elements* such as houses, roads, dams, canals and bridges. The result is the so-called "modified landscape".

MOSAICATURE. A set of operations that allows a series of individual frames to be joined together, creating an organic continuity.

MOSAICKED (images). See *Photomosaic.*

NATURAL LANDSCAPE. The "natural" landscape takes into consideration elements before human intervention that does not depend on human action. Natural causes can provoke violent and sudden changes known as natural catastrophes such as floods, earthquakes, volcanic eruptions and tornadoes or changes very slow and not noticeable such as the flowing of water that provokes the rocks' erosion and the transportation of sand and debris that pile up in a valley.

OPTICAL OR GEOMETRIC RESOLUTION. Minimal distance where two objects appear distinct and separate in the image.

OPTICAL RESOLUTION. The ability of the camera to resolve into two or more surface objects bodies which - only apparently - appear to form a unitary element. An example of this are the eyes of a subject observed at a certain distance, which can only be distinguished when observed from very close up, while from a distance they appear to be united.

ORBIT. The curve taken either by a natural celestial body (*planet* see, *satellite* see) or an artificial one (*unmanned spacecraft*—see). The natural celestial bodies of the solar system normally have elliptical orbits.

ORBITAL ALTITUDE. It is the altitude of a man-made satellite from the surface at the time when the picture was taken.

ORBITER. The Orbiting Module of a spacecraft.

ORBITING MODULE. Orbiting Module (or Orbiter) of an automatic spacecraft (see also *Unmanned Spacecraft*) that is without a human crew.

PAREIDOLIA. Subconscious illusion that tends to connect unknown objects of uncertain shape to known objects with known shapes and/or profiles. It is an instinctive trend of our brain that deciphers everything by connecting to familiar shapes a series of images that are, somehow, "confused" and "confusing" for both our eyes and mind. This phenomenon has nothing to do with the close observation (see *satellite photographic image*) of structures present on the planets where the principle that rules them is the *satellite photogrammetry* (see).

PHOTOGRAMMETRY. Metric technique that allows the acquisition of metric information of an object (its shape and position) through the acquisition and study of a couple of photograms analysed using the so-called solid geometry (see *satellite Aero-photogrammetry*).

PHOTOGRAPHY. Photography is the transfer of a real object in an image ("still"). The result is a bi-dimensional image that contains information related to the investigated phenomenon or process. Photography makes a clear distinction between *Analogue* (see) and *Digital Photos* (see).

PHOTO INTERPRETATION. It is a technique that allows experts in *Planetography* (see) to detect, determine and classify territorial phenomena and natural or artificial structures located on a celestial body just by watching and studying aerial photos or images taken from a satellite.

PHOTOMETRY. It is a "branch" of *Radio Astronomy* (see) that works measuring the amount (quantity) and *Spectral Class* of radiations emitted by a star.

PHOTOMOSAIC. Assembling technique that consists in the assemblage of two or more satellite photos in different panes where each photo is joint to the other/s. The result is a mosaic composed of many panes and it is called "photomosaic". This technique provides for an overlapping that creates a composed image of the whole area covered by the portion of the surface detected by the satellite orbiting.

The photomosaics give a global and continuous vision of the territory. The procedure is called "mosaicking".

PIXEL. A "pixel" is the single points of a built digital image (*digital photo*, see) and represents the smallest part of an image that can be detected. Every pixel can be considered in relation to three types of information: the position, the dimension and the brightness.

As regards the *position* in an image, two coordinates are taken into consideration. The first one indicates the line number and the second the number of the matrix' column. The *dimension* is the cell's side that composes the pixel. The dimensions depend on the footage instruments' characteristics but also on the distance between the footage camera and the detected structure and on the dimensions of the structure. The dimensions of the cell indicate the *geometrical resolution*. Every pixel can record a variation in brightness (see *radiometric resolution*) that adheres to the medium radiance observed in the particular interval spectral where the sensor detects the resolution cell on the ground related to a given pixel.

PLANET. A spinning body (made of rock or gases—so-called "gas giants") orbiting around the sun or any other single or multiple star system.

PLANETARY CARTOGRAPHY. The planetary cartography is a plane and bi-dimensional representation that regards the general characteristics (natural or artificial) of the total surface or portions of a planet, based on stereographic projections (see). in the planetary cartography, the photomosaic technique is widely used.

PLANETARY CHARACTERISTICS. There are elements of a planet that can be detected from very far away, thanks to Earth-based (or even orbital) optical instruments (spyglass, telescopes, space telescopes, et cet.) named "Albedo Features" (see) or from a closer distance, thanks to cameras located on board artificial satellites (i.e., satellite cameras) and named Geomorphological or *Topographic* (see) reliefs.

PLANETARY IMAGES. With the term planetary images, we basically mean the structures detected by orbiting satellites (see orbiter). In this case they are also known as orbital images; on the other hand, as regards structures detected by Descent Modules (see lander) we refer to them as images obtained from the surface or terrestrial images.

PLANETARY SCIENCE. Subject that studies the characteristics (for example: atmospheric, mineralogical, morphological et cet.) of any given celestial body. *Planetology* (see) and *Planetography* (see).

PLANETOGRAPHIC COORDINATES. The coordinates on every celestial body regard the longitude on meridian "0" (zero). It is a conventional index with the same function of the Greenwich Meridian on Earth. Conventionally, we distinguish the coordinates that indicate the "pointing" (data of longitude and latitude considered as fixed points) and the coordinates that indicate a huge area (usually an ellipse) located around a certain (given) point.

PLANETOGRAPHY. It is the planetary equivalent of *Cosmography* (see), totally separated from *planetology* (see). It is a study related to the exterior forms and characteristics of a celestial body. This word, very popular until the 19th century, was later abandoned because of the reactionary wave of Antoniadi and Co. Antoniadi refused the direct observations (*Observative Study of the Planets*—see) using optical instruments (*Spyglasses* see, *Telescopes* see) unnecessarily favouring indirect physical studies (the modern Astrophysics, see *Astrophysics—Astrophysical Techniques*).

Recently, the word *Planetography* has been redefined by the Author and appeared on an article published by the magazine *L'Astronomia* in 2005. This ruled the birth of "Modern Planetography" as an independent subject separated from the former *Planetography* (the one before the Space Age). *Planetography* is the first attempt of investigating a planet according to procedures already used to study the planet Earth, just through the observation of the "landscape" (see *Landscape Study*).

PLANETOLOGY. Planetary equivalent of *cosmology* (see). Study of the substance of the chemical and physical characteristics of the internal part of a planet (both gaseous and rocky) internal or external to the solar system. Apart from this meaning that is proper for it, "planetology" has been also used inappropriately to designate the study and the shape of a planet that should instead been investigated by "Planetography" (see).

PLANETS' STUDY. It takes place using the visual observations (Observative Studies), regarding the shape (Morphology) and the Astrophysical Observations (Physical Studies) regarding the substance (Chemistry). Both the Observative and Physical Studies can be carried out in a direct or even an indirect way.

The *direct observative studies* are made using telescopes (see), refractive or reflectors located at the astronomical observatories (see); on the other hand, the *indirect observative studies* are carried out through the *space observatories (*see) that are huge telescopes orbiting around the Earth (i.e. Hubble) or *artificial satellites* (see) placed in the orbit around the planet to be investigated, duly equipped with cameras that detect the surface (*satellite detection*, see—*satellite cameras*, see).

The *direct physical studies* are carried out using the *radioastronomy* (see), *spectroscopy* (see) and radiotelescopes (*linear* or *parabolic antennas*) located at the *astrophysical observatories* (see). The *indirect physical studies* are carried out using the *laboratories*, if and when available, placed on the lander in order to perform a few (and simple) *biological experiments* on the spot.

POSITIONAL ASTRONOMY. Properly named *Astrometry*. It is used to determine the position (coordinates) of the celestial bodies, their movement (orbit and/or trajectory—for example, the trajectory of a comet) and their variations in time.

PRISTINE (images). Unprocessed (i.e., original, raw) image/frame/picture/photogram, et cet.

QUALITY OF AN IMAGE. Level of photographic adhesion (see *Photography*) of an image in comparison with the real scene. Only considering the technical aspect, the intrinsic quality of a Satellite Image can be determined using the *Geometric* and *Radiometric Resolution* (see). An image can be modified (see *Modification of an Image*) or deleted (see *Cancellation of an Image*).

QUALITY INDEX (IQ). The quality index places a ratio between the defined resolution (v.) – which is a theoretical, mathematically calculated geometric resolution – and the geometric resolution values actually obtained. From this ratio, the so-called quality index is deduced, which is a value relating to the degree of adherence to reality of a given photographic image. The formula to be applied is RD: RG, where "RD" is the defined resolution (mentioned above), "RG" is the geometric resolution: the operation is a percentage division of the minor datum, in relation to the major datum, and it is used to measure the difference in plus or minus. The percentage obtained represents the quality index of a given image.

RADIOASTRONOMY. Detection of radiations coming from a celestial body (see also Astrophysics).

RADIOMETRIC RESOLUTION. A numeric digital image is composed of a series of lines and columns; the small squares related to them are called pixels and they are marked by a number comprised between 0 and 255. Every square represents a grade of grey, where 0 represents the colour black and 255 the colour white.

SATELLITE. Natural or artificial body spinning around a planet or orbiting synchronously with it. in this second case, we talk about, for instance, of *geostationary satellites.*

SATELLITE AERO-PHOTOGRAMMETRY. This is the photogrammetry (see) carried out using an artificial satellite, on board of which one or more observation cameras are mounted to cover a large surface area. This may vary from a few dozen to hundreds or thousands of square kilometres. The greater the height of the satellite with respect to the Datum, or Zero Altitude, of the imaged object, the lower the resolving capacity of the details on the ground.

SATELLITE CAMERA. Part of a television system (on board of a satellite) that transforms the optical images into, among other things, electronic signals.

SATELLITE DETECTION. Detection of a planet using an orbital satellite that spins around a celestial body to investigate it.

SATELLITE PHOTOGRAPHIC IMAGE. It is an image obtained by a camera mounted on board of an observer satellite. An image's details are relative above all to the height of the footage system that determines the so-called "Instantaneous Field Of View" (IFOV) (see *Optical* and *Geometrical Resolution*). The images originated from Satellite photographic detections operated by the Viking Orbiters are available in two different versions that is the *pristine* (see) and the *mosaicked* (see).

The satellite images can present themselves in black and white (grades of grey) or in colours (chromatic varieties). in the first case, the images are technically valid and can be functionally studied. in the second case, even though the images are authentic, the study is even more difficult because of the lack of contrast differences. The third variety of satellite images are the so-called *monochromatic images.*

These are not exactly black and white frames, but tone on tone images (in other words they are photographs modified by a coloured filter that exalts the different tones). It is not established that *monochromatic* images are fake, but a few space agencies use this technique to conveniently hide (*melting*) some details caught in the satellite images.

SCALE. It is the relationship existing between the reproduced image (such as the picture) and the real dimensions of the captured scene. in a nutshell, it represents the numerical value telling us that a certain portion of a planet's surface has been (dramatically) reduced, so to be clearly represented on a simple sheet of paper. The mathematic expression is 1:1.000; 1:1.000.000 et cet.

SCEPTICISM. Scepticism is properly a mental attitude that does not consider an assertion to be established until it has been proven (and so far, nothing wrong with that, on the contrary...), Conservatism on the other hand is the attachment to power that passes itself off as scepticism and that, through this screen, would like to assume a guise and role that does not belong to it and does not suit it. Conservatism, masquerading as Scepticism, in truth, refuses to consider a claim even after it has been proven.

A classical definition considers scepticism to be 'any doctrine which, affirming the non-existence of a valid criterion for distinguishing the true from the false, regards doubt as insuperable for man', and a traditional distinction asserts the existence of a traditional scepticism (which we call conservatism here) and a methodical one, which we call scepticism in the proper sense. For the latter, doubt is not a definitive result, but only a means to the search for truth.

SCIENCE AND PSEUDO-SCIENCES. The difference between Science and Pseudo-Sciences is that the Scientific Method allows, through the repetition of experiments, the acceptance of new theories and the defeat of 'sceptics' (i.e. conservatives), while Pseudo-Sciences, which do not feel obliged to demonstrate what they claim, can remain unchanged even for millennia despite the fact that in the meantime the field they claim to study has completely changed. In this sense it can be said that Planetology, (only for that part concerned with the study of Intelligent Life, possibly present on other planets) is undoubtedly a discipline that errs on the side of approximation, if it is true, as it is true, that the results of the analyses conducted by it in that context have never been (that is to say, at an official level) examined and verified according to the terms of the Experimental Scientific Method.

SHADED RELIEF. in general, it is a Surface Relief projecting a shadow (or *umbra*) on the terrain. It is also a topographic, cartographic and photographic technique (see *Topographic Structures, Planetary Cartography and Photography*) to reproduce a landscape of interest. Such a technique resolves itself in a configuration (which is only supposedly/theoretically three-dimensional!) of the chosen surface, definitely improved through the use of fake shadows that would have been projected onto the ground by hills of medium and high height, with incidental light coming from the north/west and/or the south/east.

SHADOW. Darkness generated by a dark body that intercepts and stops the rays of light emitted by a bright source.

SPACE OBSERVERS. They are satellites designed by humans and put in orbit around the planets that need to be observed. The space observers replaced the astronomical observers, such as the telescopes (see) located on the ground (in the so-called *astronomical observatories*, see). in the middle of these two elements there are the space telescopes—that orbit around the Earth (i.e., the HST, such as the Hubble Space Telescope)—or at an intermediate distance between the Earth and the observed object (i.e., Soho Solar Observer, Parker Solar Probe, et cet.).

SPECTROSCOPY. Branch of *Radio Astronomy* (see) that works using the measurement of the "quality" of radiations emitted (primarily but not exclusively) by a star.

SPYGLASS. Optical instrument to observe distant objects. Definition now replaced by t*elescope* (see).

STEREOSCOPIC PHOTOGRAPHY. Image obtained from two frames taken at the same time and related to the same object, by using two different cameras that have been put at about 6–7 cm from one another (such as approx. the same distance that exists between the eyes of a human being). The obtained image gives us the possibility to appreciate the plasticity (shape, relief, depth and dimensions) of an observed and detected object.

STELLAR STATISTICS. It represents the union of *Classical Astronomy* (see) and *Astrophysics* (see); it studies the distribution and movement of the stars using the *statistical method.*

STEREOGRAPHIC PROJECTION. In geometry and cartography (see Planetary Cartography) it is a geometric projection of the points of a structure on the surface of a sphere, starting from a point (N) on the sphere itself (often identified as the North Pole of the sphere), above a plane that is the Equatorial Plane (see topographic map).

STUDY OF THE LANDSCAPE. The observed *raw landscape* becomes a *sensitive* geographic landscape when some of its objective elements (*reliefs* and *structures*, for example) are visible through a simple and direct observation, when they are considered for what they appear to be, and, in the end, when their spatial relationships are coherent. Any landscape undergoes some modifications in time and the causes of such a transformation could be *natural* (see *Natural Landscape*) or *artificial* (see *Artificial Landscape*).

TELESCOPE. Instrument that allows the observation of objects very distant from the observer. It is the optic instrument to observe celestial bodies. Its focus could be a lens or a system of lenses (dioptric telescope or refractor) or a concave mirror with very high capacity of reflection along with a magnifying lens (catoptrical telescope or reflector).

TOPOGRAPHIC MAP. The USGS (United States Geological Survey) built a digital map of the surface of Mars using the *pristine* images (see) and arranging them according to a Cartographic Projection's Model. They used a sinusoidal equivalent (*Planetary Cartography Working Group*) where it was possible to establish a correspondence between a latitude parallel to every line compressing the length and the cosine of the same latitude.

In this way, every pixel of the image can be identified using the so-called absolute geographic coordinates of latitude and longitude, thus obtaining a Global Cartographic Reference Frame. As far as the polar regions are concerned, a Stereographic Cartographic Polar Projection was used.

TOPOGRAPHIC RESOLUTION. The *geometrical resolution* (see) makes possible for an image to show some *structures* of a certain (meaning "conspicuous") dimension. in the topographic resolution, they are considered the dimension of a *structure* and the relationship between said *structure* and the level of the considered geometrical resolution. This means that the "objective" images are not necessarily HR (High-Resolution) satellite pictures, but rather photos that possess an excellent (or just favourable) relationship between the "footage distance" (see) and the "dimensions" (see) of a given structure.

TOPOGRAPHIC STRUCTURE. Part of the surface of a celestial body observed at a close distance through an artificial *satellite* (see) put in orbit around it.

UNMANNED SPACECRAFT. Automatic spacecraft or a probe without human crew.

VIEW ANGLE. Visual angle formed by the relationship existing between the ideal vertical line connecting the satellite from the exact point of the ground that the satellite itself is overflying and the tilted line connecting the footage camera located onboard the spacecraft to a given point of the surface that must be photographed (a.k.a.: "tilt").

VISUAL OBSERVATIONS (Planetography). Direct observations of a certain celestial body that is carried out by using optic Earth-placed instruments (see *Spyglasses* and *Telescopes*) or on board of a man-made satellite (see *satellite cameras*).

Fig. 5a. Photograph of the surface of Mars at the VL-2 site on sol 161 at 1100 hours local lander time.

Tab. 169A	Viking Lander 2 – Image 21C198/161; From: Scientific results of the Viking Project – Reprinted from Journal of Geophysical Research – NASA 1977 – p. 4483 – Fig. 5a.

Acknowledgments and Sources

Behind this book are various scientific institutions which, to varying degrees, have helped and made this research possible. We would like to thank the institutions and individuals listed below:

United States Geological Survey - Branch of Astrogeology (Department of the Interior - USA): Ms Jody Swann (data facility manager), for her invaluable and irreplaceable collaboration; Ms Adrienne Wassermann (data facility manager), Ms Kathy Hoyt and Ms Karen Zanter (planetary data facility).

NASA Headquarters (USA): Mr Joseph L. Headlee (Chief - Imaging Branch - Public Affairs Division Broadcast and Audio-Visual Branch), who provided the initial address list of the US space agency, which was useful for all subsequent contacts; Mr David J. Bohlin (Office of Space Science), Ms Cynthia M. 'O Carroll and Mr Bert Ulrich for a series of planetary information.

Jet Propulsion Laboratory (JPL-NASA), California Institute of Technology (USA): Mr John Bluth (archives and records office), who provided useful information (including some addresses) and bibliographical references, and also successfully forwarded a request for information to JPL-RPIF; Miss Elizabeth Moorthy for providing an official NASA text on satellite photographic surveys; Miss Amy Culver for information on geometric resolution and a text on Mariner 9 image conversions.

Regional Planetary Image Facility - NASA - Lunar and Planetary Institute Centre for Information and Research Services (USA): Miss Mary Noel-Black (curatorial assistant) who provided information regarding the various NASA photo archives (RPIF).

Deutsches Zentrum für Luft und Raumfahrt and V. (German Aerospace Centre, Institute of Planetary Research, RPIF, Berlin, Germany): Ms Susanne Pieth (head), for providing data and information useful for contacting other people.

RPIF-NASA – Photothéque planétaire d'Orsay – Université Paris-Sud (France): Ms Daniéle Gaspard (manager) for providing essential information regarding some Mariner 9 and Viking photomosaics.

Goddard Space Flight Center - NASA World Data on Rockets and Satellites – Hughes Stx Corp. (USA): Mr Joseph King (director) and Mr David R. Williams (head of Planetary Image Archives) who provided fundamental information regarding certain satellite images of Mars; Miss Barbara Pope for certain information regarding the Mariner 9 images.

National Air and Space Museum - NASA (USA): Miss Diane Barbe (volunteer - Mail Response Museum), who provided planetographic material.

Geosciences Node (Washington University – St. Louis – MO - USA): Miss Susan Slavney and Mr Margo Mueller (Wunder. Wustl. Edu), for providing important data and facilitating the retrieval of useful information.

Armagh Planetarium (NASA - Northern Ireland): Mr Ian Griffin (director), who provided astronomical material concerning the planet Mars.

Institute of Space Astrophysics and Planetology (IAPS-Rome): Dr Marco Feroci (director), for allowing the use of some images from the RPIF photo library in Rome.

European Space Agency - European Space Research Institute (ESA-ESRIN - Italy): Dr Angelo Bodini, who provided some fundamental clarifications regarding satellite observation and geometric resolution.

National Research Council (CNR-Italy): Dr Ruggero Casacchia, for providing a text on planetary cartography.

National Antarctic Museum of the University of Siena (Italy): Dr Antonio Zeoli (geologist) for allowing us to use some elements of his study on Martian geomorphology.

Institute of Human Geography of the University of Milan (Italy): Dr Maristella Bergaglio, for providing bibliographical references on aerial photogrammetry.

Department of Astronomy, University of Bologna (Italy): Prof. Pierluigi Battistini, for providing useful bibliographical references on satellite images and general topographical information about the planet Mars.

Arizona State University (ASU-USA): Mr Charles Hewett (director) and Mr Daniel Ball, Ms Sara Santiago and Ms Veronica Ann Zabala Aliberto, for providing a series of images from the Mariner 9 mission.

The research documents and photographic material used in this volume come from the following centres (in addition to the centres mentioned in the acknowledgements):

1.. Aerospace Education Services Program, Oklahoma State University, Educational and Research Foundation Incorporation (USA, Oklahoma);
2.. Allied Signal Technical Services Corp, NASA (USA, Maryland);
3.. American Society for Photogrammetry and Remote Sensing (USA, Maryland);
4.. Ancient Astronaut Society (USA, Illinois);
5.. Anglo-American Publishing (Diffusione Edizioni Anglo, Americane) (Italy, Bologna);
6.. Arcturus Book Inc. (USA, Florida);
7.. Astronomical Society for the Pacific (USA, Maryland);
8.. Bara-King Photographic Inc. (USA, Maryland);
9.. Earth Data Analysis Centre, University of New Mexico (USA, New Mexico);
10.. Earth Observation Satellite Company (USA, Maryland);
11.. Eros Data Centre (The), Department of the Interior, USGS (USA, South Dakota);
12.. Finley Holiday Film Corporation (USA, California);
13.. GAF, International Publisher (USA, California);
14.. George C. Marshall Space Flight Centre, NASA (USA, Alabama);
15.. Gruppo Astrofili Catanesi "Guido Ruggier" (Italia, Catania);
16.. John F. Kennedy Space Centre (USA, Florida);
17.. Langley Research Centre, NASA (USA, Virginia);
18.. Lyndon B. Johnson Space Centre, NASA (USA, Texas);
19.. Mars Research (USA, Maryland);
20.. Media Service Corporation (USA, Texas);
21.. MMI Corporation, Teaching Materials for Tomorrow's Scientist (USA, Maryland);
22.. NASA, Central Management (USA, Washington, DC);
23.. National and Air Space Museum, Smithsonian Institution (USA, Washington, DC);
24.. National Oceanic and Atmospheric Administration (USA, North Carolina);
25.. National Space Society (USA, Washington, DC);
26.. Photographic Reflection Incorporation (USA, Virginia);
27.. Planetary Society (USA, California);
28.. Science Service, Science Youth Program (USA, Washington, DC);
29.. Sky Publishing Corporation (USA, Massachusetts);

30.. Smithsonian Institution, Astrophysical Observatory (USA, Massachusetts);
31.. Stennis John C. Space Centre, NASA (USA, Massachusetts);
32.. Technology Application Centre, The University of New Mexico (USA, New Mexico).

Biographical Notes

Gianni Viola (182)

He studied sociology at Carlo Bo University of Urbino (Italy). As a photo reporter for *Stampa Alternativa* in Rome, he published historiography works. The most relevant is *Polizia. Controinchiesta su abusi ed eccidi delle forze dell'ordine in Italia dal 1943 al 1976* (trad. *Police. A counter enquiry about abuses and massacres of Italian law enforcement from 1943 to 1976* printed in London in 1976 by Lewis&Cann and published by Stampa Alternativa in Rome), followed by *Police—Chronicles and documents of the repression in Italy 1860–1977* (published by Stampa Alternativa in co-edition with Bertani Editore of Verona in 1978).

In addition, about the public order, he contributed with the magazine *Nuova Polizia e Riforma dello Stato* (New Police and reformation of the State) in Rome, body of the Union for the Police Workers (the first police union in Italy).

Researcher for the "Ecumenical Russian Centre" in Rome, was awarded from this centre (along with the International Association for the Human Rights and the Italian Committee of Helsinki), of Premio Padre Ulisse Floridi S.J. (Award Father Ulisse Floridi S.J.) for the work *Enciclopedia delle Religioni dell'Unione Sovietica* (Roma 1989) (USSR Religions Encyclopaedia).

As regards the Russian question, he recently published an essay titled "La verità sulla fine dell' URSS" (The truth about the end of the USSR) (Prospettiva Editrice, Civitavecchia 2010).

Currently, he is responsible for the technical-scientific commission of *FLIP* (*Free Lance International Press* of Rome).

In the scientific field, he dealt with Planetography and he published *La Civiltà di Marte* (Osservazioni, geografia, esseri intelligenti, Edizioni Mediterranee, Roma 2002) (*The Civilisation on Mars. Observations, geography, intelligent beings*, Edizioni Mediterranee, Rome 2002 (**Tab. 184**)[156] and he cooperates with *L'Astronomia*, (The Astronomy) **(Tabs. 10, 13)** the most important astronomy magazine in Italy. He took charge of NASA photographic library and of Regional Planetary Image Facility, allocated in the

[156] A pamphlet apeared in 1995 (**Table 183**), wich was a preview of the Work later printed in 2002 (Table 184).

Research Area 2 in Tor Vergata at the premises of the Institute of Astrophysics and Space Planetology (IAPS) (**Tabs E,F**). The IAPS is part of the National Institute of Astrophysics.

In 1995, the writer published a CD - which was distributed free of charge - where the various programmes for decoding the satellite images of Mars and Venus concerning Mariner mission (Mariner 6,7,9,10) and Viking (Viking 1&2) missions were collected. The CD also contains the complete list of Mariner 9 and Viking Orbiter 1&2 images (**Tab.D**).

In 2010, Prague-based Youth Time magazine (Czech Republic), part of the Youth Time International Movement, published an interview with the author by journalist Angela Curatolo entitled "The Other Truth About the Red Planet", which appeared in both Russian and English. The article was among the 30 winners of the Gagarin 2011 competition in Baikonur (Kazakhstan), a former Soviet missile base.

He was involved in international politics and he was a political-military adviser at the Embassy. He recently published an essay on the end of Jugoslavia, "Il soave profumo dell'imperialismo" Kimerik Edizioni, Patti-ME, 2010 (The soothing aroma of Imperialism). The text was presented on 24 March 2011 on the TV channel *Tele Ambiente* in Rome, by the GAMADI group in Rome. The writer Miriam Pellegrini Ferri conducted in the studio.

In 2019, the author was interviewed by Serbian journalist Ivan Maksimovich of the *Tabloid Magazine* (Kosovska Mitrovica Kosovo/ Serbia). The interview was also published in the magazine *FB Reporter* e KM Novine—di Ponedeljak—Serbia, the latter in English. in Italy, it appeared in the Salerno periodical *Cronache Italiane* (December 2019).

Journalistic investigations: "La Scienza di Cartapesta", started in 1995 and finished in 2009. It is an *open investigation* with question referred to satellite images. The representative sample consists of physics and astronomy teachers, astronomers, astrophysics and researchers involved in space agencies. The text 'La Scienza di Cartapesta' (The Science of Papier-Mâché) was printed in November 2025 by 'Pagine Editore' in Rome (Italy).

La "Scienza di Cartapesta" is an Italian investigation that provides a first-hand account of the tragic situation facing scientific research, which is at the mercy of operators who are completely ignorant of the basic principles of the scientific method. The subjects contacted confused terrestrial structures with Martian structures, having no knowledge of either. The results of the investigation are a warning sign – certainly not against certain individuals (to whom we express our gratitude) – but rather highlight a very dangerous situation where, from positions of privilege, these individuals – despite themselves – spread unfounded news, false data and incorrect deductions, within a mechanism destined to reproduce itself through the perverse school system or the deleterious medium of television. In this sense, the aim of this investigation is to affirm science, rather than scientism, which represents a belief that is in stark contrast to the revolutionary spirit inherent in every scientific discovery (Tab184A).

Participation to television programs: TV "Attualità" at Teletna (Catania), conducted by Giovanna Quasimodo, with a debate with prof. Carlo Blanco, director of the Astrophysical Observatory of Catania; *TV "Stargate"* channel La7 di Roma, conducted by Roberto Giacobbo, in a debate with Dr Enrico Flamini, in charge of planetary missions of the Italian Space Agency of Rome (Frascati).

Video production:: *La scoperta della civiltà di Marte*, 1995 (The Discovery of Mars civilisation). Partially or fully transmitted by many channels among which Teletna, Telejonica and Antenna Sicilia (Catania), Raitre (Catania).

Proposition of scientific research's methodologies: elaboration of planetography's principles as regards the planets' shape (geography), to be distinguished by planetology that regards the substance instead (geology).1: Process of reconfirmation of satellite images; 2: Calculation of the topographical resolution; 3: Method of retrieval and redefinition of modified satellite images.[157]

Registered scientific discoveries: OLAF section SIAE Rome (1994) (**Table 185**). *Paesaggio modificato di Marte* (Mars' modified landscape), about experimentations operated in Mars photographs by probes Mariner 9 (1971–1972) and Viking 1 and 2 about experimentations operated in photographs by the probes (1975–1982). Experimentations applied to studies related to the Martian regions of Capri Chasma, Iani Chaos, Nanedi Valles, Cydonia Colles, Pavonis Mons, Galaxias Chaos, Lybia Montes, Chryse Planitia, Utopia Planitia, Nirgal Vallis, Echus Fossae, Ophir Planum, Aeolis Mensae.

The Author has been mentioned in several TV and radio channels. Among these Raitre Sicilia (Catania), Teletna (Catania), Telejonica (Catania), Tele Trudy (Catania), Tele Granducato (Livorno); Radiomare (Ladispoli); Italia 7 (Bologna); ReiTV (Acireale); Radio Universal Giarre (Giarre); Tele Video Randazzo (Randazzo); Antenna 1 (Lentini); su molte testate, fra cui: L'Astronomia (Milano), Cronaca Vera (Rho-MI), Cronache Italiane (Salerno), Rinascita-quotidiano (Roma), Gazzetta del Sud (Messina), Il Resto del Carlino (Rovigo), La Sicilia (Catania), Cronaca Vera (Rho), Gioia (Milano), Il Gazzettino di Giarre (Giarre), Fatti and Voci (Giarre), Flash Magazine (Frosinone), Lògos (Acireale, CT), La Voce dei Free Lance (Roma), La Cerniera (Giarre, CT), Il Mercatino (Catania), Il Messaggero del Sud (Mascalucia, CT), Il Messaggio (Civitavecchia, Roma), Il Periodico Etneo (Aci Catena, CT), Il Tirreno (Livorno), La Procellaria (Reggio Cala-

[157] See infra, Part II, Chap 5 - Quantitative analysis.

bria), Nuovi Orientamenti (Roma), Pomezia Notizie (Pomezia, Roma), Sicilia Sera (Catania), Tribuna di Giarre (Giarre, CT).

Gianni Viola's works are present in the main libraries of many regions of Italy among them: Library of History and Art History Dpt of the University of Trieste, the Central National Library of Rome; University La Sapienza in Rome; Biblioteca della Fondazione Giangiacomo Feltrinelli in Milan; University of Pavia; Council Library Alberto Geisser in Turin; University of Messina; Council Library Zelantea in Acireale (Sicily), National Central Library of Florence; National Archive of Florence; National Library Marciana in Venice; University Ca' Foscari in Venice and many more.

Websites regarding libraries or places of collection and distribution: Biblioteca della Fondazione Querini Stampalia Onlus; Rete Bibliotecaria Bresciana (Brescia), Libraries of Romagna (Forlì), Province of Padova (Padua), Cultural Heritage Province of Lucca (Lucca), Parma Library System (Parma), Autonomous Region of Sardinia (Cagliari), Unique Catalogue of Pavia (Pavia), Integrated Documentative System of the Florentine Area (Florence), Primo Moroni Archive, Documentation Centre of Pistoia (Pistoia), Oblates' Library (Florence), Rete Documentary Web Province of Pistoia (Pistoia), Cultural Institutes in Rome (Rome), Franco Serantini Library (Pisa), Informacarcere Library (Florence), Contemporary Art, Archive of Carta Stampata, Books Online (Piedmont Region), Libnat, Bisonomy, Bibliography, Paolo Staccioli-Bibliography, Abebooks, Iperlibro.com, Archives Net non dimenticare, PoloMaceratese, Macrolibrarsi, Unilibro, IBS, Libraccio.it, Il GiardinodeiLibri, Mondadori Store, LaFeltrinelli, Amazon, Free.Lance.International.Press, Libreriagrandangolo, Aseq, Ebay, Hoepli, Keej, LibreriaUniversitaria, LibreriaPuntoEinaudi, Picclick, Pde, Librarythyng, BookDepository, Lucca Sapiens **bookshop**.

Abroad: *Canada:* University of Toronto; *France*: RPIF-NASA – Photothéque Planétaire d'Orsay – Université Paris-Sud; Libreria "Le Libre Penseur", Paris; *Portugal*: Universitade de Lisboa, Instituto de Ciências Sociais; *Germania*: Biblioteca Sonomy – University of Kassel; Staats – und Universitätsbibliothek Bremen; *Netherlands*: Bibliotheek Universiteit van Amsterdam; International Institute of Social History –IISG- Amsterdam; *Great Britain*: The British Library, St. Pancras Reading Rooms - London; University of London Research Library Services, Senate House Library - London; University of Manchester, The John Ryland's University Library – Main Library - Manchester; London School of Economics and Political Science, London. *Spain:* Libreria "Perez Galdos El Galeon" of Madrid; *Australia*: National Library of Australia – Canberra; *USA*: Library of Congress – Washington; Jesuits' Library of Loyola University di Chicago (USA);

Yale University Library, New Haven, Connecticut; The Ohio State University; Stanford University Libraries, Stanford.

Some reviews on The civilisation of Mars by Gianni Viola (ed. Mediterranee—Rome 2002)
Very well-documented and methodologically compelling, this volume provides an overview of the last scientific data currently available and a plethora of photographical documentation that unequivocally demonstrate the presence of a very ancient civilisation on Mars' surface analysing everything geographically, orographically, hydrographically and highlighting the structures certainly artificial still visible in several areas of the planet.
A very rich source of information and stunning scientific counter-analysis that in addition would highlight censorship operated by NASA based on photographic material, this book will arouse a lot of enthusiasm but also some doubts for bold conjectures formulated by the Author we largely admire for giving us a work that exudes commitment in the research and a visceral passion for that Red Planet and its mysteries that from ages fascinate human beings.
(Tom Bosco—managing director of the Italian edition of the Australian periodical "Nexus New Times", number 63, 2005).

After ten years of research, the author acquired a fundamental truth about the mysterious Red Planet. Gianni Viola dedicates "The civilisation of Mars" both to those who believe we are alone in the Universe and to those who believe there are other celestial bodies inhabited by forms of intelligent lives. These pages contain the answers to those entire question that had not a valid explanation before thanks to shocking documents and unpublished photographs coming from the archives of NASA. (Daphne Diorio—"Leggo", Rome, 23/01/2003)
This is a book that summarises ten-year-old studies carried out by the authors regards the Red Planet. According to Viola a successful journalist and researcher, the hypothesis of a past intelligent life on Mars is possible thanks to a huge number of documents. The book—a very useful tool for curious people but also for students and University's teachers—offers objective truths coming from a rigorous, logical-mathematical reasoning. (Valentina Ivana Chiarappa, Stargate Magazine, Rome, year II, n. 9, April 2003).

This book finally goes beyond the four photos of Cydonia and the Martian sphinx that are usually shown and presents an overwhelming amount of data to support the hypothesis of ancient or recent anthropomorphic structures on the Red Planet. The author is motivated by the question, "Does life exist on other planets in our solar system?" And he does it with great skill. Gianni Viola started his journalistic career at the age of 17 and it shows.
What would you think about a book that presents you with 200 photographs, many of which are unpublished? And has a whole section on Martian faces? Lots more than just the original "Face on Mars"! Finally, he felt the need to present a lot of consistent data, like a breath of fresh air, into an environment that had become stale. This is excellent work by Viola, who from 1974 to 1981 collaborated with the press agency Stampa Alternativa, in Rome (with which he published Polizia in 1976).

This was followed by an alternative history of Italy from 1943 to 1976, with a second volume in 1978, covering the period 1860–1977, a co-edition published with Bertani in Verona). And the hand of the historian can be clearly discerned. Viola has studied a huge number of photographs released by American institutions but scattered (intentionally?) between the IAS—CNR, JPL-NASA, ESA-ESRIN, GSFC-NASA, RPIF-NASA and USGS.

With great constancy, he has been able to reconstruct this enormous puzzle, drawing on some astonishing data, discovering hidden images or examining more closely those that have not received enough attention, to draw up a systematic and consistent theory on the history of Mars. His book "La Civiltà di Marte (Martian civilisation)" (Edizioni Mediterranee, Rome 2002, 383 pages), is packed with information. Better yet, it overwhelms you with the sheer volume of data it contains. It is impossible to refute or contest it, remain sceptical or turn a blind eye. Viola is known for his Carthusian works.

In 1976, he began research at the Russian Ecumenical Centre in Rome, on behalf of which he wrote a voluminous Encyclopaedia of the religions of the USSR, winning the P. Ulisse Floridi Prize). Therefore, the volume of documentation was no deterrent to him. The result is an excellent book. in 1978, Viola joined the Sindacato Nazionale Scrittori (National Writers Union) and since 1987 has carried out independent research in the field of descriptive astronomy, with data from NASA, including the Goddard Space Flight Centre, the Jet Propulsion Laboratory and the Regional Planetary Image Facility.

In 1994, he wrote his first work on the planet Mars and began to promote his theory in lectures, interviews and essays. This researcher from Acireale is a credit to his country, which has always been a cradle of illustrious ingenuity. The painstaking patience with which he has catalogued and illustrated every one of the Martian anomalies is astonishing. There are simply no words sufficient to describe his most valuable work! Let his readers be the judge. This book should definitely be a part of your library of mysteries. A real masterpiece! (Prof. Alfredo Lissoni—Milan 2003).

Paolo Catello Fienga

Born in Milan on 13 May 1963, he graduated in Law in Milan in 1987 and after a few months he started to work for SAIPEM (ENI Group), a company working in the oil and engineering field. He worked there till 1995. Meanwhile, thanks to a passion transmitted by his father (a Magistrate) and his mother (a Sociologist), he started to study Astrophysics.

He graduated in 1998 in Holland after studying in the USA, Switzerland and Holland itself. Because of the long illness of his wife, he had met in USA, he decided to assist her constantly until her death in 2013. Because of this, he also left his job as a freelance consultant.

In June 2004, Paolo founded, at the same time, the non-profit Free Association *Lunar Explorer Italia* and a blog (*True Planets*); in the AD 2011, in Loving Memory of his father, he also created the *Italian Planetary Foundation* (*IPF*): a sort of Italian answer (daily operated by Dr Paolo C. Fienga and through the heading *Astronomy Picture of the Day*, still on-line today but no longer updated) to the "Planetary Society".

He is now dedicated to study, works for a Design and Engineering Company of Milan as a freelancer and, every time he can, deals with the spreading of Planetary and Frontier Science.

Maria Grazia Lucrezia Leotta

Maria Grazia Lucrezia Leotta attended a private High School for Languages and she graduated in Modern Foreign Languages and Literature at the University of Catania. She also took a diploma in Theology. She worked as a clerk and interpreter at the Magistrates Court of Catania, as a reporter for the newspaper *La Voce dell'Jonio* in Acireale and as a foreign languages teacher in several secondary schools. She was also a teacher at the Università Popolare of Acireale and a private tutor.

For more than ten years, she devoted herself to choral music and pop music working as a singer in a piano bar and local television channels, anchor-woman and events planner. Ten years ago, she moved to England where she took an MA Translation Studies at the University of Sheffield.

Currently, she works at the Minster of York, she is a translator and interpreter and an active member for *The Missing Princes Project* led by Philippa Langley for the Richard III Society where she is a researcher and an expert of medieval history and palaeography. She writes for several bulletins and magazines published by the Richard III Society and she has just published the first Italian biography of Richard III titled *Riccardo III il re piú amato* published in November 2022 by Edizioni Efesto.

She is also a poet. Two of her poems were published in England for the anthology *White Rose Bards*. She lives in York and she is fond of antiques, music and craft.

Jody Swann

Jody Swann (secretary; B.S., Eastern New Mexico University) [The following information was abstracted from an interview with Jody Swann by Gerald G. Schaber on 16 February 2001.] Jody (Loman) Swann, was born in Texas, but was reared primarily in northeastern New Mexico where her mother was high school teacher and her father owned his own water-well drilling company. She received her Bachelor's Degree at Eastern New Mexico University with a double major in secondary education and secretarial science. When she graduated she got a job as a secretary, and in a few months got a job working for the Civil Service. Her first husband [Sam Loman] had been transferred to Flagstaff, Arizona, and Jody was anxious to keep her Civil Service continuity, and so ended up getting a job out at the Navajo Army Depot (NAD) [at Bellmont, just west of Flagstaff] as a secretary in the personnel office. While Jody was working at NAD in 1963, Don Elston and Wanda Wilkinson (Administrative Officer for the Branch of Astrogeology at that time) came out to NAD to look at the register of competent people to be considered for recruitment as secretary to Gene Shoemaker. Elston asked Jody if she would like to come into town and talk to them at Astrogeology [at Astrogeology's wing at the Museum of Northern Arizona.] Jody Swann: "So I did. I went in, talked to Don and looked the place over a bit–and I called him [Don] back the next day and said no, I didn't think I could do that. Then, Don called me one Sunday and asked if I would reconsider coming to work for Astrogeology. And I said no again. Then, he called me at my office at NAD and said I know you don't want the job–you said that–but I really want you to just come and meet Gene Shoemaker. I [finally] said okay, I'd come out. This was November 1963. 306 I arrived at MNA and it was a hubbub; everybody was still working; the lights were still on and everybody was doing their job–and I was amazed! I had come from the Department of Defense. I had worked for the Department of Agriculture in New Mexico, for the Air Force; and I was currently working for the Army. I was used to people working until 4:30 PM. You leave work at 4:30 PM; that is what you did–and you don't stop or you're going to get run over [by the people scrambling to leave the building]. I remember thinking that we should get on with this [Don was not there at MNA to meet her when she arrived] because there was a young man there in ragged Levi's and a flannel shirt that I feared wanted to clean up the place. I was soon informed that he was Gene Shoemaker! I sat down and talked to Gene–and I was hooked! I quit my job [at NAD] the next day and went to work for Astrogeology. Gene was the most charismatic personality-plus person I have ever known. He could sell ice to Eskimos! So that was when I started to work for Astrogeology. That was December 2, 1963 out at the Museum of northern Arizona. Gene was different from anybody. I came from a very small town and a relatively small university, and I was incredibly naive and unworldly. I had never flown on a commercial aircraft in my life, and Gene just assumed that I could make travel reservations, get tickets and hotel reservations, and take dictation. I could take dictation because that is something I learned to do in college. Gene wanted me to listen in on every phone conversation he had–from beginning to end. And we had telephone receivers that had a little button on them that was sort of a press to talk button. Otherwise, I could pick up the receiver and keep it tucked between my ear and my neck–and my neck is still a piece of trash! I would listen to every word he said on the phone. And for good reason for Gene–that was to keep track of any commitments that he may have made whether on the calendar, a trip, something he said that he would send to someone–which I would just take care of. He wouldn't have to write it down and remember it–think about it. It would just get done! And that makes a lot of sense. I couldn't possibly absorb all of that [Gene's phone conversations]. I didn't have the background to absorb it–but even if I did, you would still have to make a selection of what's important to your mind and what you're going to write down–what you're going to do–what you're going to remember to do. Anyway, if they only had headsets then! I remember when Jack Schmitt first came to Astrogeology in Flagstaff. Like a lot of brilliant and energetic people, Jack had written something for Gene–and he wanted it typed. I was the only secretary then–and when Gene was in town, Gene was my priority. Gene was gone so much, and always had a pile of stuff on his desk–he had to get things done. I had to do it then. I recall onetime when Jack Schmitt brought me something to type; and I guess I was just naive enough to be kind of rude. No, that's not true–but anyway I was frantic trying to get things done for Gene– and transcribing. Jack later came up and said, have you got that typed yet, and I said if you don't get your ugly face out of here, you're never going to get it typed! Now go away and I'll let you know when it's done. Jack said, yes maam! After that, we got along just fine. But sometimes under the pressure of trying to do too many things at once–you're not always at your best." 307 The first time I actually flew in a commercial airliner was to go out to Pasadena [CA] for Surveyor 1 to type reports and take dictation, among other things." Schaber: "People tell me that first Surveyor landing made things quite hectic at JPL, as they never expected it to land—much less take over 11,000 pictures." Jody Swann: "Yes, and we were there. I was out there over one weekend for which Gene [Shoemaker] was going to be out of town. My sister lived in the San Bernardino area, so she came up to Pasadena to pick me up. She sat in the parking lot and waited to me for six hours because Gene was dictating, and he said it'd just be a little bit longer–just a little more. I ended up taking a complete steno pad–front and back–in one session for a Surveyor-1 report. But NASA was big on these reports–interim reports; but that was good because it got things out to the public. But from then on–from Surveyor and on through Apollo–there were so many real-time things we had to do. Everything got done [because it had to]. A lot of things that got done–and got done well–were done on the fly. It was sort of what-if-let's try! There was a lot of imagination and a lot of youth." Schaber: "And really old technology compared to today's standards-right?" Jody Swann: "Well, that's true! Oh, my goodness–if I'd had a personal computer in those days gosh!" Schaber; "Or even a word processor!" Jody Swann: "Yes! We just typed, and re-typed [with a typewriter]. You couldn't make typographical errors–it just took way too much time." Schaber: "Actually, we scientists were doing the science—and thoroughly enjoying it, although it too was hectic. But you know, many of us never thought about the pressure that was put on the secretaries at that time–it was amazing. You all absorbed some of the excitement of course, but then again, you were doing the hard work like the drudgery of typing and taking dictation [and putting up with the scientists demands 12 to 16 hours a day]." Jody Swann: "One day [in 1964] I was working over at Gene's house. It was a cold and snowy day and Gene had been sick. People were coming over to their house to keep appointments with Gene, or to see Gene, and I was being the secretary over there. Gene had an appointment that morning with Jack Schmitt. Gene wanted Jack to apply to be an astronaut. And Jack said, Gene, I don't think I can make it because several years ago I had to have major surgery and they took out several [extra] feet of my colon or large intestine, and I don't think I could pass the physical. Gene said, I don't see why that would be a problem. So, that morning, Gene and Jack dictated his application for me to type up for him to apply to be a scientist astronaut–and the rest is history." Schaber: "Gene already knew he had physical problems by that time [I am here referring to the fact that Gene had earlier been diagnosed as having Addison's disease; see main text, spring of 1963]" 308 Jody Swann: "Gene couldn't even get a pilot's license. He had to quit flying on a private airplane by that time." Also, in those early days [of the Branch] we had only the old telephone switchboard that was in the Forest Service Building [now the Federal Building on San Francisco Street in downtown Flagstaff]–and everything had to go to the switchboard. There were only so many trunk lines. Dialing numbers to get out- meant getting many re-orders. There were no touch-tone phones then; you dialed every number all the way through. It could take a while. You'd dial this whole long number, and you'd get a re-order and you would start all over. It could be a challenge." [Author's Note: The name Bob Wildey [astrophysicist] came up during our conversation, and Gordon Swann, who was sitting in on my interview with Jody, made the following comment]: Gordon: "Ken Watson hired him. Bob Wildey is the guy that Hal James, the Chief Geologist at the time, commented on during a visit to Flagstaff. He was looking over the Branch roster and said, what the hell is the Geological Division of the Geological Survey doing with an Astrophysicist on its staff?"] Jody Swann: "Bob Wildey really irritated me one time. Gene got a lot of correspondence from people interested in what we were doing. Well, one of the letters was in German and I didn't have any idea what it said, so I asked Bob Wildey to translate it for me. And he did–he translated it into Spanish. [Gordon Swann's comment: "But that was Bob Wildey's sense of humor!]" Well, I didn't see the humor in it." Schaber: What were you doing for our Apollo Lunar Surface Geology Team in Houston during Apollo 11?" Jody Swann: "I did stuff for Gene; and for the Geology team when we were actually in the Science Support room [at Houston's Mission Control]. I was either typing up notes during the EVA's or typing up sets of questions and getting them circulated through the higher-ups so they could be transmitted up to the surface [of the Moon]. That is the kind of thing I was doing. Then in late 1970, early 1971, Sam and I moved to Flagstaff; and Sam and I were divorced soon thereafter. Sam went to work for the State and eventually moved out of Flagstaff–and I stayed working for the Branch of Astrogeology."

The Society for Planetary SETI Research (SPSR)

The Society for Planetary SETI Research (SPSR)[158] is an organization of scientists and scholars from a variety of disciplines formed around their common interest in anomalies on planets and their satellites whose origins may be the result of intelligent activity. The focus of SPSR research is primarily the surfaces of Mars and the Moon as revealed by orbiter and lander investigation.

Because they believe the Planetary SETI question is of great potential importance to humanity, SPSR's members donate their research time freely despite the challenges currently posed by the controversial implications of the subject matter. These have included not only lack of research funding but prejudicial barriers to the publication of research papers, difficulties in obtaining access to raw data on a timely basis and a general attitude of ridicule in the halls of mainstream science.

Despite these challenges, SPSR is committed to keeping the Mars investigation in particular as much in the scientific mainstream as possible. To this end, members are encouraged to submit their papers to peer review within the group and to seek publication in appropriate peer-reviewed science journals. Despite the difficulties, SPSR members have succeeded in having a number of peer-reviewed articles published as well as papers presented at professional conferences.

Because SPSR members represent such a broad diversity of interests, areas of expertise, and opinions, the organization as a whole does not generally endorse any particular conclusions drawn by its members, and no claim of endorsement by SPSR as an organization would be appropriate. Some of the position statements by individual members reflect this diversity of viewpoints.

Nevertheless we all agree that anomalous features we have seen in images from Mars convince us that:

- the issue of artificial structures on Mars should be openly debated in the scientific community,
- that mainstream journals should be open to review of qualified technical papers on this subject matter and.,
- that follow-up images to test the predictions of the hypotheses of natural versus artificial origin for these anomalous features should be a priority for the remaining lifetime of the MGS spacecraft and its successors.

[158] The Author has been a member of the SPSR since 10 February 2025.

Tab. 182	The Author

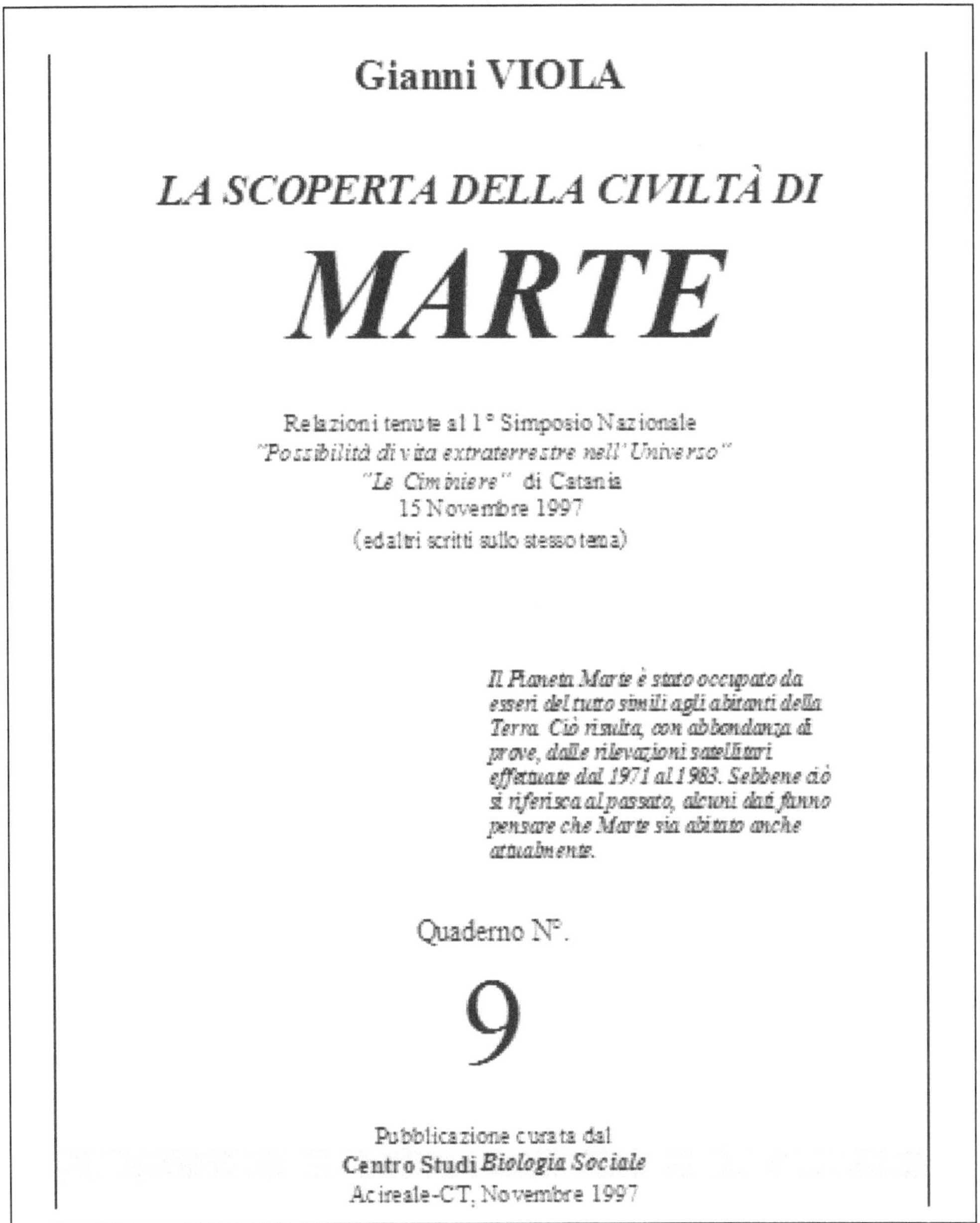

Gianni VIOLA

LA SCOPERTA DELLA CIVILTÀ DI

MARTE

Relazioni tenute al 1° Simposio Nazionale
"Possibilità di vita extraterrestre nell'Universo"
"Le Ciminiere" di Catania
15 Novembre 1997
(ed altri scritti sullo stesso tema)

Il Pianeta Marte è stato occupato da esseri del tutto simili agli abitanti della Terra. Ciò risulta, con abbondanza di prove, dalle rilevazioni satellitari effettuate dal 1971 al 1983. Sebbene ciò si riferisca al passato, alcuni dati fanno pensare che Marte sia abitato anche attualmente.

Quaderno N°.

9

Pubblicazione curata dal
Centro Studi *Biologia Sociale*
Acireale-CT, Novembre 1997

Tab 183	Gianni Viola, The Discovery of Civilisation on Mars, published by the Centre for Social Biology Studies in Acireale (Sicily). – Acireale, 1997.

Tab. 184	Author's text "La Civiltà di Marte" (Mediterranean Editions, Rome 2002).

SOCIETA' ITALIANA AUTORI ED EDITORI (S.I.A.E.)
Direzione Generale - 00100 - Viale della Letteratura, 30 (E.U.R.)

SEZIONE OLAF - SERVIZIO DEPOSITO OPERE INEDITE

ROMA,28-04-94

Si attesta l'avvenuto deposito dell'opera:

TITOLO:
MARTE

AUTORE:
VIOLA GIOVANNI

VIOLA GIOVANNI

VIA P. ALMERICO, 21

95018 RIPOSTO (CT)

accettato ai sensi dell'art. 75 lett. b) del Regolamento della S.I.A.E. e alle condizioni previste per l'espletamento del servizio, riportati a tergo.
Il deposito, contrassegnato dal n. **9401108** di repertorio, decorre dal **28-04-94** con scadenza il **27-04-99**

S.I.A.E. - SEZIONE OLAF

Tab. 185	Certificate of deposit of the work "Marte" by Gianni Viola, dated 28 April 1994, with the "Società Italiana Autori ed Editori" (S.I.A.E.) – Sezione OLAF – Servizio deposito opere inedite.

Tab. 185A	The boss checks over report with Girl Friday. Gene Shoemaker, researcher on the Ranger and Surveyor programs, and secretary Jody Loman Swann in new office (USGS). From: Arizona Daily Sun of 1965.

The Society for Planetary SETI Research

SPSR - Members

SPSR
News
Members
Members Websites
Peer Reviewed Journal Publications and Other Articles
Books by SPSR members on anomaly research
Film and Video Archive
Related Links of Interest

Mr. Gianni Viola

Posted by Administrator on Feb 10 2025

Germinal280@libero.it

Italian journalist writer, Embassy military political consultant.

Researcher with scientific discoveries registered in 1994 code 9401108 Società Italiana Autori ed Editori Italia. Author La Civiltà di Marte Edizioni Mediterranee Rome 2002.

Curator Photo Library Nasa Southern Europe Regional Planetary Imaging Facility RPIF Rome 2004-2019. Since 2006 Head of Scientific Commission Free Lance International Press Rome (investigative journalism and participatory interviews).

Author operational project Laboratory of Planetography - RPIF 2012. Refoundation of Planetography (complementary to Planetology) in L'Astronomia di Milano, 2005-2006.

Unpublished elements of aero-photogrammetric planetography: reconfirmation process of satellite images, calculation of topographic resolution, recognition of modified landscapes, recovery and redefinition of modified images.

Last changed: Feb 10 2025 at 5:21 PM

Back to Overview

Tab.185B	Author's membership card for "The Society for Planetary SETI Research – SPSR" (USA).

Bibliography

ASTRONOMY

*Norman, L. J. (1895) *Astronomia*, with notes and additions by Giovanni Celoria, Manuali Hoepli, II edition, Milan: Ulrico Hoepli.

*Ricci, E. (1992) *Telescope no problem*, Milan: Biblioteca Antares.

*Astrophysics-INAF—Astronomical Observatory of Brera; from the website "Scienzainrete".

*"There are billions of living planets in the Milky Way", on *La Repubblica* 29 March 2012, by Emanuela Di Pasqua.

PLANETOLOGY

*Dick, S. J. (2002) *Vita nel Cosmo—Esistono gli extraterrestri?*, Milan: Raffaello Cortina Editore.

*Conrad, B. A. (1989) *Le Chiavi del Cosmo*, Padua: Franco Muzzio Editore.

*Sindoni, E. (1997) *Esistono gli extraterrestri?*, Edizioni Il Saggiatore, Milan: Flammarion.

*Introduzione alla Planetologia, Nemesis Project Research.

*Zubrin, R. (2009) *The Case for Mars*, Curno (BG): Aldebaran Editions.

*Kuiper, G. R. and Barbara M. (1986) Middlehurst in "Pianeti e satelliti". Cit. in V. N. Žarkov, Struttura interna della Terra e dei Pianeti, Editori Riuniti, Rome: Edizioni Mir.

*NASA SP-469—*The geology of the terrestrial planets*—Carr e Wilhelms (USGS), Saunders (JPL), Strom (Arizona State University), NASA 1984 Topographical maps of Mars, Photo lander, Atmosphere composition, Bibliography.

*Nasa sp-469—The Geology of the Terrestrial Planets—Carr e Wilhelms (USGS), Saunders (JPL), Strom (Arizona University), NASA 1984. Topographical maps of Mars; Viking photo lander; Atmosphere, composition; Bibliography.

*The Geology of Mars, Mutch, Arvidson, Head, Jones, Saunders, Princeton University Press, Princeton, New Jersey (USA). Used data: Mars satellites (p.20); Mariner 4,6,7 (pp.29); Shooting from Mariner 9 (p.38); Biological experiments by Viking Lander (p.45).

PLANETOGRAPHY

*Viola, G. (2002) *La Civilta di Marte*, Rome: Edizioni Mediterranee, pp. 384, 132 ill. b/w and colours (16 not in the text).

*Viola, G. (2005) "Contraddizioni areologiche", on *L'Astronomia, n. 268;* it contains the original formula of "planetography".

*Viola, G. (2006) "Marte: fisionomia di un pianeta", on *L'Astronomia, n. 272;* it contains a first mention of "Geography of Mars".

*Ruggieri, G. (1971) "La Scoperta del pianeta Marte", Milan: Arnoldo Mondadori Editore.

*Schiaparelli, G. V. (2002) *La vita sul pianeta Marte*, Milan: Mimesis Edizioni. Pasquale Tucci, *Fu solo illusione? I Canali di Marte tra il 1877 e il 1910.*

* Herschel, F.W. (1985) *Outlines of Astronomy*, New Canaan, London: Readex Micropint.

*Schiaparelli, G. *La vita sul pianeta Marte,* series of published writings on the magazine "Natura ed Arte", Milan 1893–1909, see *Le Opere di G. V. Schiaparelli*, XI vol., Rome 1929–1943.

*G. V. Schiaparelli, *Osservazioni astronomiche e fisiche sull'asse di rotazione e sulla topografia e costituzione del pianeta Marte fatte nella reale Specola di Brera in Milano coll'equatoriale di Merz-Repsold durante l'opposizione del 1890*, "Atti della R. Accademia dei Lincei", serie V, 1910, "Memorie della classe di scienze fisiche matematiche e naturali", in *Le Opere di G. V. Schiaparelli*, XI voll., Rome 1929–1943, Tome III.

*G. V. Schiaparelli, *Gli abitanti di altri mondi*, in *Le Opere di G. V. Schiaparelli*, XI vol., Rome 1929–1943, Tome X.

*The correspondence Lowell—Schiaparelli—by Alessandro Manaro and Acts of the XXI National Meeting of history and physics and astronomy. Department of Physics, University of Calabria, Arcavacata di Rende (CS)—6, 7 and 8 June 2001. p.14.

*Maggini, M. (1939) *Il Pianeta Marte*, Milan: Ulrico Hoepli Editore.

*(edited by) Maraschi, Manfrini, Sandrelli (2003) *Schiaparelli e Marte: Un sogno scientifico*, Milan: Edizioni Scientifiche Coelum.

*Raffaella Simili, *Giovanni Schiaparelli astronomo e uomo di scienza*, in: "Giovanni Schiaparelli: storico della Astronomia e uomo di cultura", edited by Antonio Panainoe and Guido Pellegrini, Editore Mimesis, Milan 1999.

*Erol Torun, "The Geomorphology and Geometry of the D&M Pyramid", unpublished, available through Compuserve ISSUES forum, section 10, file's name PYRAMI.RSH. Extracted from: Graham Hancock, Robert Bauval, John Grigsby, L'Enigma di Marte, Transl. by Lucia Corradini—TEA Tascabili degli Editori Associati—Milan 1998.

* Baker, V. R. (1982) *The Channels of Mars*, Austin (USA): University of Texas Press.

* Slovskij, J. S. and Sagan, C. (1980) *La vita intelligente nell'universo*, Milan: Feltrinelli.

ATLAS AND CARTOGRAPHY

*Batson R.M. (1987) *Digital cartography of the planets: New methods, its status and its future. Photogrammetric Engineering and Remote Sensing.*

*Nicola, F. (1950) *Elementi di cartografia*, Florence: Istituto Geografico Militare.

* Batson, R. M., Bridges, P.M. and Inge, J.L. (1979) *Atlas of Mars* (scale 1:5.000.000), NASA-438.

*"Cambiamenti climatici globali su Marte" in *Le Scienze*, n. 341, January 1997.

GEOMATICS AND OPTICAL PHYSICS

*Mario A. (2004) Gomarasca, Elementi di Geomatica, Milan: Associazione Italiana di Telerilevamento.

* Paliaga Gian Paolo (2008), *I vizi di rifrazione,* Torino Edizioni Minerva Medica.

EPISTEMOLOGY

*Ernest, N. (1989) *La logica dell'ipotesi*, Milan: Rusconi Editore.

*Ernest, N. *Les Fondateurs de L'Astronomie Moderne*, III édition.

*Arsena, A. (2016) *"La filosofia e la religione" di Ernest Naville*, Rome: editions Stamen.

SOURCE MATERIAL FOR THE PLANETOGRAPHIC ANALYSIS AND THE SCIENTIFIC EXPERIMENTATION

Mission Mariner 9 (1971–1972)

*Nasa, Technical Memorandum 33-585 volume I, Mariner Mars 1971 Television Picture Catalogue, Experiment Design and Picture Data, Cutts, JPL California Institute of Technology, Pasadena, California (USA), 30 June 1974. Used data: Pavonis Mons (p.126); Capri Chasma (p.156).

*Nasa, Technical Memorandum 33-585 volume I, Addendum 1, Mariner Mars 1971 Television Picture Catalogue, Experiment Design and Picture Catalogue, Crosby, JPL California Institute of Technology, Pasadena, California (USA), 15 August 1974. Used data: Photomosaic 211-4294 (p. 9).

*Nasa, Technical Memorandum 33-585 volume II, Mariner Mars 1971 Television Picture Catalogue, Sequence Design and Picture Coverage, Koskela, Helton, Seeley e Zawacki, JPL California Institute of Technology, Pasadena, California (USA), 15 December 1972. Used data: Pavonis Mons (6-145, 6-171, 6-212, 6-257); Nanedi Valles (6-363); Capri Chasma (6-371); Aurorae Planum (6-367); Iani Chaos (6-379).

*Nasa, Technical Memorandum 33-585 volume II, Addendum 1, Mariner Mars 1971 Television Picture Catalogue, Sequence Design and Picture Coverage, Koskela, JPL California Institute of Technology, Pasadena, California (USA), 1 July 1973.

*Nasa, Technical Memorandum 33-595, Mariner 9 Television Pictures: microfiche library user's guide, mtc/mtvs realtime pictures, Becker, JPL, California Institute of Technology, Pasadena, California (USA), 15 February 1973. Table of correspondence "roll number" and orbitals (pp.8, 9).

*Nasa, Technical Memorandum 33-723, Guide to the use of Mariner images, Saunders, Mutch e Jones, JPL California Institute of Technology, Pasadena, California (USA), 15 October 1975. Guide to the use of the images from missions Mariner.

*Nasa, Technical Memorandum 33-628, A user's guide to the Mariner 9 television reduced data record, Seidma, Green, Jepsen, Ruiz, Thorpe, JPL California Institute of Technology, Pasadena, California (USA), 1 September 1973. Guide to the television shooting done by mission Mariner 9.

*Mariner Mars 1971 Science Data Team Library, Final Report: JPL Report 610, 243, JPL Pasadena USA, 1973.

*Mariner Mars 1971 Science Experimenter Teams, Mariner Mars 1971 Project Final Report, v.4, *Science Results:* JPL, California Institute of Technology Technical Report 32 1550, 1973.

*Mars as viewed by Mariner 9, NASA 1974. Used data: Pavonis Mons (p. 11); Aram Chaos, Iani Chaos (p.86); Hydaspis Chaos (pp. 88, 89); Apollinaris Patera (p. 175).

*The New Mars, The Discoveries of Mariner 9, NASA sp-337, 1974. Used data: Telescopic Map of Mars (II cover); Pavonis Mons (p. 85); First mapping of Mars (p. 87); Valles Marineris (p. 89); Apollinaris Patera (p. 105); Maps (p. 146); Territorial division (p.147); Mars satellite mapping (III cover).

*The Many Faces of Mars, NASA JPL/California Institute of Technology, Pasadena, California (USA). Used data: General topographic map Mercator (p. 8); Pavonis Mons (pp. 10, 23); Valles Marineris (p.20); Mars first mapping (p. 25).

*Mariner to Mars, Collection of Papers from Science, 21 January 1972, American Association for the Advancement of Science, Washington DC (USA). Telescopic topographic maps.

**Mars as viewed by Mariner 9*, NASA 1974.

*Mariner 9 Television Team 1974, *Mars as Viewed by Mariner 9*, Washington, NASA SP 329.

* *The New Mars*, The Discoveries of Mariner 9, NASA SP-337, 1974.

Mission Viking 1 and 2 (1975–1982)

*Scientific Results of the Viking project, American Geophysical Union, Washington, DC (USA), 1977. Orbiter Science Platform; Used data: Camera Viking (p.3977); Drop-shape islands in Ares Vallis (p.4020); Galaxias Chaos (p.4044); Fluvial structures (p.4100); Cydonia Mensae (p.4113); Utopia Planitia (p.4483).

*Nasa Contractor Report 3326—The Mosaic of Mars—As Seen by the Viking Lander Cameras—Elliott C. Levinthal and Kenneth L. Jones, September 1980. Mosaics realised on the findings of descent modules of Viking mission.

*Nasa Contractor Report 3277, Viking Orbiter Stereo Imaging Catalogue—Karl R. Blasius, Amelia V. Vetrone and Michael D. Martin. June1980. Catalogue of the Images from Viking mission.

*Nasa Contractor Report 3496—The Viking Mosaic Catalogue—Nancy Evans. March 1982. Vols. 1–2. Catalogue of photomosaics built on the data by the satellite findings of missions Viking Orbiter 1 and 2.

*Liebes Sidney Jr., *Viking Lander Atlas of Mars*, NASA Scientific and Technical Information Office, NASA Contractor Report 3568—1982.

*Carr M.H. and Evans Nancy, *Images of Mars: The Viking Extended Mission*, Washington DC, NASA SP-444, 1980.

*Corliss W., *The Viking Mission to Mars*, Washington DC NASA SP-334, 1974.

*Tyner R.L. e Carrol R.D., *A Catalogue of Selected Viking Orbiter Images*, Washington DC NASA Reference Publication 1093, 1983.

*AA.VV. *The Martian Landscape—By The Viking Lander Imaging Team*, Scientific and Technical Informaton Office—NASA Washington DC 1978—NASA SP-425. Images of the lander shooting.

*Viking Orbiter Imaging Team, *Viking Orbiter Views of Mars*, Washington NASA SP-441, 1980.

*Tucker Robert, Viking Lander Imaging Investigation: Picture Catalogue of Primary Mission Experiment Data Record NASA RP-1007.

*Anonymous, *Viking 1: Early Results*, Washington, NASA SP-408, 1976.

*Scientific Results of the Viking Project, American Geophysical Union, Washington, DC (USA)—1977.

Missions Mariner 9—Viking 1 and 2

*Planetary Image Conversion Task—JPL Publication 85-50—Final Report, Martin, Stanley, Laughlin, NASA, JPL California Institute of Technology, Pasadena, California (USA), 15 September 1985. Tables of conversion satellite images codes. Used data: Mars. Viking (pp.80–115); Mariner 9 (pp.123–146); Mariner 6/7 (pp.147–157).

*Colour Coded Contour Map of Mars—Scala 1:25.000.000.—US Geological Survey (USA), 2003.

*Nasa sp-469—The Geology of the Terrestrial Planets—Carr and Wilhelms (USGS), Saunders (JPL), Strom (Arizona University), NASA 1984. Mars topographic maps; lander photo from Viking; Atmosphere, composition; Bibliography.

*The Geology of Mars, Mutch, Arvidson, Head, Jones, Saunders, Princeton University Press, Princeton, New Jersey (USA). Used data: Mars satellites (p.20); Mariner 4, 6, 7 (p.29); Shooting from Mariner 9 (p.38); Biological experiments from Viking Lander (p.45).

*Greeley R., *Planetary Landscapes*, 2^{a} ed. London: Allen und Unwin, 1987.

*Morrison D., *Exploring Planetary Worlds*, New York: W.H. Freeman, 1993. JPL Publication 85-50; Planetary Image Conversion Task, Final Report. Martin, Stanley, Laughlin, NASA, JPL California Institute of Technology—Pasadena, California (USA), 15 September 1985. Conversion's tables of satellite images codes—Mars.

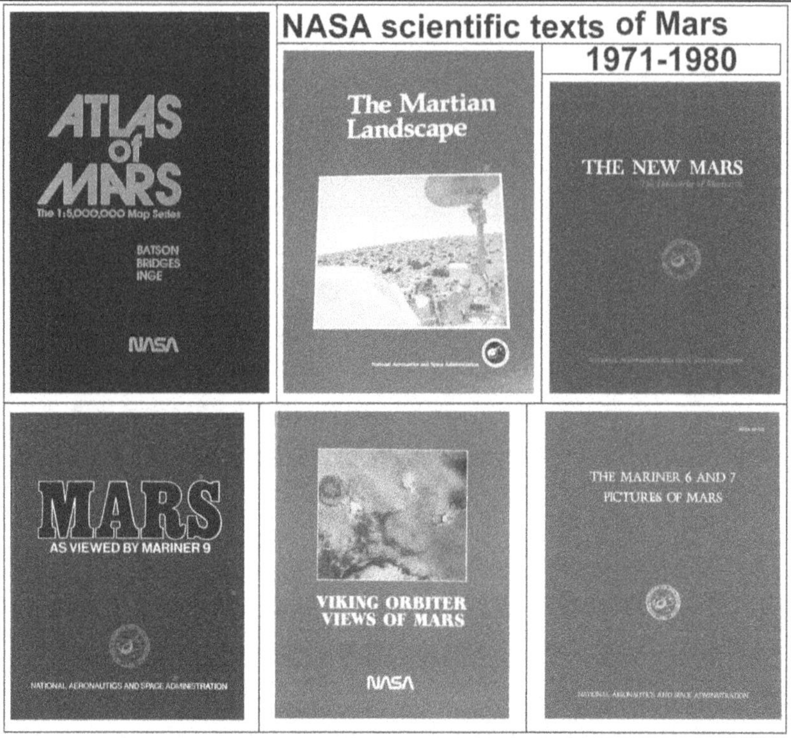

Tab. 186	The fundamental scientific texts, published by NASA, summarising the space exploration of Mars, in the period 1969-1983: Atlas of Mars (1979) – Mariner 9 and Viking 1&2 Missions; The Martian Landscape (1978) – Viking Lander 1&2 Missions; The New Mars (1974) – Mariner 9 Mission; Mars as viewed by Mariner 9 (1974); Viking Obiter views of Mars (1980); The Mariner 6 and 7 pictures of Mars (1971).

EDITORIAL DISCAIMERS
RELEASES RELATING TO THE USE OF TEXTS, ARTICLES AND PHOTOGRAPHIC IMAGES AND BOOKS

Text: - Vita nel Cosmo (Life in the Cosmos) - by Steven J. Dick
Raffaello Cortina Publisher - Milan 2002
pp. 41,45,49,57,65,70,71,73,76,77,84,86
Release received from Rights Office - Laura Maccagni

Text: C'è vita nell'Universo (Is there life in the Universe?) - by Pippo Battaglia and Walter Ferreri
Lindau Editions - Turin 2008 (p.7)
Release received from Rights Office - Davide Platzer Ferrero

Text: Elementi di Geomatica (Elements of Geomatics) - by Mario Gomarasca Italian Association of Remote Sensing Editions - Milan 2004 Headings: geography, cosmography.
Release received from the author of the text.

Other releases have been requested (and we are awaiting replies) from the following publishers:

Edizioni Scientifiche Coelum (A..VV., Schiaparelli e Marte: un sogno scientifico, 2003.

Edizioni Antares (E.Ricci, Telescope no probleme, 1992).

Franco Muzzio Editore (Bohm Conrad, Le Chiavi del Cosmo, 1989).

Minerva Medica (Gianpaolo Paliaga, I vizi di rifrazione, 2008.

Arnoldo Mondadori Editore (Guido Ruggieri, La scoperrta del pianeta Marte, 1971);

Feltrinelli Editore (Slovslij-Sagan, La vita intelligente nell'universo, 1980).

Aldebaran Edizioni (Robert Zubrin, The Case for Mars, 2009).

Editori Riuniti-Edizioni Mir (Zarkov, Struttura interna della Terra e dei pianeti, 1986).

Rusconi Editore (Ernest Naville, La logica dell'ipotesi, 1989).

Editori Associati (Graham Hancock, L'Enigma di Marte, 1998).

PHOTOGRAPHIC TABLES:

1.. Image 2 is in the public domain. Photographic portrait of Giovanni Schiaparelli by painter Paul D. Stewart (1870);

2.. Image 3 is in the public domain. Italian commemorative stamp dedicated to Giovanni Virginio Schiaparelli on the 100th anniversary of his death (1910-2010).

3. Images V, 1, 4, 5, 6, 10, 11, 12, 13, 14, 17, 19, 20, 21, 22, 31, 32, 34, 35, 36, 39, 40, 41, 45, 47, 49A, 52, 54, 54A, 56, 57, 58, 59, 60, 62, 64, 66, 71, 81, 82, 83, 84, 85, 85A, 97, 98, 99, 119, 120, 123, 124, 132, 152, 159, 160, 161, 162, 163, 164, 165, 166, 182, 183, 184, 185, 186A are the property of
the Author;

4.. Images 4, 6, 9A, 16, 19, 39, 41, 46, 48, 49, 50, 51, 65, 66, 67, 77, 80A, 80B, 86A, 87, 88, 89B, 90, 92, 93, 94, 95, 102A, 102B, 103, 104, 105B, 106, 107, 109, 110, 111, 112B, 113, 114, 116A, 116B, 117, 118, 121, 122B, 127, 128, 133, 134, 135, 136, 138, 139, 140, 141, 142, 147A, 147B, 148, 149, 150, 151B, 154A, 154B, 155, 156, 157, 158 are graphic modifications by the author, who used images provided by NASA (USA);

5.. Images 78 and 79 are graphic modifications by the author, who used images provided by the USGS – United States Geological Survey (USA);

6. Images 143, 144 and 145 are graphic modifications by the author, who used images provided by ESA – European Space Agency (Germany, Berlin) and NASA (USA);

7. Images 23, 24, 25, 26, 27, 28, 29, 30, 42, 43, 44, 61, 67, 68, 69, 70A, 70B, 72, 72A, 73, 74, 76, 86B, 89A, 91, 96, 96B, 100, 101, 105A, 108, 112, 115, 122A125, 126, 129, 130, 131, 137, 146, 151A, 153, 167, 168, 169, 169A, 170, 171, 172, 173, 174, 175, 176, 177, 178, 179, 180, 181, 186 are used under licence from NASA (USA);

8.. Photographic plates nos. I, II and 187, owned by IAPS - Institute of Space Astrophysics and Planetology (Rome), are published by kind permission of IAPS. Release received from the Director of the IAPS, Dr. Marco Feroci.
Document received on 12 September 2025.

9.. Image 75 is used under licence from the USGS (USA);

10.. Images 33, 37 and 38 are the property of Edizioni Minerva Medica. Pending authorisation;

11.. Images 53 and 55 are used under licence from Dr Mark Carlotto (USA);

12. Image 18 is the property of EDP Sciences (France). Pending authorisation;

13.. Image 51 is the property of Kynthia Lynn (USA). Pending authorisation;

14.. Image 63B is the property of Thomas Mutch (USA). Text: The Geology of Mars. Pending authorisation.

15. Image 65 is a remake of a topographical map by Victor Baker (USA). Text: The Channels of Mars. Pending authorisation.

16. Images 8 and 9 are the property of A. Manara and F. Chlistovskiy (Italy). Text: Lowell-Schiaparelli sign. Pending authorisation.

17.. Images 71 and 185A are in the public domain. Arizona Daily Sun newspaper (USA), 30 August 1972 and 1965.

18.. Image 15 is the property of the website 'Valutazione delle distanze' (Distance Assessment). Pending authorisation.

19.. Image 7 is used with the authorisation of the Centre d'Iconographie Genevoise - Geneva (Switzerland).

20.. Image 67 was created by a NASA official for my scientific text.

21.. Image137.A2 are in the public demain; Slovskij-Sagan, *La Vita intelligente nell'Universo,* Feltrinelli Editore Milano 1980.

22.. The image on page 377 is published with the permission of the author, Prof. Yulia Bondarenko (Russia, Omsk).

Index

Mars was the national god of the Romans. 'The Roman Empire was the most powerful empire after that of Heaven. The Roman people claim that their father, the father of the Empire, is none other than Mars.'

(Titus Livius. Ab Urbe Condita, Book I)

Caption for the image on page 377.

Counterclockwise: Colosseum (detail), Roman construction begun in 70 or 72 and inaugurated in 80 AD; Ponte Sant'Angelo, Roman construction inaugurated in 136 AD; in the background, a detail of the Fortezza Adriana (Castel Sant'Angelo), Roman construction begun in 135 and inaugurated in 139 AD; Capitoline Wolf, Etruscan sculpture from the 5th century BC; in later times, in the 15th century AD, the two children, possibly Romulus and Remus, were added. The photographs are by Yulia Bondarenko (Omsk, Russia).